THE FAST FOURIER
TRANSFORM
AND ITS APPLICATIONS

D0783278

PRENTICE HALL SIGNAL PROCESSING SERIES

Alan V. Oppenheim, Editor

THE FAST FOURIER TRANSFORM AND ITS APPLICATIONS

E. Oran Brigham

Avantek, Inc.

 Prentice-Hall International, Inc.

ISBN   0-13-307547-8

Printed in Great Britain by BPCC Wheatons Ltd, Exeter

10 9 8 7 6 5 4

ISBN   0-13-307547-8

Prentice-Hall International (UK) Limited, *London*
Prentice-Hall of Australia Pty. Limited, *Sydney*
Prentice-Hall Canada Inc., *Toronto*
Prentice-Hall Hispanoamericana, S.A., *Mexico*
Prentice-Hall of India Private Limited, *New Delhi*
Prentice-Hall of Japan, Inc., *Tokyo*
Simon & Schuster Asia Pte. Ltd., *Singapore*
Editora Prentice-Hall do Brasil, Ltda., *Rio de Janeiro*
Prentice-Hall, Inc., *Englewood Cliffs, New Jersey*

To Cami

A very special daughter

CONTENTS

PREFACE

The fast Fourier transform (FFT) is a widely used signal-processing and analysis concept. Availability of special-purpose hardware in both the commercial and military sectors has led to sophisticated signal-processing systems based on the features of the FFT. The implementation of FFT algorithms on large mainframe computers has made unprecedented solution techniques readily achievable. Personal computers have generated yet a further proliferation of FFT applications. To the student, the professional at home, engineers, computer scientists, and research analysts, the FFT has become an invaluable problem-solving tool.

Popularity of the FFT is evidenced by the wide variety of application areas. In addition to conventional radar, communications, sonar, and speech signal-processing applications, current fields of FFT usage include biomedical engineering, imaging, analysis of stock market data, spectroscopy, metallurgical analysis, nonlinear systems analysis, mechanical analysis, geophysical analysis, simulation, music synthesis, and the determination of weight variation in the production of paper from pulp. Clearly, an applications text cannot address in depth such a breadth of technology. The objective of this book is to provide the foundation from which one can acquire the fundamental knowledge to apply the FFT to problems of interest.

The book is designed to be *user friendly*. We stress a pictorial, intuitive approach supported by mathematics, rather than an elegant exposition that is difficult to read. Every major concept is developed by a three-stage sequential process. First, the concept is introduced by an intuitive graphical development. Second, a nonsophisticated (but theoretically sound) mathematical treatment is developed to support the intuitive arguments. The third stage consists of practical examples designed to review and expand the concept. This three-step procedure, with an emphasis on graphical techniques, gives *meaning* as well as mathematical substance to the basic properties and applications of the FFT. Readers should expect a high efficiency in transferring the development of the text into practical applications.

This book is a sequel to *The Fast Fourier Transform*. The focus of the original volume was on the Fourier transform, the discrete Fourier transform, and the FFT. Only a cursory examination of FFT applications was presented. This text extends the original volume with the incorporation of extensive developments of fundamental FFT applications. *Applications* of the FFT are based on its unique property to rapidly compute the Fourier, inverse Fourier, or Laplace transforms. For this reason, we develop in detail the methods for applying the FFT to transform analysis and interpreting results. We then extend the development and apply the FFT to the computation of convolution and correlation integrals. All developments employ a rich use of graphical techniques and examples to insure clarity of the presentation. We then build on these fundamentals and expand the basic FFT uses to a higher level of application topics. Topical areas include two-dimensional FFT analysis, FFT digital filter design, FFT multichannel bandpass filtering, FFT signal processing, and FFT systems applications.

The text should provide an excellent basis for a senior level or introductory graduate course on digital signal processing. Course outlines emphasizing a thorough examination of the Fourier Transform will find the text particularly appealing. The added applications material allows students to develop the experience necessary to apply the FFT to problems spanning a wide variety of disciplines. Students are expected to have access to a digital computer. The text should serve equally well as a supplementary text for a course with broad systems analysis and signal-processing objectives. The book should also be very attractive as a reference to the practicing signal-processing community because it offers not only a readable introduction to the FFT, but a thorough and unified reference for applying the FFT to any field of interest. Readers should also find that the material provides an excellent self-study text.

The text is divided into five major subject areas:

1. *The Fourier Transform*. In Chapters 2 through 6, we lay the foundation for the entire book. We investigate the Fourier transform, its inversion formula, and its basic properties; graphical explanation of each discussion lends physical insight to the concept. The transform properties of the convolution and correlation integrals are explored in detail. Numerous examples are presented to facilitate understanding. For reference in later chapters, Fourier series and waveform sampling of baseband signals are developed in terms of Fourier transform theory.

2. *The Discrete Fourier Transform.* Chapters 6 and 7 develop the discrete Fourier transform. A graphical presentation develops the discrete transform from the continuous Fourier transform. This graphical presentation is substantiated by a theoretical development. Discrete transform properties are derived. The relationship between the discrete and continuous Fourier transform is explored in detail; numerous waveform classes are considered by illustrative examples. Discrete convolution and correlation are defined and compared with continuous equivalents by illustrative examples.

3. *The Fast Fourier Transform.* In Chapter 8, we develop the FFT algorithm. A simplified explanation of why the FFT is efficient is presented. We follow with the development of a signal flow graph, a graphical procedure for examining the FFT. Based on this flow graph, we describe sufficient generalities to develop a computer flowchart and computer programs. Theoretical developments of the various forms of the FFT are presented.

4. *Basic Applications of the FFT.* Chapters 9 through 11 focus on an investigation of the basic applications of the FFT. Application of the FFT to the computation of discrete and inverse discrete Fourier transforms is presented with emphasis on a graphical examination of resolution and common FFT user mistakes (aliasing, time-domain truncation, noncausal time functions, and periodic functions). FFT data-weighting functions are examined in depth. Laplace transform computation using the FFT is presented with graphical examples. FFT implementation of discrete convolution and correlation is developed by extensive graphical presentations. Computational procedures are carefully defined and a computer program is provided. Two-dimensional Fourier transforms, convolution, and correlation are developed (graphically and by example), as in the one-dimensional case. Application of the FFT to two-dimensional Fourier transform and convolution computation are described and computer programs are provided.

5. *Signal Processing and System FFT Applications.* The design and application of digital filters using the FFT is explored from a practical usage perspective. A novel application of the FFT to multichannel band-pass filtering is developed in a manner from which the reader can readily expand the results.

Because waveform sampling is fundamental to FFT signal-processing applications, band-pass and quadrature waveform sampling is addressed in detail. The philosophy underlying the remaining discussions is to address a range of FFT techniques that are applicable to sonar, seismic, radar, communications, medical, optical, system analysis, and antenna applications. Specific FFT application areas addressed include signal-to-noise enhancement, matched filtering, deconvolution filtering, time-difference-of-arrival measurements, phase interferometry measurements, antenna analysis, system simulation, power spectrum analysis, and array beamforming.

I would like to take this opportunity to express my sincere appreciation to the many individuals who have contributed to the contents of this book. A special note of thanks goes to Dr. Patty Patterson, who contributed significantly in correcting and improving the manuscript. Charlene Rushing and Neil Ishman contributed to the computer subroutines.

To my wife Vangee, I am indebted for her patience and understanding for the many hours I have stolen from her life while preparing the manuscript. To my daughter Cami, I thank you for your efforts, dedication, and enthusiasm towards a *commitment to excellence*; I hope some of your ideals are incorporated in this book.

1

INTRODUCTION

The fast Fourier transform (FFT) is a fundamental problem-solving tool in the educational, industrial, and military sectors. Since 1965 [1], FFT usage has rapidly expanded and personal computers fuel an explosion of additional FFT applications. **The single focus of this book is the FFT and its applications.**

In this chapter, we survey briefly the broad application areas of the FFT to give the reader a perspective for its seemingly universal appeal. We will establish the FFT as one of the major developments in signal-processing technology. The diverse applications of the FFT follow from the roots of the FFT: the discrete Fourier transform and hence the Fourier transform. Our overview of the Fourier transform and its interpretation with respect to the time and frequency domains is presented.

1.1 THE UBIQUITOUS FFT

Ubiquitous is defined as being everywhere at the same time. The FFT is certainly ubiquitous because of the great variety of *apparent* unrelated fields of application. However, we know that the proliferation of applications across broad and diverse areas is because they are united by a common entity, the Fourier transform. For years only the *elitist* theoretical mathematician was capable of staying abreast of such a broad spectrum of technologies. However, with the FFT, Fourier analysis has been reduced to a readily available and practical procedure that can be applied effectively with-

Applied Mechanics

- structural dynamics
- aircraft wing-flutter suppression
- machinery dynamics diagnostics
- nuclear power plant modeling
- vibration analysis

Sonics and Acoustics

- acoustic imaging
- passive sonar
- ultrasonic transducers
- array processing
- architecture acoustic measurement
- music synthesis

Biomedical Engineering

- diagnosis of airways obstruction
- muscle fatigue monitoring
- assessing heart valve damage
- tissue structure characterization
- gastric disturbances investigation
- cardiac patients diagnosis
- ECG data compression
- artery dynamics investigation

Numerical Methods

- high-speed interpolation
- conjugate gradient method
- boundary value problems
- Riccati and Dirichlet equations
- Rayleigh's integral

- Wiener-Hopf integral equation
- diffusion equation
- numerical integration
- Karhunen-Loeve transform
- elliptic differential equations

Signal Processing

- matched filters
- deconvolution
- real-time spectral analysis
- cepstrum analysis
- coherence function estimation
- speech synthesis and recognition
- random process generation
- transfer function estimation
- echo/reverbation removal

Instrumentation

- chromatography
- microscopy
- spectroscopy
- x-ray diffraction
- electrochronography

Radar

- cross-section measurement
- moving target indicator
- synthetic aperture
- doppler processor
- pulse compression
- clutter rejection

Electromagnetics

- microstrip line propagation
- conducting bodies scattering

Figure 1.1   Summary of FFT Applications.

- antenna   radiation   patterns
- dielectric   substrate
  capacitance
- phased-array   antenna
  analysis
- time-domain   reflectometry
- waveguide   analysis
- network   analysis

Communications

- systems   analysis
- transmultiplexers
- demodulators
- speech   scrambler   system
- multichannel   filtering
- M-ary   signaling

- signal   detection
- high-speed   digital   filters
- voice   coding   systems
- video   bandwidth
  compression

Miscellaneous

- magnetotellurics
- metallurgy
- electrical   power   systems
- image   restoration
- nonlinear   system   analysis
- geophysics
- GaAs   FET   transient
  response
- integrated   circuit   modeling
- quality   control

Figure 1.1   (cont.)

out sophisticated training or years of experience. The FFT has become a *standard analysis module* because of its usefulness and availability.

The FFT is no longer a textbook novelty. In Fig. 1.1, we show an abbreviated listing of typical application areas of the FFT. Key reference materials in the FFT application fields shown are included in the bibliography. The FFT, once the province of engineers and scientists, has become a technique used in areas ranging from the analysis of stock market trends to the determination of weight variations in the production of paper from pulp. Computer technology evolution, particularly that of the personal computer, has positioned the FFT as a handy and powerful analysis tool whose availability is no longer limited only to the signal-processing specialist. As shown in Fig. 1.1, the application fields of the FFT are extremely diverse. In an age where it is virtually impossible to stay abreast of technology, it is stimulating to find an analysis concept that enables one to approach an unfamiliar field with familiar tools. Certainly, the FFT has become one of the major developments in digital signal-processing technology.

As stated previously, the common bond throughout the varied application of the FFT is the Fourier transform. A key property of the Fourier transform is its ability to allow one to examine a function or waveform from the perspective of both the time and frequency domains. The Fourier transform is the cornerstone of this text.

1.2 INTERPRETING THE FOURIER TRANSFORM

A simplified interpretation of the Fourier transform is illustrated in Fig. 1.2. As shown, the essence of the Fourier transform of a waveform is to decompose or separate the waveform into a sum of sinusoids of different frequencies. If these sinusoids sum to the original waveform, then we have determined the Fourier transform of the waveform. The pictorial representation of the Fourier transform is a diagram that displays the amplitude and frequency of each of the determined sinusoids.

Figure 1.2 also illustrates an example of the Fourier transform of a simple waveform. The Fourier transform is the two sinusoids that add to yield the waveform. As shown, the Fourier transform diagram displays both the amplitude and frequency of each sinusoid. We have followed the usual convention and displayed both positive and negative frequency sinusoids for each frequency; the amplitude has been halved accordingly. The Fourier transform then decomposes the example waveform into its two individual sinusoidal components.

The Fourier transform identifies or distinguishes the different frequency sinusoids (and their respective amplitudes) that combine to form an arbitrary waveform. Mathematically, this relationship is stated as

$$S(f) = \int_{-\infty}^{\infty} s(t)e^{-j2\pi ft}\, dt \tag{1.1}$$

where $s(t)$ is the waveform to be decomposed into a sum of sinusoids, $S(f)$ is the Fourier transform of $s(t)$, and $j = \sqrt{-1}$. An example of the Fourier transform of a square-wave function is illustrated in Fig. 1.3(a). An intuitive justification that a square waveform can be decomposed into the set of sinusoids determined by the Fourier transform is shown in Fig. 1.3(b).

We normally associate the analysis of periodic functions such as a square wave with Fourier series rather than Fourier transforms. However, as we will show in Chapter 5, the Fourier series is a special case of the Fourier transform.

If the waveform $s(t)$ is not periodic, then the Fourier transform will be a continuous function of frequency, that is, $s(t)$ is represented by the summation of sinusoids of all frequencies. For illustration, consider the pulse waveform and its Fourier transform, as shown in Fig. 1.4. In this example, the Fourier transform indicates that one sinusoid frequency becomes indistinguishable from the next and, as a result, all frequencies must be considered.

The Fourier transform is then a frequency-domain representation of a function. As illustrated in both Figs. 1.3(a) and 1.4, the Fourier transform frequency domain contains exactly the same information as that of the orig-

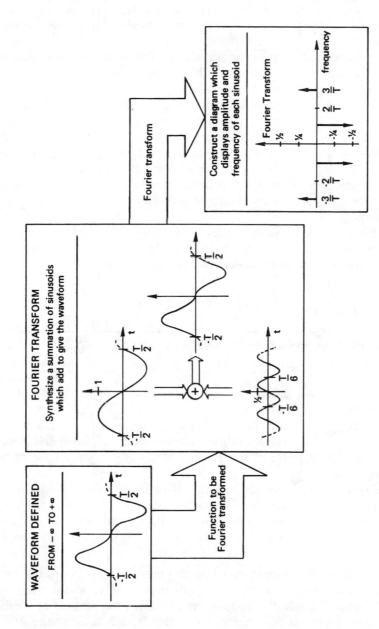

Figure 1.2 Interpretation of the Fourier transform.

5

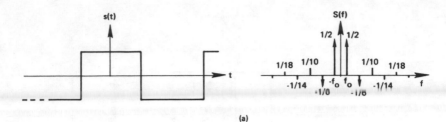

(a)

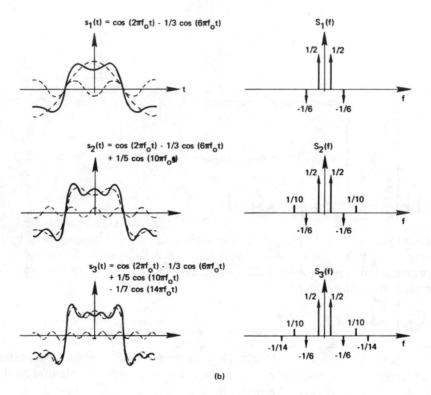

(b)

Figure 1.3 Fourier transform of a square-wave function.

inal function; they differ only in the manner of presentation. Fourier analysis allows one to examine a function from another point of view, the transform domain. As we will see in the discussions to follow, the method of Fourier transform analysis employed, as illustrated in Fig. 1.2, is often the key to problem-solving success.

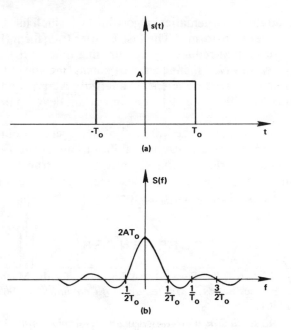

Figure 1.4   Fourier transform of a pulse waveform.

1.3 DIGITAL FOURIER ANALYSIS

Because of the wide range of problems that are susceptible to attack by the Fourier transform, we would expect the logical extension of Fourier transform analysis to the digital computer. Numerical integration of Eq. (1.1) implies the relationship:

$$S(f_k) = \sum_{i=0}^{N-1} s(t_i)e^{-j2\pi f_k t_i}(t_{i+1} - t_i) \qquad k = 0, 1, \ldots, N-1 \qquad (1.2)$$

For those problems that do not yield to a closed-form Fourier transform solution, the discrete Fourier transform of Eq. (1.2) offers a potential method of attack. However, careful inspection of Eq. (1.2) reveals that if there are N data points of the function $s(t_i)$ and if we desire to determine the amplitude of N separate sinusoids, then computation time is proportional to N^2, the number of multiplications. Even with high-speed computers, computation of the discrete Fourier transform requires excessive machine time for large N.

An obvious requirement existed for the development of techniques to reduce the computing time of the discrete Fourier transform; however, the scientific community met with little success. Then, in 1965, Cooley and

Tukey published their mathematical algorithm [1], which has become known as the "fast Fourier transform." The fast Fourier transform (FFT) is a computational algorithm that reduces the computing time of Eq. (1.2) to a time proportional to $N \log_2 N$. This increase in computing speed has completely revolutionized many facets of scientific analysis. A historical review of the discovery of the FFT illustrates that this important development was almost ignored [4, 5].

The FFT has revolutionized the use of the discrete Fourier transform. It is important to recognize that one's ability to apply the FFT relies principally on an understanding of the discrete Fourier transform and not the FFT algorithm. For this reason, this text emphasizes the fundamentals of the Fourier and discrete Fourier transforms.

REFERENCES

1. COOLEY, J. W., AND J. W. TUKEY. "An Algorithm for the Machine Calculation of Complex Fourier Series." *Mathematics of Computation* (1965), Vol. 19, No. 90, pp. 297–301.

2. BRACEWELL, RON. *The Fourier Transform and Its Applications*, 2d Rev. Ed. New York: McGraw-Hill, 1986.

3. PAPOULIS, A. *Probability, Random Variables, and Stochastic Processes*, 2d Ed. New York: McGraw-Hill, 1984.

4. COOLEY, J. W., R. L. GARWIN, C. M. RADER, B. P. BOGERT, AND T. C. STOCKHAM. "The 1968 Arden House Workshop of Fast Fourier Transform Processing." *IEEE Trans. on Audio and Electroacoustics* (June 1969), Vol. AU-17, No. 2, pp. 66–75.

5. COOLEY, J. W., P. W. LEWIS, AND P. D. WELCH. "Historical Notes on the Fast Fourier Transform." *IEEE Trans. on Audio and Electroacoustics* (June 1967), Vol. AU-15, No. 2, pp. 76–79.

6. *IEEE Trans. on Audio and Electroacoustics, Special Issue on the Fast Fourier Transform* (June 1969), Vol. AU-17, No. 2.

7. BRIGHAM, E. O. *The Fast Fourier Transform*. Englewood-Cliffs, NJ: Prentice Hall, 1974.

8. RAMIREZ, R. W. *The FFT: Fundamentals and Concepts*. Englewood-Cliffs, NJ: Prentice Hall, 1985.

9. BURRIS, C. S., AND T. W. PARKS. *DFT-FFT & Convolution Algorithms & Implementation*. New York, Wiley, 1985.

10. ELLIOT, D. F., AND K. R. RAO. *Fast Transforms, Algorithms, Analyses, Applications*. Orlando, FL: Academic Press, 1982.

2

THE FOURIER TRANSFORM

A principal analysis tool in many of today's scientific challenges is the Fourier transform. Possibly the most well-known application of this mathematical technique is the analysis of linear time-invariant systems. But, as emphasized in Chapter 1, the Fourier transform is essentially a universal problem-solving technique. Its importance is based on the fundamental property that one can examine a particular relationship from an entirely different viewpoint. Simultaneous visualization of a function and its Fourier transform is often the key to successful problem solving.

2.1 THE FOURIER INTEGRAL

The Fourier integral is defined by the expression

$$H(f) = \int_{-\infty}^{\infty} h(t)e^{-j2\pi ft}\, dt \qquad (2.1)$$

If the integral exists for every value of the parameter f, then Eq. (2.1) defines $H(f)$, the Fourier transform of $h(t)$. Typically, $h(t)$ is termed a function of the variable time and $H(f)$ is termed a function of the variable frequency. We use this terminology throughout the book: t is *time* and f is *frequency*. Further, a lowercase symbol represents a function of time; the Fourier transform of this time function is represented by the same uppercase symbol as a function of frequency.

In general, the Fourier transform is a complex quantity:

$$H(f) = R(f) + jI(f) = |H(f)| e^{j\theta(f)} \qquad (2.2)$$

where $R(f)$ is the real part of the Fourier transform,

$I(f)$ is the imaginary part of the Fourier transform,

$|H(f)|$ is the *amplitude* or *Fourier spectrum* of $h(t)$ and is given by $\sqrt{R^2(f) + I^2(f)}$,

$\theta(f)$ is the phase angle of the Fourier transform and is given by $\tan^{-1}[I(f)/R(f)]$.

Example 2.1   Exponential Waveform

To illustrate the various defining terms of the Fourier transform, consider the function of time

$$h(t) = \beta e^{-\alpha t} \qquad t > 0$$
$$= 0 \qquad t < 0 \qquad (2.3)$$

From Eq. (2.1),

$$H(f) = \int_0^\infty \beta e^{-\alpha t} e^{-j2\pi f t}\, dt = \beta \int_0^\infty e^{-(\alpha + j2\pi f)t}\, dt$$

$$= \frac{-\beta}{\alpha + j2\pi f} e^{-(\alpha + j2\pi f)t} \Big|_0^\infty = \frac{\beta}{\alpha + j2\pi f} \qquad (2.4)$$

$$= \frac{\beta\alpha}{\alpha^2 + (2\pi f)^2} - j\frac{2\pi f\beta}{\alpha^2 + (2\pi f)^2}$$

$$= \frac{\beta}{\sqrt{\alpha^2 + (2\pi f)^2}} e^{j\tan^{-1}[-2\pi f/\alpha]}$$

Hence,

$$R(f) = \frac{\beta\alpha}{\alpha^2 + (2\pi f)^2}$$

$$I(f) = \frac{-2\pi f\beta}{\alpha^2 + (2\pi f)^2}$$

$$|H(f)| = \frac{\beta}{\sqrt{\alpha^2 + (2\pi f)^2}}$$

$$\theta(f) = \tan^{-1}\left[\frac{-2\pi f}{\alpha}\right]$$

Each of these functions is plotted in Fig. 2.1 to illustrate the various forms of Fourier transform presentation.

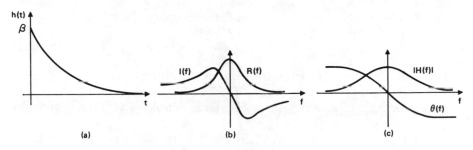

Figure 2.1 (a) Example of a time-domain function, (b) real and imaginary presentations of the Fourier transform, and (c) magnitude and phase presentations of the Fourier transform.

2.2 THE INVERSE FOURIER TRANSFORM

The inverse Fourier transform is defined as

$$h(t) = \int_{-\infty}^{\infty} H(f)e^{j2\pi ft}\, df \tag{2.5}$$

Inversion transformation, Eq. (2.5), allows the determination of a function of time from its Fourier transform. If the functions $h(t)$ and $H(f)$ are related by Eqs. (2.1) and (2.5), the two functions are termed a *Fourier transform pair*, and we indicate this relationship by the notation

$$h(t) \Longleftrightarrow H(f) \tag{2.6}$$

Example 2.2   Inverse Fourier Transform of Example 2.1

Consider the frequency function determined in the previous example:

$$H(f) = \frac{\beta}{\alpha + j2\pi f} = \frac{\beta\alpha}{\alpha^2 + (2\pi f)^2} - j\frac{2\pi f\beta}{\alpha^2 + (2\pi f)^2}$$

From Eq. (2.5),

$$h(t) = \int_{-\infty}^{\infty} \left[\frac{\beta\alpha}{\alpha^2 + (2\pi f)^2} - j\frac{2\pi f\beta}{\alpha^2 + (2\pi f)^2} \right] e^{j2\pi ft}\, df$$

Because $e^{j2\pi ft} = \cos(2\pi ft) + j\sin(2\pi ft)$, then

$$\begin{aligned}
h(t) = &\int_{-\infty}^{\infty} \left[\frac{\beta\alpha\,\cos(2\pi ft)}{\alpha^2 + (2\pi f)^2} + \frac{2\pi f\beta\,\sin(2\pi ft)}{\alpha^2 + (2\pi f)^2} \right] df \\
&+ j\int_{-\infty}^{\infty} \left[\frac{\beta\alpha\,\sin(2\pi ft)}{\alpha^2 + (2\pi f)^2} - \frac{2\pi f\beta\,\cos(2\pi ft)}{\alpha^2 + (2\pi f)^2} \right] df
\end{aligned} \tag{2.7}$$

The second integral of Eq. (2.7) is zero because each integrand term is an odd function. This point is clarified by examination of Fig. 2.2; the first integrand term in the second integral of Eq. (2.7) is illustrated. Note that the function is odd, that is,

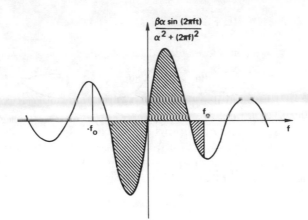

Figure 2.2   Integration of an odd function.

$g(t) = -g(-t)$. Consequently, the area under the function from $-f_0$ to f_0 is zero. Therefore, in the limit as f_0 approaches infinity, the integral of the function remains zero; the infinite integral of any odd function is zero.

Equation (2.7) becomes

$$h(t) = \frac{\beta\alpha}{(2\pi)^2} \int_{-\infty}^{\infty} \frac{\cos(2\pi t f)}{(\alpha/2\pi)^2 + f^2} \, df + \frac{2\pi\beta}{(2\pi)^2} \int_{-\infty}^{\infty} \frac{f \sin(2\pi t f)}{(\alpha/2\pi)^2 + f^2} \, df \qquad (2.8)$$

From a standard table of integrals,

$$\int_{-\infty}^{\infty} \frac{\cos(ax)}{b^2 + x^2} \, dx = \frac{\pi}{b} e^{-ab} \qquad a > 0$$

$$\int_{-\infty}^{\infty} \frac{x \sin(ax)}{b^2 + x^2} \, dx = \pi e^{-ab} \qquad a > 0$$

Hence, Eq. (2.8) can be written as

$$h(t) = \frac{\beta\alpha}{(2\pi)^2} \left[\frac{\pi}{(\alpha/2\pi)} e^{-(2\pi t)(\alpha/2\pi)} \right] + \frac{2\pi\beta}{(2\pi)^2} [\pi e^{-(2\pi t)(\alpha/2\pi)}]$$

$$= \frac{\beta}{2} e^{-\alpha t} + \frac{\beta}{2} e^{-\alpha t} = \beta e^{-\alpha t} \qquad t > 0$$

(2.9)

The time function

$$h(t) = \beta e^{-\alpha t} \qquad t > 0$$

and the frequency function

$$H(f) = \frac{\beta}{\alpha + j(2\pi f)}$$

are related by both Eqs. (2.1) and (2.5) and hence are a Fourier transform pair:

$$\beta e^{-\alpha t} \quad (t > 0) \quad \Longleftrightarrow \quad \frac{\beta}{\alpha + j(2\pi f)} \qquad (2.10)$$

2.3 EXISTENCE OF THE FOURIER INTEGRAL

To this point, we have not considered the validity of Eqs. (2.1) and (2.5); the integral equations have been assumed to be well-defined for all functions. In general, for most functions encountered in practical scientific analysis, the Fourier transform and its inverse are well-defined. We do not intend to present a highly theoretical discussion of the existence of the Fourier transform but rather to point out conditions for its existence and to give examples of these conditions. Our discussion follows that of Papoulis [3].

Condition 1.   If $h(t)$ is integrable in the sense

$$\int_{-\infty}^{\infty} |h(t)| \, dt < \infty \tag{2.11}$$

then its Fourier transform $H(f)$ exists and satisfies the inverse Fourier transform of Eq. (2.5).

It is important to note that Condition 1 is a sufficient but not necessary condition for the existence of a Fourier transform. There are functions that do not satisfy Condition 1 but have a transform satisfying Eq. (2.5). This class of functions is covered by Condition 2.

Example 2.3   Symmetrical Pulse Waveform

To illustrate Condition 1, consider the pulse time waveform

$$\begin{aligned}
h(t) &= A & |t| < T_0 \\
&= \frac{A}{2} & t = \pm T_0 \\
&= 0 & |t| > T_0
\end{aligned} \tag{2.12}$$

which is shown in Fig. 2.3. Equation (2.11) is satisfied for this function; therefore, the Fourier transform exists and is given by

$$H(f) = \int_{-T_0}^{T_0} A e^{-j2\pi ft} \, dt$$

$$= A \int_{-T_0}^{T_0} \cos(2\pi ft) \, dt - jA \int_{-T_0}^{T_0} \sin(2\pi ft) \, dt$$

The second integral is equal to zero because the integrand is odd:

$$\begin{aligned}
H(f) &= \frac{A}{2\pi f} \sin(2\pi ft) \Big|_{-T_0}^{T_0} \\
&= 2AT_0 \frac{\sin(2\pi T_0 f)}{2\pi T_0 f}
\end{aligned} \tag{2.13}$$

Those terms that obviously can be canceled are retained to emphasize the $[\sin(af)]/af$ characteristic of the Fourier transform of a pulse waveform, as shown in Fig. 2.3.

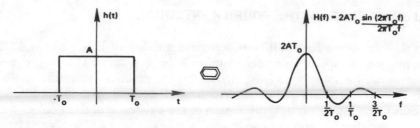

Figure 2.3 Fourier transform of a symmetrical-pulse time-domain waveform.

Because this example satisfies Condition 1, then $H(f)$ as given by Eq. (2.13) must satisfy Eq. (2.5).

$$h(t) = \int_{-\infty}^{\infty} 2AT_0 \frac{\sin(2\pi T_0 f)}{2\pi T_0 f} e^{j2\pi ft} \, df$$

$$= 2AT_0 \int_{-\infty}^{\infty} \frac{\sin(2\pi T_0 f)}{2\pi T_0 f} [\cos(2\pi ft) + j \sin(2\pi ft)] \, df$$

(2.14)

The imaginary integrand term is odd; therefore,

$$h(t) = \frac{A}{\pi} \int_{-\infty}^{\infty} \frac{\sin(2\pi T_0 f) \cos(2\pi ft)}{f} \, df \tag{2.15}$$

From the trigonometric identity

$$\sin(x) \cos(y) = \tfrac{1}{2}[\sin(x + y) + \sin(x - y)] \tag{2.16}$$

$h(t)$ becomes

$$h(t) = \frac{A}{2\pi} \int_{-\infty}^{\infty} \frac{\sin[2\pi f(T_0 + t)]}{f} \, df + \frac{A}{2\pi} \int_{-\infty}^{\infty} \frac{\sin[2\pi f(T_0 - t)]}{f} \, df$$

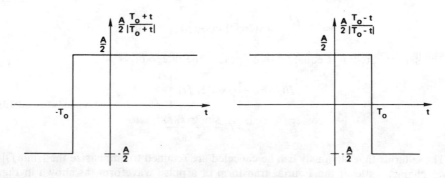

Figure 2.4 Graphical evaluation of Eq. (2.19).

and can be rewritten as

$$h(t) = A(T_0 + t) \int_{-\infty}^{\infty} \frac{\sin[2\pi f(T_0 + t)]}{2\pi f(T_0 + t)} \, df$$

$$+ A(T_0 - t) \int_{-\infty}^{\infty} \frac{\sin[2\pi f(T_0 - t)]}{2\pi f(T_0 - t)} \, df \qquad (2.17)$$

Because

$$\int_{-\infty}^{\infty} \frac{\sin(2\pi ax)}{2\pi ax} \, dx = \frac{1}{2|a|} \qquad (2.18)$$

($|\ |$ denotes magnitude or absolute value), then

$$h(t) = \frac{A}{2} \frac{T_0 + t}{|T_0 + t|} + \frac{A}{2} \frac{T_0 - t}{|T_0 - t|} \qquad (2.19)$$

Each term of Eq. (2.19) is illustrated in Fig. 2.4; by inspection, these terms add to yield

$$h(t) = A \qquad |t| < T_0$$

$$= \frac{A}{2} \qquad t = \pm T_0 \qquad (2.20)$$

$$= 0 \qquad |t| > T_0$$

The existence of the Fourier transform and the inverse Fourier transform has been demonstrated for a function satisfying Condition 1. We have established the Fourier transform pair (Fig. 2.3):

$$h(t) = A \quad (|t| < T_0) \quad \Longleftrightarrow \quad 2AT_0 \frac{\sin(2\pi T_0 f)}{2\pi T_0 f} \qquad (2.21)$$

Example 2.4   General Pulse Time Waveform

Consider the pulse time waveform

$$h(t) = A \qquad 0 < t < 2T_0$$

$$= \frac{A}{2} \qquad t = 0; t = 2T_0 \qquad (2.22)$$

$$= 0 \qquad \text{otherwise}$$

which is shown in Fig. 2.5(a). The Fourier transform is given by

$$H(f) = \int_0^{2T_0} Ae^{-j2\pi ft} \, dt$$

$$= A \int_0^{2T_0} \cos(2\pi ft) \, dt - jA \int_0^{2T_0} \sin(2\pi ft) \, dt$$

$$= (A/2\pi f) \sin(2\pi ft) \big|_0^{2T_0} + j(A/2\pi f) \cos(2\pi ft) \big|_0^{2T_0} \qquad (2.23)$$

$$= \frac{2AT_0 \sin[2\pi(2T_0)f]}{2\pi(2T_0)f} + j \left\{ 2AT_0 \frac{\cos[2\pi(2T_0)f]}{2\pi(2T_0)f} - \frac{2AT_0}{2\pi(2T_0)f} \right\}$$

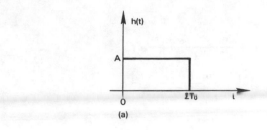

(a)

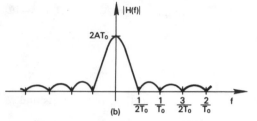

(b)

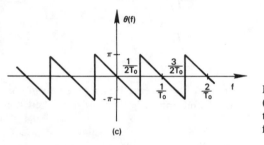

(c)

Figure 2.5 (a) General pulse waveform, (b) Fourier transform amplitude function, and (c) Fourier transform phase function.

The amplitude spectrum is given by

$$| H(f) | = \frac{2AT_0}{2\pi(2T_0)f} \{\sin^2[2\pi(2T_0)f] + \cos^2[2\pi(2T_0)f]$$

$$- 2 \cos[2\pi(2T_0)f] + 1\}^{1/2} \tag{2.24}$$

$$= \frac{2AT_0}{2\pi(2T_0)f} \{2 - 2 \cos[2\pi(2T_0)f]\}^{1/2}$$

$$= 2AT_0 \left| \frac{\sin[2\pi T_0 f]}{2\pi T_0 f} \right|$$

Because $\cos(x) - 1 = -\sin^2(x)$ and $\sin(x) = 2 \sin(x/2) \cos(x/2)$, then the phase angle is given by

$$\theta(f) = \tan^{-1} \left\{ \frac{\cos[2\pi(2T_0)f] - 1}{\sin[2\pi(2T_0)f]} \right\}$$

$$= \tan^{-1} \left\{ \frac{-\sin[2\pi T_0 f]}{\cos[2\pi T_0 f]} \right\} \tag{2.25}$$

$$= -2\pi T_0 f$$

The amplitude spectrum $|H(f)|$ and phase angle $\theta(f)$ of the Fourier transform of $h(t)$ are shown in Figs. 2.5(b) and (c), respectively. Note that the $\tan^{-1}(x)$ function is restricted to $-\pi < \theta < \pi$ by normal convention.

Condition 2.   If $h(t) = \beta(t) \sin(2\pi ft + \alpha)$ (where f and α are arbitrary constants), if $\beta(t + k) < \beta(t)$, and if for $|t| > \lambda > 0$, the function $h(t)/t$ is absolutely integrable in the sense of Eq. (2.11), then $H(f)$ exists and satisfies the inverse Fourier transform, Eq. (2.5).

An important example is the function $[\sin(at)]/at$, which does not satisfy the integrability requirements of Condition 1.

Example 2.5   Pulse Frequency Waveform

Consider the function

$$h(t) = 2Af_0 \frac{\sin(2\pi f_0 t)}{2\pi f_0 t} \tag{2.26}$$

illustrated in Fig. 2.6. From Condition 2, the Fourier transform of $h(t)$ exists and is given by

$$H(f) = \int_{-\infty}^{\infty} 2Af_0 \frac{\sin(2\pi f_0 t)}{2\pi f_0 t} e^{-j2\pi ft} \, dt$$

$$= \frac{A}{\pi} \int_{-\infty}^{\infty} \frac{\sin(2\pi f_0 t)}{t} [\cos(2\pi ft) - j \sin(2\pi ft)] \, dt \tag{2.27}$$

$$= \frac{A}{\pi} \int_{-\infty}^{\infty} \frac{\sin(2\pi f_0 t) \cos(2\pi ft)}{t} \, dt$$

The imaginary term integrates to zero because the integrand term is an odd function. Substitution of the trigonometric identity of Eq. (2.16) gives

$$H(f) = \frac{A}{2\pi} \int_{-\infty}^{\infty} \frac{\sin[2\pi t(f_0 + f)]}{t} \, dt + \frac{A}{2\pi} \int_{-\infty}^{\infty} \frac{\sin[2\pi t(f_0 - f)]}{t} \, dt$$

$$= A(f_0 + f) \int_{-\infty}^{\infty} \frac{\sin[2\pi t(f_0 + f)]}{2\pi t(f_0 + f)} \, dt \tag{2.28}$$

$$+ A(f_0 - f) \int_{-\infty}^{\infty} \frac{\sin[2\pi t(f_0 - f)]}{2\pi t(f_0 - f)} \, dt$$

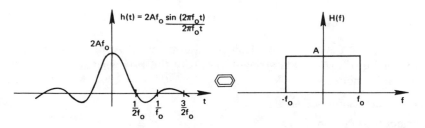

Figure 2.6   Fourier transform of $A [\sin(at)/at]$.

Equation (2.28) is of the same form as Eq. (2.17); identical analysis techniques yield

$$H(f) = A \qquad |f| < f_0$$

$$= \frac{A}{2} \qquad f = \pm f_0 \tag{2.29}$$

$$= 0 \qquad |f| > f_0$$

Because this example satisfies Condition 2, $H(f)$ [Eq. (2.29)], must satisfy the inverse Fourier transform relationship, Eq. (2.5):

$$h(t) = \int_{-f_0}^{f_0} Ae^{j2\pi ft} \, df$$

$$= A \int_{-f_0}^{f_0} \cos(2\pi ft) \, df = A \left. \frac{\sin(2\pi ft)}{2\pi t} \right|_{-f_0}^{f_0} \tag{2.30}$$

$$= 2Af_0 \frac{\sin(2\pi f_0 t)}{2\pi f_0 t}$$

By means of Condition 2, the Fourier transform pair

$$2Af_0 \frac{\sin(2\pi f_0 t)}{2\pi f_0 t} \iff H(f) = A \qquad |f| < f_0 \tag{2.31}$$

has been established and is illustrated in Fig. 2.6.

Condition 3. Although not specifically stated, all functions for which Conditions 1 and 2 hold are assumed to be of *bounded variation*, that is, they can be represented by a curve of finite height in any finite time interval. By means of Condition 3, we extend the theory to include singular (impulse) functions.

If $h(t)$ is a periodic or impulse function, then $H(f)$ exists only if one introduces the theory of distributions. Appendix A has an elementary discussion of distribution theory; with the aid of this development, the Fourier transform of singular functions can be defined. It is important to develop the Fourier transform of impulse functions because their use greatly simplifies the derivation of many transform pairs.

Impulse function $\delta(t)$ is defined as [Eq. (A.8)]

$$\int_{-\infty}^{\infty} \delta(t - t_0)x(t) \, dt = x(t_0) \tag{2.32}$$

where $x(t)$ is an arbitrary function continuous at t_0. Application of the definition of Eq. (2.32) yields straightforwardly the Fourier transform of many important functions.

Example 2.6   Impulse Function

Consider the function

$$h(t) = K\delta(t) \tag{2.33}$$

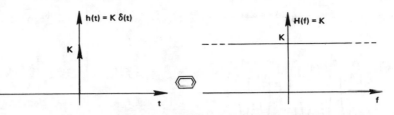

Figure 2.7 Fourier transform of an impulse function.

The Fourier transform of $h(t)$ is easily derived using the definition of Eq. (2.32):

$$H(f) = \int_{-\infty}^{\infty} K\delta(t)e^{-j2\pi ft}\, dt = Ke^0 = K \tag{2.34}$$

The inverse Fourier transform of $H(f)$ is given by

$$h(t) = \int_{-\infty}^{\infty} [K]e^{j2\pi ft}\, df = \int_{-\infty}^{\infty} K\cos(2\pi ft)\, df + j\int_{-\infty}^{\infty} K\sin(2\pi ft)\, df \tag{2.35}$$

Because the integrand of the second integral is an odd function, the integral is zero; the first integral is meaningless unless it is interpreted in the sense of distribution theory. From Eq. (A.21), Eq. (2.35) exists and can be rewritten as

$$h(t) = K \int_{-\infty}^{\infty} e^{j2\pi ft}\, df = K\int_{-\infty}^{\infty} \cos(2\pi ft)\, df = K\delta(t) \tag{2.36}$$

These results establish the Fourier transform pair

$$K\delta(t) \quad \diamondsuit \quad H(f) = K \tag{2.37}$$

which is illustrated in Fig. 2.7.

Similarly, the Fourier transform pair, as shown in Fig. 2.8,

$$h(t) = K \quad \diamondsuit \quad K\delta(f) \tag{2.38}$$

can be established where the reasoning process concerning existence is exactly as argued previously.

Example 2.7   Periodic Functions

To illustrate the Fourier transform of periodic functions, consider

$$h(t) = A\cos(2\pi f_0 t) \tag{2.39}$$

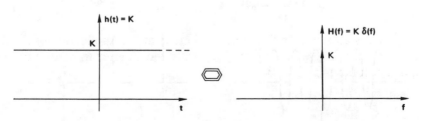

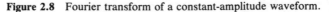

Figure 2.8 Fourier transform of a constant-amplitude waveform.

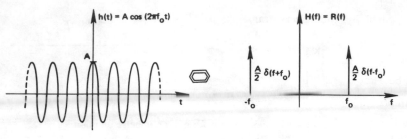

Figure 2.9   Fourier transform of $A \cos(at)$.

The Fourier transform is given by

$$
\begin{aligned}
H(f) &= \int_{-\infty}^{\infty} A \cos(2\pi f_0 t) e^{-j2\pi ft} \, dt \\
&= \frac{A}{2} \int_{-\infty}^{\infty} [e^{j2\pi f_0 t} + e^{-j2\pi f_0 t}] e^{-j2\pi ft} \, dt \\
&= \frac{A}{2} \int_{-\infty}^{\infty} [e^{-j2\pi t(f - f_0)} + e^{-j2\pi t(f_0 + f)}] \, dt \\
&= \frac{A}{2}\delta(f - f_0) + \frac{A}{2}\delta(f + f_0)
\end{aligned}
\tag{2.40}
$$

where arguments identical to those leading to Eq. (2.36) have been employed. The inversion formula yields

$$
\begin{aligned}
h(t) &= \int_{-\infty}^{\infty} \left[\frac{A}{2}\delta(f + f_0) + \frac{A}{2}\delta(f - f_0) \right] e^{j2\pi ft} \, df \\
&= \frac{A}{2} e^{j2\pi f_0 t} + \frac{A}{2} e^{-j2\pi f_0 t} \\
&= A \cos(2\pi f_0 t)
\end{aligned}
\tag{2.41}
$$

The Fourier transform pair

$$
A \cos(2\pi f_0 t) \quad \Longleftrightarrow \quad \frac{A}{2}\delta(f - f_0) + \frac{A}{2}\delta(f + f_0)
\tag{2.42}
$$

is illustrated in Fig. 2.9.

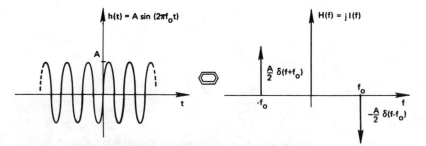

Figure 2.10   Fourier transform of $A \sin(at)$.

Similarly, the Fourier transform pair (Fig. 2.10)

$$A \sin(2\pi f_0 t) \quad \Longleftrightarrow \quad j\frac{A}{2}\delta(f + f_0) - j\frac{A}{2}\delta(f - f_0) \qquad (2.43)$$

can be established. Note that the Fourier transform is imaginary.

Example 2.8   Sequence of Impulse Functions

The Fourier transform of a sequence of equidistant impulse functions is another sequence of equidistant impulses [3].

$$h(t) = \sum_{n=-\infty}^{\infty} \delta(t - nT) \quad \Longleftrightarrow \quad H(f) = \frac{1}{T} \sum_{n=-\infty}^{\infty} \delta\left(f - \frac{n}{T}\right) \qquad (2.44)$$

A graphical development of this Fourier transform pair is illustrated in Fig. 2.11.

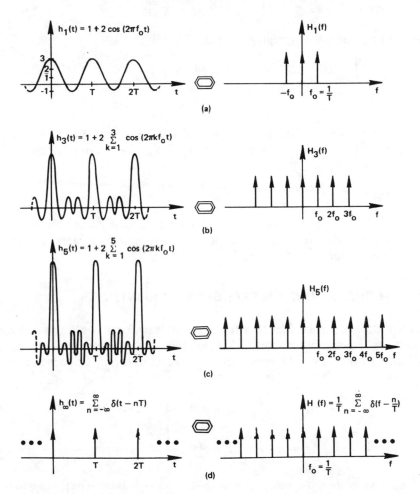

Figure 2.11 Graphical development of the Fourier transform of a sequence of equidistant impulse functions.

The importance of the Fourier transform pair of Eq. (2.44) becomes obvious in future discussions of discrete Fourier transforms.

Inversion Formula Proof

By means of distribution theory concepts, it is possible to derive a simple formal proof of the inversion formula of Eq. (2.5).

Substitution of $H(f)$ [Eq. (2.1)] into the inverse Fourier transform of Eq. (2.5) yields

$$\int_{-\infty}^{\infty} H(f)e^{j2\pi ft}\, df = \int_{-\infty}^{\infty} e^{j2\pi ft}\, df \int_{-\infty}^{\infty} h(x)e^{-j2\pi fx}\, dx \qquad (2.45)$$

Because [Eq. (A.21)]

$$\int_{-\infty}^{\infty} e^{j2\pi ft}\, dt = \delta(t)$$

then an interchange of integration in Eq. (2.45) gives

$$\int_{-\infty}^{\infty} H(f)e^{j2\pi ft}\, df = \int_{-\infty}^{\infty} h(x)\, dx \int_{-\infty}^{\infty} e^{j2\pi f(t-x)}\, df$$

$$= \int_{-\infty}^{\infty} h(x)\delta(t - x)\, dx \qquad (2.46)$$

But by the definition of the impulse function of Eq. (2.32), Eq. (2.46) simply equals $h(t)$. This statement is valid only if $h(t)$ is continuous.[1] However, if it is assumed that

$$h(t) = \frac{h(t^+) + h(t^-)}{2} \qquad (2.47)$$

that is, if $h(t)$ is defined as the midvalue at a discontinuity, then the inversion formula still holds. Note that in the previous examples we carefully defined each discontinuous function consistent with Eq. (2.47).

2.4 ALTERNATE FOURIER TRANSFORM DEFINITIONS

It is a well-established fact that the Fourier transform is a universally accepted tool of modern analysis. Yet, to this day, there is not a common definition of the Fourier integral and its inversion formula. To be specific, the Fourier transform pair is defined as

$$H(\omega) = a_1 \int_{-\infty}^{\infty} h(t)e^{-j\omega t}\, dt \qquad \omega = 2\pi f \qquad (2.48)$$

$$h(t) = a_2 \int_{-\infty}^{\infty} H(\omega)e^{j\omega t}\, d\omega \qquad (2.49)$$

---

[1] See Appendix A. The definition of the impulse response is based on the continuity of the testing function $h(t)$.

where the coefficients a_1 and a_2 assume different values depending on the user. Some set $a_1 = 1$ and $a_2 = 1/2\pi$; others set $a_1 = a_2 = 1/\sqrt{2\pi}$; or set $a_1 = 1/2\pi$ and $a_2 = 1$. Equations (2.48) and (2.49) impose the requirement that $a_1 a_2 = 1/2\pi$. Various users are then concerned with the splitting of the product $a_1 a_2$.

To resolve this question, we must define the relationship desired between the Fourier transform and the Laplace transform and the definition we wish to assume for the relationship between the total energy computed in the time domain and the total energy computed in ω, the radian frequency domain. For example, Parseval's Theorem states:

$$\int_{-\infty}^{\infty} h^2(t)\, dt = 2\pi a_1^2 \int_{-\infty}^{\infty} |H(\omega)|^2\, d\omega \qquad (2.50)$$

If the energy computed in t is required to be equal to the energy computed in ω, then $a_1 = 1/\sqrt{2\pi}$. However, if the requirement is made that the Laplace transform, universally defined as

$$L[h(t)] = \int_{-\infty}^{\infty} h(t)e^{-st}\, dt = \int_{-\infty}^{\infty} h(t)e^{-(\alpha+j\omega)t}\, dt \qquad (2.51)$$

shall reduce to the Fourier transform when the real part of s is set to zero, then a comparison of Eqs. (2.48) and (2.51) requires $a_1 = 1$, i.e., $a_2 = 1/2\pi$, which is in contradiction to the previous hypothesis.

A logical way to resolve this conflict is to define the Fourier transform pair as follows:

$$H(f) = \int_{-\infty}^{\infty} h(t)e^{-j2\pi ft}\, dt \qquad (2.52)$$

$$h(t) = \int_{-\infty}^{\infty} H(f)e^{j2\pi ft}\, df \qquad (2.53)$$

With this definition, Parseval's Theorem becomes

$$\int_{-\infty}^{\infty} h^2(t)\, dt = \int_{-\infty}^{\infty} |H(f)|^2\, df$$

and Eq. (2.52) is consistent with the definition of the Laplace transform. Note that as long as integration is with respect to f, the scale factor $1/2\pi$ never appears. For this reason, the latter definition of the Fourier transform pair was chosen for this book.

2.5 FOURIER TRANSFORM PAIRS

A pictorial table of Fourier transform pairs is given in Fig. 2.12. This graphical and analytical catalog is by no means complete, but does contain the most frequently encountered transform pairs.

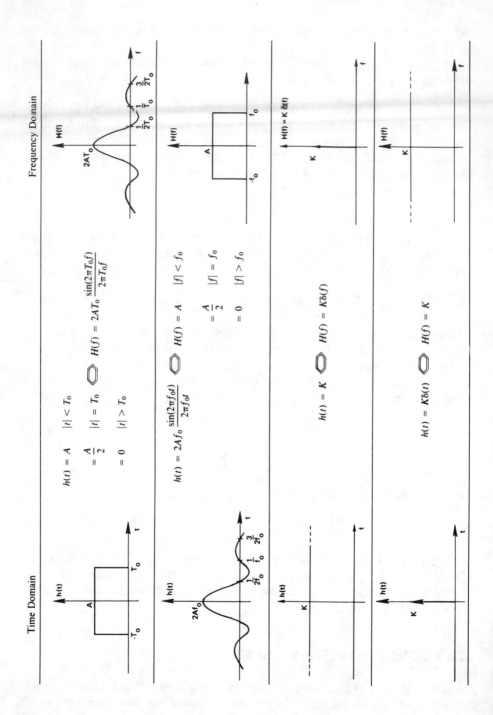

Time Domain

$h(t) = A \quad |t| < T_0$

$\quad = \dfrac{A}{2} \quad |t| = T_0$

$\quad = 0 \quad |t| > T_0$

Frequency Domain

$H(f) = 2AT_0 \dfrac{\sin(2\pi T_0 f)}{2\pi T_0 f}$

$h(t) = 2Af_0 \dfrac{\sin(2\pi f_0 t)}{2\pi f_0 t}$

$H(f) = A \quad |f| < f_0$

$\quad = \dfrac{A}{2} \quad |f| = f_0$

$\quad = 0 \quad |f| > f_0$

$h(t) = K \qquad H(f) = K\delta(f)$

$h(t) = K\delta(t) \qquad H(f) = K$

24

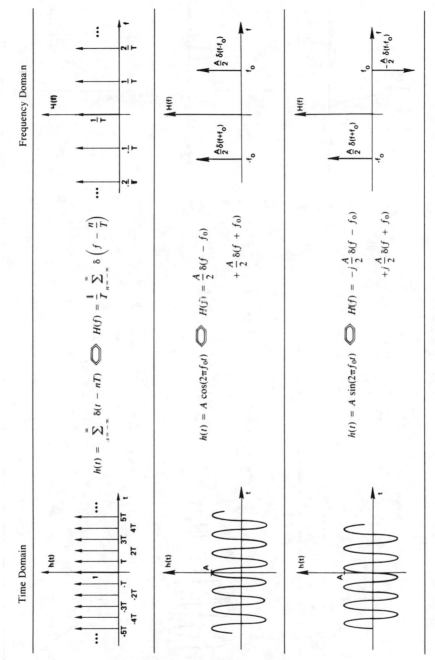

Figure 2.12 Catalog of Fourier transform pairs.

$$h(t) = \sum_{n=-\infty}^{\infty} \delta(t - nT) \Longleftrightarrow H(f) = \frac{1}{T} \sum_{n=-\infty}^{\infty} \delta\left(f - \frac{n}{T}\right)$$

$$h(t) = A\cos(2\pi f_0 t) \Longleftrightarrow H(f) = \frac{A}{2}\delta(f - f_0) + \frac{A}{2}\delta(f + f_0)$$

$$h(t) = A\sin(2\pi f_0 t) \Longleftrightarrow H(f) = -j\frac{A}{2}\delta(f - f_0) + j\frac{A}{2}\delta(f + f_0)$$

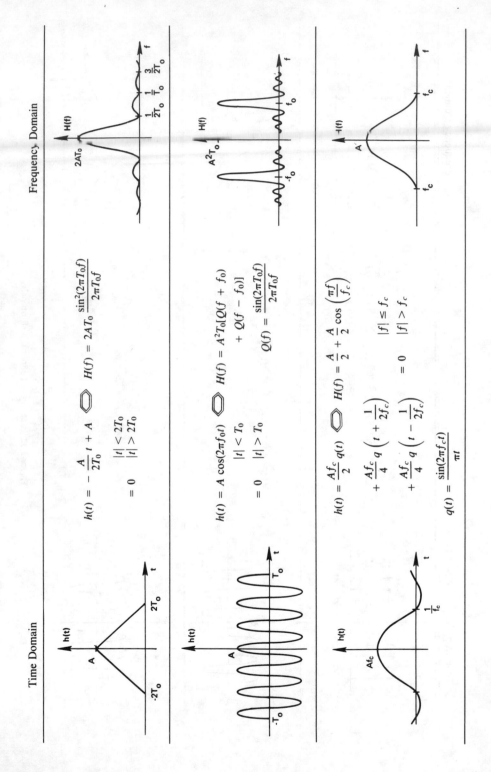

$$h(t) = -\frac{A}{2T_0}t + A$$
$$= 0 \qquad \begin{array}{l}|t| < 2T_0 \\ |t| > 2T_0\end{array}$$
$$H(f) = 2AT_0\frac{\sin^2(2\pi T_0 f)}{2\pi T_0 f}$$

$$h(t) = A\cos(2\pi f_0 t) \qquad |t| < T_0$$
$$= 0 \qquad |t| > T_0$$
$$H(f) = A^2 T_0[Q(f+f_0) + Q(f-f_0)]$$
$$Q(f) = \frac{\sin(2\pi T_0 f)}{2\pi T_0 f}$$

$$h(t) = \frac{Af_c}{2}q(t)$$
$$+ \frac{Af_c}{4}q\left(t+\frac{1}{2f_c}\right)$$
$$+ \frac{Af_c}{4}q\left(t-\frac{1}{2f_c}\right)$$
$$q(t) = \frac{\sin(2\pi f_c t)}{\pi t}$$
$$H(f) = \frac{A}{2} + \frac{A}{2}\cos\left(\frac{\pi f}{f_c}\right) \qquad |f| \le f_c$$
$$= 0 \qquad |f| > f_c$$

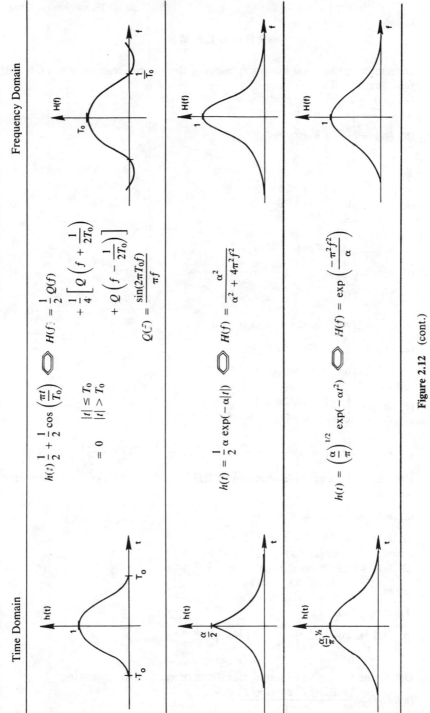

Figure 2.12 (cont.)

27

PROBLEMS

2.1. Determine the real and imaginary parts of the Fourier transform of each of the following functions:

(a) $h(t) = e^{-a|t|} \quad -\infty < t < \infty$

(b) $h(t) = \begin{cases} k & t > 0 \\ \dfrac{k}{2} & t = 0 \\ 0 & t < 0 \end{cases}$

(c) $h(t) = \begin{cases} -A & t < 0 \\ 0 & t = 0 \\ A & t > 0 \end{cases}$

(d) $h(t) = \begin{cases} A \cos(2\pi f_0 t) & t > 0 \\ \dfrac{A}{2} & t = 0 \\ 0 & t < 0 \end{cases}$

(e) $h(t) = \begin{cases} A & a < t < b;\ a,\ b > 0 \\ \dfrac{A}{2} & t = a;\ t = b \\ 0 & \text{elsewhere} \end{cases}$

(f) $h(t) = \begin{cases} Ae^{-\alpha t} \sin(2\pi f_0 t) & t \geq 0 \\ 0 & t < 0 \end{cases}$

(g) $h(t) = \dfrac{1}{2}\left[\delta(t + a) + \delta(t - a) + \delta\left(t + \dfrac{a}{2}\right) + \delta\left(t - \dfrac{a}{2}\right) \right]$

2.2. Determine the amplitude spectrum $|H(f)|$ and phase $\theta(f)$ of the Fourier transform of $h(t)$:

(a) $h(t) = \dfrac{1}{t} \quad -\infty < t < \infty$

(b) $h(t) = e^{-\pi t^2} \quad -\infty < t < \infty$

(c) $h(t) = A \sin(2\pi f_0 t) \quad 0 \leq t < \infty$

(d) $h(t) = Ae^{-\alpha t} \cos(2\pi f_0 t) \quad 0 \leq t < \infty$

(e) $h(t) = \begin{cases} At & 0 < t < T_0 \\ 0 & \text{elsewhere} \end{cases}$

(f) $h(t) = \cos^2(2\pi f_0 t)$

(g) $h(t) = \cos(2\pi f_0 t) \qquad |t| \leq \dfrac{4}{f_0}$

$\qquad = 0 \qquad\qquad\qquad \text{otherwise}$

2.3. Determine the inverse Fourier transform of each of the following:

(a) $H(f) = \dfrac{\sin(2\pi fT)\cos(2\pi fT)}{2\pi f}$

(b) $H(f) = (1 - f^2)^2 \qquad |f| < 1$

$\qquad = 0 \qquad\qquad\qquad \text{otherwise}$

(c) $H(f) = \dfrac{f}{(f^2 + \alpha)(f^2 + 4\alpha)}$

(d) $H(f) = A \cos(2\pi f t_0)$

REFERENCES

1. ARSAC, J. *Fourier Transforms and the Theory of Distributions*. Englewood Cliffs, NJ: Prentice Hall, 1966.

2. BRACEWELL, R. *The Fourier Transform and its Applications*, 2d Ed. New York: McGraw-Hill, 1986.

3. PAPOULIS, A. *The Fourier Integral and Its Applications*, 2d Ed. New York: McGraw-Hill, 1984.

4. CHAMPENEY, D. C. *Fourier Transforms and Their Physical Application*. New York: Academic Press, 1973.

3

FOURIER TRANSFORM
PROPERTIES

In dealing with Fourier transforms, there are a few properties that are basic to a thorough understanding. A visual interpretation of these fundamental properties is of equal importance to knowledge of their mathematical relationships. The purpose of this chapter is to develop not only the theoretical concepts of the basic Fourier transform pairs, but also the *meaning* of these properties. For this reason, we use ample analytical and graphical examples.

3.1 LINEARITY

If $x(t)$ and $y(t)$ have the Fourier transforms $X(f)$ and $Y(f)$, respectively, then the sum $x(t) + y(t)$ has the Fourier transform $X(f) + Y(f)$. This property is established as follows:

$$\int_{-\infty}^{\infty} [x(t) + y(t)]e^{-j2\pi ft}\, dt = \int_{-\infty}^{\infty} x(t)e^{-j2\pi ft}\, dt$$

$$+ \int_{-\infty}^{\infty} y(t)e^{-j2\pi ft}\, dt \qquad (3.1)$$

$$= X(f) + Y(f)$$

The Fourier transform pair

$$x(t) + y(t) \quad \Longleftrightarrow \quad X(f) + Y(f) \qquad (3.2)$$

is of considerable importance because it reflects the applicability of the Fourier transform to linear-system analysis.

Example 3.1   Addition of a Constant and a Sinusoid

To illustrate the linearity property, consider the Fourier transform pairs

$$x(t) = K \quad \Longleftrightarrow \quad X(f) = K\delta(f) \tag{3.3}$$

$$y(t) = A \cos(2\pi f_0 t) \quad \Longleftrightarrow \quad Y(f) = \frac{A}{2}\delta(f - f_0) + \frac{A}{2}\delta(f + f_0) \tag{3.4}$$

By the linearity theorem,

$$x(t) + y(t) = K + A\cos(2\pi f_0 t) \quad \Longleftrightarrow \quad X(f) + Y(f) = K\delta(f) + \frac{A}{2}\delta(f - f_0)$$
$$+ \frac{A}{2}\delta(f + f_0) \tag{3.5}$$

Figures 3.1(a), (b), and (c) illustrate each of the Fourier transform pairs, respectively.

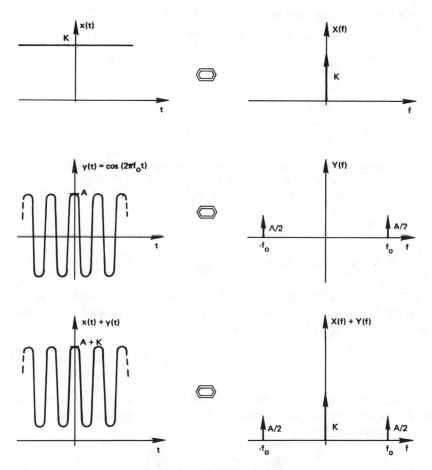

Figure 3.1   The linearity property.

3.2 SYMMETRY

If $h(t)$ and $H(f)$ are a Fourier transform pair, then

$$H(t) \iff h(-f) \tag{3.6}$$

The Fourier transform pair of Eq. (3.6) is established by rewriting Eq. (2.5).

$$h(-t) = \int_{-\infty}^{\infty} H(f)e^{-j2\pi ft} \, df \tag{3.7}$$

and by interchanging the parameters t and f:

$$h(-f) = \int_{-\infty}^{\infty} H(t)e^{-j2\pi ft} \, dt \tag{3.8}$$

Example 3.2 Pulse Time and Frequency Waveforms

To illustrate this property, consider the Fourier transform pair:

$$h(t) = A \quad (\,|\,t\,|\, < T_0) \iff \frac{2AT_0 \sin(2\pi T_0 f)}{2\pi T_0 f} \tag{3.9}$$

illustrated previously in Fig. 2.3. By the symmetry theorem,

$$2AT_0 \frac{\sin(2\pi T_0 t)}{2\pi T_0 t} \iff h(-f) = h(f) = A \quad |\,f\,| < T_0 \tag{3.10}$$

which is identical to the Fourier transform pair of Eq. (2.31) illustrated in Fig. 2.6. Utilization of the symmetry theorem can eliminate many complicated mathematical developments; a case in point is the development of the Fourier transform pair of Eq. (2.31).

3.3 TIME AND FREQUENCY SCALING

If the Fourier transform of $h(t)$ is $H(f)$, then the Fourier transform of $h(kt)$, where k is a real constant greater than zero, is determined by substituting $t' = kt$ in the Fourier integral equation:

$$\int_{-\infty}^{\infty} h(kt)e^{-j2\pi ft} \, dt = \int_{-\infty}^{\infty} h(t')e^{-j2\pi t'\,(f/k)} \frac{dt'}{k} = \frac{1}{k} H\left(\frac{f}{k}\right) \tag{3.11}$$

For k negative, the term on the right-hand side changes sign because the limits of integration are interchanged. Therefore, time scaling results in the Fourier transform pair:

$$h(kt) \iff \frac{1}{|\,k\,|} H\left(\frac{f}{k}\right) \tag{3.12}$$

When dealing with time scaling of impulses, extra care must be exercised; from Eq. (A.10),

$$\delta(at) = \frac{1}{|a|}\delta(t) \tag{3.13}$$

Example 3.3   Time Scale Expansion

The time-scaling Fourier transform property is well-known in many fields of scientific endeavor. As shown in Fig. 3.2, time scale expansion corresponds to frequency scale compression. Note that as the time scale expands, the frequency scale not only contracts, but the amplitude increases vertically in such a way as to keep the area constant. This is a well-known concept in radar and antenna theory.

Frequency Scaling

If the inverse Fourier transform of $H(f)$ is $h(t)$, the inverse Fourier transform of $H(kf)$, where k is a real constant, is given by the Fourier transform pair:

$$\frac{1}{|k|}h\left(\frac{t}{k}\right) \iff H(kf) \tag{3.14}$$

The relationship of Eq. (3.14) is established by substituting $f' = kf$ into the inversion formula:

$$\int_{-\infty}^{\infty} H(kf)e^{j2\pi ft}\, df = \int_{-\infty}^{\infty} H(f')e^{j2\pi f'(t/k)}\frac{df'}{k} = \frac{1}{|k|}h\left(\frac{t}{k}\right) \tag{3.15}$$

Frequency scaling of impulse functions is given by

$$\delta(af) = \frac{1}{|a|}\delta(f) \tag{3.16}$$

Example 3.4   Frequency-Scale Expansion

Analogous to time scaling, frequency-scale expansion results in a contraction of the time scale. This effect is illustrated in Fig. 3.3. Note that as the frequency scale expands, the amplitude of the time function increases. This is simply a reflection of the symmetry property of Eq. (3.6) and the time-scaling relationship of Eq. (3.12).

Example 3.5   Infinite Sequence of Impulse Functions

Many texts state Fourier transform pairs in terms of the radian frequency ω. For example, Papoulis [2] gives

$$h(t) = \sum_{n=-\infty}^{\infty} \delta(t - nT) \iff H(\omega) = \frac{2\pi}{T}\sum_{n=-\infty}^{\infty}\delta\left(\omega - \frac{2n\pi}{T}\right) \tag{3.17}$$

By the frequency-scaling relationship of Eq. (3.16), we know that

$$\frac{2\pi}{T}\sum_{n=-\infty}^{\infty}\delta\left[2\pi\left(f - \frac{n}{T}\right)\right] = \frac{1}{T}\sum_{n=-\infty}^{\infty}\delta\left(f - \frac{n}{T}\right) \tag{3.18}$$

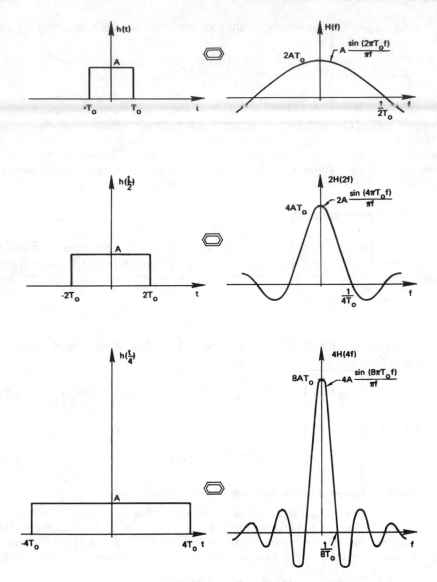

Figure 3.2 Time-scaling property.

and Eq. (3.17) can be rewritten in terms of the frequency variable f:

$$h(t) = \sum_{n=-\infty}^{\infty} \delta(t - nT) \quad\Longleftrightarrow\quad H(f) = \frac{1}{T} \sum_{n=-\infty}^{\infty} \delta\left(f - \frac{n}{T}\right) \tag{3.19}$$

which is Eq. (2.44).

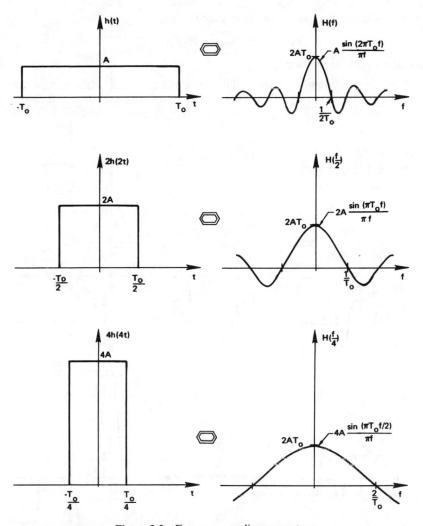

Figure 3.3 Frequency-scaling property.

3.4 TIME AND FREQUENCY SHIFTING

If $h(t)$ is shifted by a constant t_0, then by substituting $s = t - t_0$, the Fourier transform becomes

$$\int_{-\infty}^{\infty} h(t - t_0)e^{-j2\pi ft}\, dt = \int_{-\infty}^{\infty} h(s)e^{-j2\pi f(s + t_0)}\, ds$$

$$= e^{-j2\pi f t_0} \int_{-\infty}^{\infty} h(s)e^{-j2\pi fs}\, ds \qquad (3.20)$$

$$= e^{-j2\pi f t_0}H(f)$$

The time-shifted Fourier transform pair is

$$h(t - t_0) \Longleftrightarrow H(f)e^{-j2\pi f t_0} \tag{3.21}$$

Example 3.6   Phase Shifting

A pictorial description of this pair is illustrated in Fig. 3.4. As shown, time shifting results in a change in the phase angle $\theta(f) = \tan^{-1}[I(f)/R(f)]$. Note that time shifting

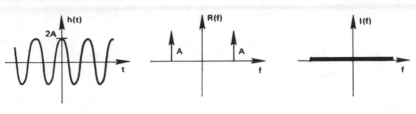

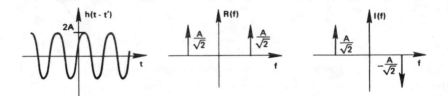

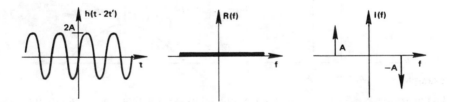

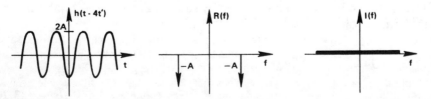

Figure 3.4   Time-shifting property.

does not alter the magnitude of the Fourier transform. This follows because

$$H(f)e^{-j2\pi ft_0} = H(f)[\cos(2\pi ft_0) - j\sin(2\pi ft_0)]$$

and hence the magnitude is given by

$$| H(f)e^{-j2\pi ft_0} | = \sqrt{H^2(f)[\cos^2(2\pi ft_0) + \sin^2(2\pi ft_0)]} = \sqrt{H^2(f)} \qquad (3.22)$$

where $H(f)$ has been assumed to be real for simplicity. These results are easily extended to the case of $H(f)$, a complex function.

Frequency Shifting

If $H(f)$ is shifted by a constant f_0, its inverse transform is multiplied by $e^{j2\pi tf_0}$

$$h(t)e^{j2\pi tf_0} \iff H(f - f_0) \qquad (3.23)$$

This Fourier transform pair is established by substituting $s = f - f_0$ into the inverse Fourier transform-defining relationship:

$$\int_{-\infty}^{\infty} H(f - f_0)e^{j2\pi ft}\, df = \int_{-\infty}^{\infty} H(s)e^{j2\pi t(s+f_0)}\, ds$$

$$= e^{j2\pi tf_0} \int_{-\infty}^{\infty} H(s)e^{j2\pi st}\, ds \qquad (3.24)$$

$$= e^{j2\pi tf_0}\, h(t)$$

Example 3.7   Modulation

To illustrate the effect of frequency shifting, let us assume that the frequency function $H(f)$ is real. For this case, frequency shifting results in a multiplication of the time function $h(t)$ by a cosine whose frequency is determined by the frequency shift f_0 (Fig. 3.5). This process is commonly known as *modulation*.

Example 3.8   Down Conversion by Frequency Multiplication

A practical application of frequency shifting is illustrated in Fig. 3.6. Multiplication of a sinusoid of frequency $2f_0$ with another sinusoid of frequency $3f_0$ results in two sinusoids. One sinusoid has a frequency that is the sum of the frequencies of the multiplied sinusoids, that is, $5f_0$. The second sinusoid has a frequency determined by the difference of the two frequencies, f_0. This difference-frequency sinusoid is commonly referred to as the *down conversion* term or component.

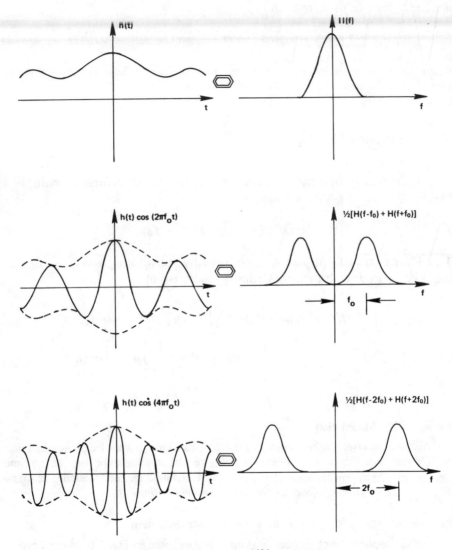

Figure 3.5 Frequency-shifting property.

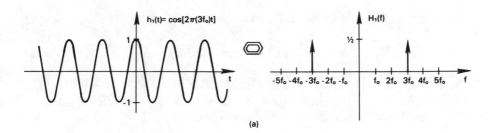

(a)

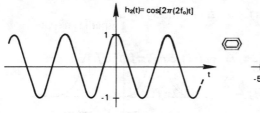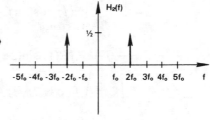

(b)

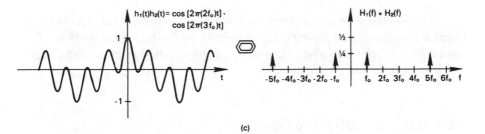

(c)

Figure 3.6 Examples of sum and difference frequencies produced by frequency multiplication.

3.5 ALTERNATE INVERSION FORMULA

The inversion formula of Eq. (2.5) can also be written as

$$h(t) = \left[\int_{-\infty}^{\infty} H^*(f) e^{-j2\pi ft} \, df \right]^* \tag{3.25}$$

where $H^*(f)$ is the conjugate of $H(f)$; that is, if $H(f) = R(f) + jI(f)$, then $H^*(f) = R(f) - jI(f)$. The relationship of Eq. (3.25) is verified by simply performing the conjugation operations indicated.

$$
\begin{aligned}
h(t) &= \left[\int_{-\infty}^{\infty} H^*(f) e^{-j2\pi ft} \, df \right]^* \\
&= \left[\int_{-\infty}^{\infty} R(f) e^{-j2\pi ft} \, df - j \int_{-\infty}^{\infty} I(f) e^{-j2\pi ft} \, df \right]^* \\
&= \left[\int_{-\infty}^{\infty} [R(f) \cos(2\pi ft) - I(f) \sin(2\pi ft)] \, df \right. \\
&\qquad \left. - j \int_{-\infty}^{\infty} [R(f) \sin(2\pi ft) + I(f) \cos(2\pi ft)] \, df \right]^* \\
&= \int_{-\infty}^{\infty} [R(f) \cos(2\pi ft) - I(f) \sin(2\pi ft)] \, df \\
&\qquad + j \int_{-\infty}^{\infty} [R(f) \sin(2\pi ft) + I(f) \cos(2\pi ft)] \, df \\
&= \int_{-\infty}^{\infty} [R(f) + jI(f)][\cos(2\pi ft) + j \sin(2\pi ft)] \, df \\
&= \int_{-\infty}^{\infty} H(f) e^{j2\pi ft} \, df
\end{aligned}
\tag{3.26}
$$

The significance of the alternate inversion formula is that now both the Fourier transform and its inverse contain the common term $e^{-j2\pi ft}$. This similarity is of considerable importance in the development of fast Fourier transform computer programs.

3.6 EVEN AND ODD FUNCTIONS

If $h_e(t)$ is an even function, that is, $h_e(t) = h_e(-t)$, then the Fourier transform of $h_e(t)$ is an even function and is real:

$$h_e(t) \quad \Diamond \quad R_e(f) = \int_{-\infty}^{\infty} h_e(t) \cos(2\pi ft) \, dt \tag{3.27}$$

This pair is established by manipulating the defining relationships:

$$H(f) = \int_{-\infty}^{\infty} h_e(t)e^{-j2\pi ft}\, dt$$

$$= \int_{-\infty}^{\infty} h_e(t)\cos(2\pi ft)\, dt - j\int_{-\infty}^{\infty} h_e(t)\sin(2\pi ft)\, dt \qquad (3.28)$$

$$= \int_{-\infty}^{\infty} h_e(t)\cos(2\pi ft)\, dt = R_e(f)$$

The imaginary term is zero because the integrand is an odd function. Because $\cos(2\pi ft)$ is an even function, then $h_e(t)\cos(2\pi ft) = h_e(t)\cos[2\pi(-f)t]$ and $H_e(f) = H_e(-f)$; the frequency function is even. Similarly, if $H(f)$ is given as a real and even frequency function, the inversion formula yields

$$h(t) = \int_{-\infty}^{\infty} H_e(f)e^{j2\pi ft}\, dt = \int_{-\infty}^{\infty} R_e(f)e^{j2\pi ft}\, df$$

$$= \int_{-\infty}^{\infty} R_e(f)\cos(2\pi ft)\, df + j\int_{-\infty}^{\infty} R_e(f)\sin(2\pi ft)\, df \qquad (3.29)$$

$$= \int_{-\infty}^{\infty} R_e(f)\cos(2\pi ft)\, df = h_e(t)$$

Example 3.9   Even Time and Frequency Functions
As shown in Fig. 3.7, the Fourier transform of an even time function is a real and even frequency function; conversely, the inverse Fourier transform of a real and even frequency function is an even function of time.

Odd Functions

If $h_0(t) = -h_0(-t)$, then $h_0(t)$ is an odd function, and its Fourier transform is an odd and imaginary function,

$$H(f) = \int_{-\infty}^{\infty} h_0(t)e^{-j2\pi ft}\, dt$$

$$= \int_{-\infty}^{\infty} h_0(t)\cos(2\pi ft)\, dt - j\int_{-\infty}^{\infty} h_0(t)\sin(2\pi ft)\, dt \qquad (3.30)$$

$$= -j\int_{-\infty}^{\infty} h_0(t)\sin(2\pi ft)\, dt = jI_0(f)$$

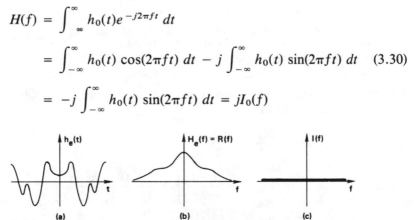

Figure 3.7   Fourier transform of an even function.

The real integral is zero because the multiplication of an odd and an even function is an odd function. Because $\sin(2\pi ft)$ is an odd function, then $h_0(t) \sin(2\pi ft) = -h_0(t) \sin[2\pi(-f)t]$ and $H_0(f) = -H_0(-f)$; the frequency function is odd. For $H(f)$ given as an odd and imaginary function, then

$$h(t) - \int_{-\infty}^{\infty} H(f)e^{j2\pi ft}\, dt - j \int_{-\infty}^{\infty} I_0(f)e^{j2\pi ft}\, df$$

$$= j \int_{-\infty}^{\infty} I_0(f) \cos(2\pi ft)\, df + j^2 \int_{-\infty}^{\infty} I_0(f) \sin(2\pi ft)\, df \qquad (3.31)$$

$$= - \int_{-\infty}^{\infty} I_0(f) \sin(2\pi ft)\, df = h_0(t)$$

and the resulting $h_0(t)$ is an odd function. The Fourier transform pair is thus established:

$$h_0(t) \iff jI_0(f) = -j \int_{-\infty}^{\infty} h_0(t) \sin(2\pi ft)\, dt \qquad (3.32)$$

Example 3.10   Odd Time and Frequency Functions

An illustrative example of this transform pair is shown in Fig. 3.8. The function $h(t)$ depicted is odd; therefore, the Fourier transform is an odd and imaginary function of frequency. If a frequency function is odd and imaginary, then its inverse transform is an odd function of time.

Figure 3.8   Fourier transform of an odd function.

3.7 WAVEFORM DECOMPOSITION

An arbitrary function can always be decomposed or separated into the sum of an even and an odd function:

$$h(t) = \frac{h(t)}{2} + \frac{h(t)}{2}$$

$$= \left[\frac{h(t)}{2} + \frac{h(-t)}{2} \right] + \left[\frac{h(t)}{2} - \frac{h(-t)}{2} \right] \qquad (3.33)$$

$$= h_e(t) + h_0(t)$$

The terms in brackets satisfy the definitions of even and odd functions, respectively. From Eqs. (3.27) and (3.32), the Fourier transform of Eq. (3.33) is

$$H(f) = R(f) + jI(f) = H_e(f) + H_0(f) \tag{3.34}$$

where $H_e(f) = R(f)$ and $H_0(f) = jI(f)$. We show in Chapter 9 that decomposition can increase the speed of computation of the FFT.

Example 3.11   Exponential Waveform Decomposition

To demonstrate the concept of waveform decomposition, consider the exponential function [Fig. 3.9(a)]

$$h(t) = e^{-at} \qquad t \geq 0 \tag{3.35}$$

Following the developments leading to Eq. (3.33), we obtain

$$
\begin{aligned}
h(t) &= \left[\frac{e^{-at}}{2} \right] + \left[\frac{e^{-at}}{2} \right] \\
&= \left\{ \left[\frac{e^{-at}}{2} \right]_{t \geq 0} + \left[\frac{e^{at}}{2} \right]_{t \leq 0} \right\} + \left\{ \left[\frac{e^{-at}}{2} \right]_{t \geq 0} - \left[\frac{e^{at}}{2} \right]_{t \leq 0} \right\} \\
&= \left\{ \frac{e^{-a|t|}}{2} \right\} + \left\{ \left[\frac{e^{-at}}{2} \right]_{t \geq 0} - \left[\frac{e^{at}}{2} \right]_{t \leq 0} \right\} \\
&= \{h_e(t)\} + \{h_0(t)\}
\end{aligned}
\tag{3.36}
$$

Figures 3.9(b) and (c) illustrate the even and odd decompositions, respectively.

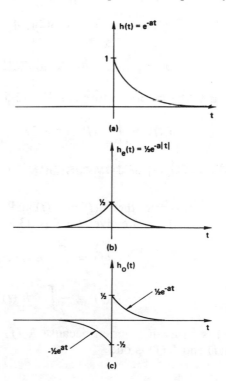

Figure 3.9 Waveform decomposition property.

3.8 COMPLEX TIME FUNCTIONS

For ease of presentation, we have to this point considered only real functions of time. The Fourier transform, Eq. (2.1), the inversion integral, Eq. (2.5), and the Fourier transform properties hold for the case of $h(t)$, a complex function of time. If

$$h(t) = h_r(t) + jh_i(t) \tag{3.37}$$

where $h_r(t)$ and $h_i(t)$ are the real and imaginary parts of the complex function $h(t)$, respectively, then the Fourier integral, Eq. (2.1), becomes

$$\begin{aligned} H(f) &= \int_{-\infty}^{\infty} [h_r(t) + jh_i(t)]e^{-j2\pi ft}\, dt \\ &= \int_{-\infty}^{\infty} [h_r(t)\cos(2\pi ft) + h_i(t)\sin(2\pi ft)]\, dt \\ &\quad -j \int_{-\infty}^{\infty} [h_r(t)\sin(2\pi ft) - h_i(t)\cos(2\pi ft)]\, dt \\ &= R(f) + jI(f) \end{aligned} \tag{3.38}$$

Therefore,

$$R(f) = \int_{-\infty}^{\infty} [h_r(t)\cos(2\pi ft) + h_i(t)\sin(2\pi ft)]\, dt \tag{3.39}$$

$$I(f) = -\int_{-\infty}^{\infty} [h_r(t)\sin(2\pi ft) - h_i(t)\cos(2\pi ft)]\, dt \tag{3.40}$$

Similarly, the inversion formula, Eq. (2.5), for complex functions yields

$$h_r(t) = \int_{-\infty}^{\infty} [R(f)\cos(2\pi ft) - I(f)\sin(2\pi ft)]\, df \tag{3.41}$$

$$h_i(t) = \int_{-\infty}^{\infty} [R(f)\sin(2\pi ft) + I(f)\cos(2\pi ft)]\, df \tag{3.42}$$

If $h(t)$ is real, then $h(t) = h_r(t)$ and the real and imaginary parts of the Fourier transform are given by Eqs. (3.39) and (3.40), respectively:

$$R_e(f) = \int_{-\infty}^{\infty} h_r(t)\cos(2\pi ft)\, dt \tag{3.43}$$

$$I_0(f) = -\int_{-\infty}^{\infty} h_r(t)\sin(2\pi ft)\, dt \tag{3.44}$$

$R_e(f)$ is an even function because $R_e(f) = R_e(-f)$. Similarly, $I_0(-f) = -I_0(f)$ and $I_0(f)$ is odd.

For $h(t)$ purely imaginary, $h(t) = jh_i(t)$ and

$$R_0(f) = \int_{-\infty}^{\infty} h_i(t) \sin(2\pi ft)\, dt \tag{3.45}$$

$$I_e(f) = \int_{-\infty}^{\infty} h_i(t) \cos(2\pi ft)\, dt \tag{3.46}$$

$R_0(f)$ is an odd function and $I_e(f)$ is an even function. Table 3.1 lists various complex time functions and their respective Fourier transforms.

Example 3.12   Simultaneous Fourier Transforms

We can employ the relationships of Eqs. (3.43), (3.44), (3.45), and (3.46) to simultaneously determine the Fourier transform of two real functions. To illustrate this point, recall the linearity property of Eq. (3.2):

$$x(t) + y(t) \Longleftrightarrow X(f) + Y(f) \tag{3.47}$$

Let $x(t) = h(t)$ and $y(t) = jg(t)$, where both $h(t)$ and $g(t)$ are real functions. It follows that $X(f) = H(f)$ and $Y(f) = jG(f)$. Because $x(t)$ is real, then from Eqs. (3.43) and (3.44)

$$x(t) = h(t) \Longleftrightarrow X(f) = H(f) = R_e(f) + jI_0(f) \tag{3.48}$$

Similarly, because $y(t)$ is imaginary, then from Eqs. (3.45) and (3.46)

$$y(t) = jg(t) \Longleftrightarrow Y(f) = jG(f) = R_0(f) + jI_e(f) \tag{3.49}$$

Hence,

$$h(t) + jg(t) \Longleftrightarrow H(f) + jG(f) \tag{3.50}$$

TABLE 3.1   Properties of the Fourier Transform for Complex Functions

Time domain $h(t)$	Frequency domain $H(f)$
Real	Real part even Imaginary part odd
Imaginary	Real part odd Imaginary part even
Real even, imaginary odd	Real
Real odd, imaginary even	Imaginary
Real and even	Real and even
Real and odd	Imaginary and odd
Imaginary and even	Imaginary and even
Imaginary and odd	Real and odd
Complex and even	Complex and even
Complex and odd	Complex and odd

where

$$H(f) = R_e(f) + jI_0(f) \tag{3.51}$$

$$G(f) = I_e(f) - jR_0(f) \tag{3.52}$$

Thus, if

$$z(t) = h(t) + jg(t) \tag{3.53}$$

then the Fourier transform of $z(t)$ can be expressed as

$$
\begin{aligned}
Z(f) &= R(f) + jI(f) \\
&= \left[\frac{R(f)}{2} + \frac{R(-f)}{2}\right] + \left[\frac{R(f)}{2} - \frac{R(-f)}{2}\right] \\
&\quad + j\left[\frac{I(f)}{2} + \frac{I(-f)}{2}\right] + j\left[\frac{I(f)}{2} - \frac{I(-f)}{2}\right]
\end{aligned} \tag{3.54}
$$

and from Eqs. (3.51) and (3.52)

$$H(f) = \left[\frac{R(f)}{2} + \frac{R(-f)}{2}\right] + j\left[\frac{I(f)}{2} - \frac{I(-f)}{2}\right] \tag{3.55}$$

$$G(f) = \left[\frac{I(f)}{2} + \frac{I(-f)}{2}\right] - j\left[\frac{R(f)}{2} - \frac{R(-f)}{2}\right] \tag{3.56}$$

Thus, it is possible to separate the frequency function $Z(f)$ into the Fourier transforms of $h(t)$ and $g(t)$. As is demonstrated in Chapter 9, this technique is used advantageously to increase the speed of computation of the FFT.

3.9 SUMMARY TABLE OF FOURIER TRANSFORM PROPERTIES

For future reference, the basic properties of the Fourier transform are summarized in Table 3.2. These relationships are of considerable importance throughout the remainder of the book.

PROBLEMS

3.1. Let

$$h(t) = \begin{cases} A & |t| < 2 \\ \dfrac{A}{2} & t = \pm 2 \\ 0 & |t| > 2 \end{cases}$$

$$x(t) = \begin{cases} -A & |t| < 1 \\ -\dfrac{A}{2} & t = \pm 1 \\ 0 & |t| > 1 \end{cases}$$

TABLE 3.2   Properties of Fourier Transforms

Time domain	Equation number	Frequency domain
Linear addition $x(t) + y(t)$	(3.2)	Linear addition $X(f) + Y(f)$
Symmetry $H(t)$	(3.6)	Symmetry $h(-f)$
Time scaling $h(kt)$	(3.12)	Inverse scale change $\dfrac{1}{\lvert k \rvert} H\!\left(\dfrac{f}{k}\right)$
Inverse scale change $\dfrac{1}{\lvert k \rvert} h\!\left(\dfrac{t}{k}\right)$	(3.14)	Frequency scaling $H(kf)$
Time shifting $h(t - t_0)$	(3.21)	Phase shifting $H(f)e^{-j2\pi f t_0}$
Modulation $h(t)e^{j2\pi t f_0}$	(3.23)	Frequency shifting $H(f - f_0)$
Even function $h_e(t)$	(3.27)	Real function $H_e(f) = R_e(f)$
Odd function $h_0(t)$	(3.30)	Imaginary $H_0(f) = jI_0(f)$
Real function $h(t) = h_r(t)$	(3.43) (3.44)	Real part even Imaginary part odd $H(f) = R_e(f) + jI_0(f)$
Imaginary function $h(t) = jh_i(t)$	(3.45) (3.46)	Real part odd Imaginary part even $H(f) = R_0(f) + jI_e(f)$

Sketch $h(t)$, $x(t)$, and $[h(t) - x(t)]$. Use the Fourier transform pair of Eq. (2.21) and the linearity theorem to find the Fourier transform of $[h(t) - x(t)]$.

3.2. Consider the functions $h(t)$ illustrated in Fig. 3.10. Use the linearity property to derive the Fourier transform of $h(t)$.

3.3. Use the symmetry theorem and the Fourier transform pairs of Fig. 2.12 to determine the Fourier transform of the following:

(a) $h(t) = \dfrac{A^2 \sin^2(2\pi T_0 t)}{(\pi t)^2}$

(b) $h(t) = \dfrac{\alpha^2}{(\alpha^2 + 4\pi^2 t^2)}$

(c) $h(t) = \exp\!\left(\dfrac{-\pi^2 t^2}{\alpha}\right)$

3.4. Derive the frequency-scaling property from the time-scaling property by means of the symmetry theorem.

3.5. Consider

$$h(t) = \begin{cases} A^2 - \dfrac{A^2 \lvert t \rvert}{2T_0} & \lvert t \rvert < 2T_0 \\ 0 & \lvert t \rvert > 2T_0 \end{cases}$$

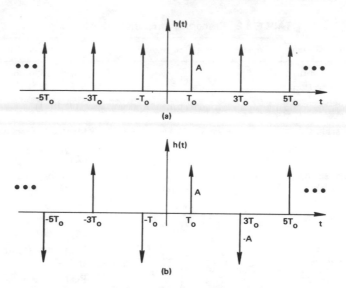

Figure 3.10   Functions for Problem 3.2.

Sketch the Fourier transform of $h(2t)$, $h(4t)$, and $h(8t)$. (The Fourier transform of $h(t)$ is given in Figure 2.12.)

3.6. Derive the time-scaling property for the case where k is negative.

3.7. By means of the shifting theorem, find the Fourier transform of the following functions:

(a) $h(t) = \dfrac{A \sin[2\pi f_0(t - t_0)]}{\pi(t - t_0)}$

(b) $h(t) = K\delta(t - t_0)$

(c) $h(t) = \begin{cases} A^2 - \dfrac{A^2}{2T_0}|t - t_0| & |t - t_0| < 2T_0 \\ 0 & |t - t_0| > 2T_0 \end{cases}$

3.8. Show that

$$h(\alpha t - \beta) \quad \text{⬡} \quad \frac{1}{|\alpha|} e^{-j2\pi\beta f/\alpha} H\left(\frac{f}{\alpha}\right)$$

3.9. Show that $|H(f)| = |e^{-j2\pi f t_0}H(f)|$, that is, the magnitude of a frequency function is independent of the time delay.

3.10. Find the inverse Fourier transform of the following functions by using the frequency-shifting theorem:

(a) $H(f) = \dfrac{A \sin[2\pi T_0(f - f_0)]}{\pi(f - f_0)}$

(b) $H(f) = \dfrac{\alpha^2}{\alpha^2 + 4\pi^2(f + f_0)^2}$

(c) $H(f) = \dfrac{A^2 \sin^2[2\pi T_0(f - f_0)]}{[\pi(f - f_0)]^2}$

3.11. Review the derivations leading to Eqs. (2.9), (2.13), (2.20), (2.29), (2.30), and (2.36). Note the mathematics that result are real for the Fourier transform of an even function.

3.12. Decompose and sketch the even and odd components of the following functions:

(a) $h(t) = \begin{cases} 1 & 1 < t < 2 \\ 0 & \text{otherwise} \end{cases}$

(b) $h(t) = \dfrac{1}{[2 - (t - 2)^2]}$

(c) $h(t) = \begin{cases} -t + 1 & 0 < t \le 1 \\ 0 & \text{otherwise} \end{cases}$

(d) $h(t) = 1 + t + t^2 + t^3$

(e) $h(t) = 1 + \sin(2\pi f t)$

3.13. Prove each of the properties listed in Table 3.1.

3.14. If $h(t)$ is real, show that $|H(f)|$ is an even function.

3.15. By making a substitution of a variable in Eq. (2.32), show that

$$\int_{-\infty}^{\infty} x(t)\delta(at - t_0)\, dt = \frac{1}{|a|} x\left(\frac{t_0}{a}\right)$$

3.16. Prove the following Fourier transform pairs:

(a) $\dfrac{dh(t)}{dt} \quad \Longleftrightarrow \quad j2\pi f H(f)$

(b) $[-j2\pi t]h(t) \quad \Longleftrightarrow \quad \dfrac{dH(f)}{df}$

3.17. Use the derivative relationship of Problem 3.16(a) to find the Fourier transform of a pulse waveform given the Fourier transform of a triangular waveform.

REFERENCES

1. BRACEWELL, R. *The Fourier Transform and its Applications*, 2d Ed. New York: McGraw-Hill, 1986.

2. PAPOULIS, A. *The Fourier Integral and Its Applications*, 2d Ed. New York: McGraw-Hill, 1984.

3. CHAMPENEY, D. C. *Fourier Transforms and Their Physical Application*. New York: Academic Press, 1973.

4

CONVOLUTION AND CORRELATION

In Chapter 3, we investigated those properties that are fundamental to the Fourier transform. However, there exists a class of Fourier transform relationships whose importance far outranks those previously considered. These properties are the convolution and correlation theorems, which are to be discussed at length in this chapter.

4.1 CONVOLUTION INTEGRAL

Convolution of two functions is a significant physical concept in many diverse scientific fields. However, as in the case of many important mathematical relationships, the convolution integral does not readily *unveil* itself as to its true implications. To be more specific, the convolution integral is given by

$$y(t) = \int_{-\infty}^{\infty} x(\tau)h(t - \tau) \, d\tau = x(t) * h(t) \tag{4.1}$$

Function $y(t)$ is said to be the convolution of the functions $x(t)$ and $h(t)$. Note that it is extremely difficult to *visualize* the mathematical operation of Eq. (4.1). We develop the true meaning of convolution by graphical analysis.

4.2 GRAPHICAL EVALUATION OF THE CONVOLUTION INTEGRAL

Let $x(t)$ and $h(t)$ be two time functions given by graphs, as represented in Figs. 4.1(a) and (b), respectively. To evaluate Eq. (4.1), functions $x(\tau)$ and $h(t - \tau)$ are required; $x(\tau)$ and $h(\tau)$ are simply $x(t)$ and $h(t)$, respectively, where the variable t has been replaced by the variable τ. $h(-\tau)$ is the image of $h(\tau)$ about the ordinate axis and $h(t - \tau)$ is simply the function $h(-\tau)$ shifted by the quantity t. Functions $x(\tau)$, $h(-\tau)$, and $h(t - \tau)$ are shown in Fig. 4.2. To compute the integral of Eq. (4.1), it is necessary to multiply and integrate the functions $x(\tau)$ [Fig. 4.2(a)] and $h(t - \tau)$ [Fig. 4.2(c)] for each value of t from $-\infty$ to $+\infty$. As illustrated in Figs. 4.3(a) and (h), this product is zero for the choice of the parameter $t = -t_1$. The product remains zero until t is reduced to zero. As illustrated in Figs. 4.3(c) and (h), the product of $x(\tau)$ and $h(t_1 - \tau)$ is the function emphasized by shading. The integral of this function is simply the shaded area beneath the curve. As t is increased to $2t_1$ and further to $3t_1$, Figs. 4.3(d), (e), and (h) illustrate the relationships of the functions to be multiplied as well as the resulting integrations. For $t = 4t_1$, the product again becomes zero, as shown by Figs. 4.3(f) and (h). This product remains zero for all t greater than $4t_1$ [Figs. 4.3(g) and (h)]. If t is allowed to be a continuum of values, then the convolution of $x(t)$ and $h(t)$ is the triangular function illustrated in Fig. 4.3(h).

The procedure described is a convenient graphical technique for evaluating convolution integrals. Summarizing the steps:

1. *Folding.* Take the mirror image of $h(\tau)$ about the ordinate axis.
2. *Displacement.* Shift $h(-\tau)$ by the amount t.
3. *Multiplication.* Multiply the shifted function $h(t - \tau)$ by $x(\tau)$.
4. *Integration.* The area under the product of $h(t - \tau)$ and $x(\tau)$ is the value of the convolution at time t.

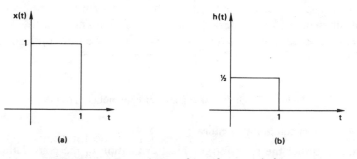

(a)   (b)

Figure 4.1   Example waveforms for convolution.

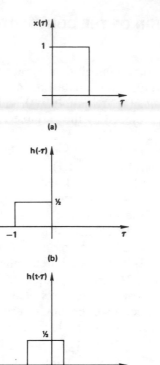

(a)

(b)

(c)

Figure 4.2 Graphical illustration of folding and displacement operations.

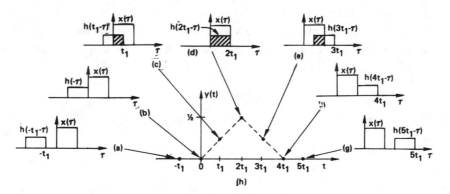

Figure 4.3 Graphical example of convolution; $t_1 = 1/2$.

Example 4.1   Convolution Procedure

To illustrate further the rules for graphical evaluation of the convolution integral, convolve the functions illustrated in Figs. 4.4(a) and (b). First, fold $h(\tau)$ to obtain $h(-\tau)$, as illustrated in Fig. 4.4(c). Next, displace or shift $h(-\tau)$ by the amount t,

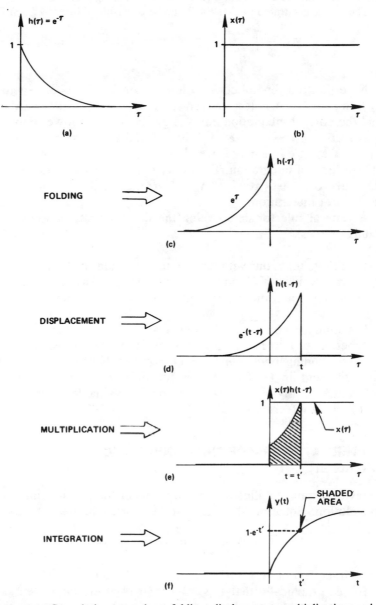

Figure 4.4 Convolution procedure: folding, displacement, multiplication, and integration.

as shown in Fig. 4.4(d). Then, multiply $h(t - \tau)$ by $x(\tau)$ [Fig. 4.4(e)] and, finally, integrate to obtain the convolution result for time t' [Fig. 4.4(f)].

The result illustrated in Fig. 4.4(f) can be determined directly from Eq. (4.1):

$$y(t) = \int_{-\infty}^{\infty} x(\tau)h(t - \tau) \, d\tau = \int_{0}^{t} (1)e^{-(t-\tau)} \, d\tau$$

$$= e^{-t}(e^{\tau} \big|_{0}^{t}) = e^{-t}[e^{t} - 1] = 1 - e^{-t}. \tag{4.2}$$

Note that the general convolution integration limits of $-\infty$ to $+\infty$ become 0 to t for Ex. 4.1. It is desired to develop a straightforward approach to find the correct integration limits. For Ex. 4.1, the lower nonzero value of the function $h(t - \tau) = e^{-(t-\tau)}$ is $-\infty$ and the lower nonzero value for $x(\tau)$ is 0. When we integrated, we chose the largest of these two values as our lower limit of integration. The upper nonzero value of $h(t - \tau)$ is t; the upper nonzero value of $x(\tau)$ is ∞. We chose the smallest of these two for our upper limit of integration.

A general rule for determining the limits of integration can then be stated as follows:

Given two functions with lower nonzero values of L_1 and L_2 and upper nonzero values of U_1 and U_2, choose the lower limit of integration as $\max[L_1, L_2]$ and the upper limit of integration as $\min[U_1, U_2]$.

It should be noted that the lower and upper nonzero values for the fixed function $x(\tau)$ do not change; however, the lower and upper nonzero values of the sliding function $h(t - \tau)$ change as t changes. Thus, it is possible to have different limits of integration for different ranges of t. A graphical sketch similar to Fig. 4.4 is also an extremely valuable aid in choosing the correct limit of integration.

4.3 ALTERNATE FORM OF THE CONVOLUTION INTEGRAL

The previous graphical illustration is but one of the possible interpretations of convolution. Equation (4.1) can also be written equivalently as

$$y(t) = \int_{-\infty}^{\infty} h(\tau)x(t - \tau) \, d\tau \tag{4.3}$$

Hence, either $h(\tau)$ or $x(\tau)$ can be folded and shifted.

To see graphically that Eqs. (4.1) and (4.3) are equivalent, consider the functions illustrated in Fig. 4.5(a). It is desired to convolve these two functions. The series of graphs on the left in Fig. 4.5 illustrate the evaluation of Eq. (4.1); the graphs on the right illustrate the evaluation of Eq. (4.3). The previously defined steps of (1) folding, (2) displacement, (3) multipli-

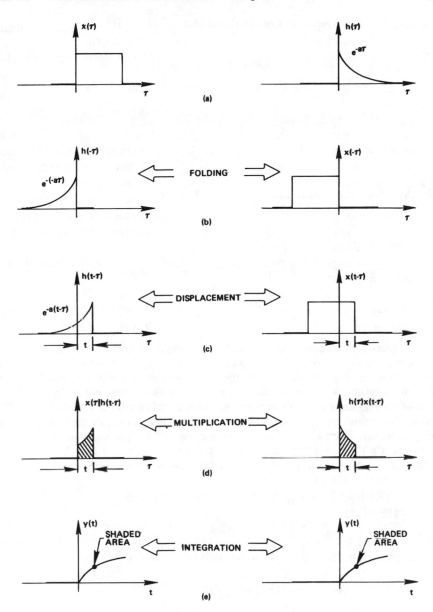

Figure 4.5 Graphical example of convolution by Eqs. (4.1) and (4.3).

cation, and (4) integration are illustrated by Figs. 4.5(b), (c), (d), and (e), respectively. As indicated by Fig. 4.5(e), the convolution of $x(\tau)$ and $h(\tau)$ is the same irrespective of which function is chosen for folding and displacement.

Example 4.2   Equivalence of Eqs. (4.1) and (4.3)

Let

$$h(t) = e^{-t} \qquad t \geq 0$$
$$= 0 \qquad t < 0 \tag{4.4}$$

and

$$x(t) = \sin t \qquad 0 \leq t \leq \frac{\pi}{2}$$
$$= 0 \qquad \text{otherwise} \tag{4.5}$$

Find $h(t) * x(t)$ using both Eqs. (4.1) and (4.3).
From Eq. (4.1)

$$y(t) = \int_{-\infty}^{\infty} x(\tau)h(t - \tau)\, d\tau$$

$$y(t) = \begin{cases} \int_0^t [\sin(\tau)]e^{-(t-\tau)}\, d\tau & 0 \leq t \leq \frac{\pi}{2} \\ \int_0^{\pi/2} [\sin(\tau)]e^{-(t-\tau)}\, d\tau & t \geq \frac{\pi}{2} \\ 0 & t \leq 0 \end{cases} \tag{4.6}$$

The integral limits are easily determined by using the procedure described previously. The lower and upper nonzero values of the function $x(\tau)$ are 0 and $\pi/2$, respectively. For the function $h(t - \tau) = e^{-(t-\tau)}$, the lower nonzero value is $-\infty$ and the upper nonzero value is t. We take the maximum of the lower nonzero values for our lower limit of integration, i.e., 0. The upper limit of integration is a function of t. For $0 \leq t \leq \pi/2$, the minimum of the upper nonzero values is t and hence the upper limit of integration is t. For $t \geq \pi/2$, the minimum of the upper nonzero values is $\pi/2$ and consequently the upper limit of integration for this range of t is $\pi/2$. A graphical sketch of the convolution process also yields these integration limits.
    Evaluating Eq. (4.6), we obtain

$$y(t) = \begin{cases} 0 & t \leq 0 \\ \frac{1}{2}(\sin t - \cos t + e^{-t}) & 0 < t \leq \frac{\pi}{2} \\ \frac{e^{-t}}{2}(1 + e^{\pi/2}) & t \geq \frac{\pi}{2} \end{cases} \tag{4.7}$$

Similarly, from Eq. (4.3), we obtain

$$y(t) = \int_{-\infty}^{\infty} h(\tau)x(t - \tau)\, d\tau$$

$$y(t) = \begin{cases} \int_0^t e^{-\tau}[\sin(t - \tau)]\, d\tau & 0 < t < \frac{\pi}{2} \\ \int_{t-\pi/2}^t e^{-\tau}[\sin(t - \tau)]\, d\tau & t \geq \frac{\pi}{2} \\ 0 & t < 0 \end{cases} \tag{4.8}$$

Although Eqs. (4.8) are different from Eqs. (4.6), evaluation yields identical results to Eq. (4.7).

4.4 CONVOLUTION INVOLVING IMPULSE FUNCTIONS

The simplest type of convolution integral to evaluate is one in which either $x(t)$ or $h(t)$ is an impulse function. To illustrate this point, let $h(t)$ be the singular function shown graphically in Fig. 4.6(a) and let $x(t)$ be the rectan-

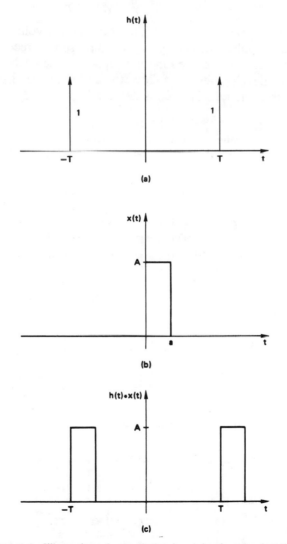

Figure 4.6   Illustration of convolution involving impulse functions.

gular function shown in Fig. 4.6(b). For these example functions, Eq. (4.1)
becomes

$$y(t) = \int_{-\infty}^{\infty} [\delta(\tau - T) + \delta(\tau + T)]x(t - \tau)\, d\tau \tag{4.9}$$

Recall from Eq. (2.28) that

$$\int_{-\infty}^{\infty} \delta(\tau - T)x(\tau)\, d\tau = x(T)$$

Hence, Eq. (4.9) can be written as

$$y(t) = x(t - T) + x(t + T) \tag{4.10}$$

Function $y(t)$ is illustrated in Fig. 4.6(c). Note that convolution of the function $x(t)$ with an impulse function is evaluated by simply reconstructing $x(t)$ with the position of the impulse function replacing the ordinate of $x(t)$. As we will see in the developments to follow, the ability to visualize convolution involving impulse functions is of considerable importance.

Example 4.3   Convolution with Impulses

Let $h(t)$ be a series of impulse functions, as illustrated in Fig. 4.7(a). To evaluate the convolution of $h(t)$ with the rectangular pulse shown in Fig. 4.7(b), we simply

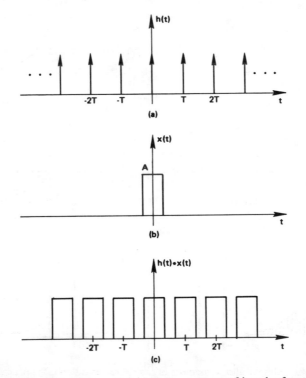

Figure 4.7 Convolution with an infinite sequence of impulse functions.

reproduce the rectangular pulse at each of the impulse functions. The resulting convolution results are illustrated in Fig. 4.7(c).

Example 4.4   Linear-System Convolution

Convolution of two functions is a significant physical concept in many diverse scientific fields. A linear system is characterized by an output that is determined by the convolution of the system input and the system impulse response. To demonstrate, consider Fig. 4.8. As shown, if the system input is an impulse, the output is the impulse response of the system. It is the impulse response of a system that allows one to express the system output in terms of the input. To illustrate this point, we assume the system is time-invariant, that is, if the impulse is delayed by a time $t =$

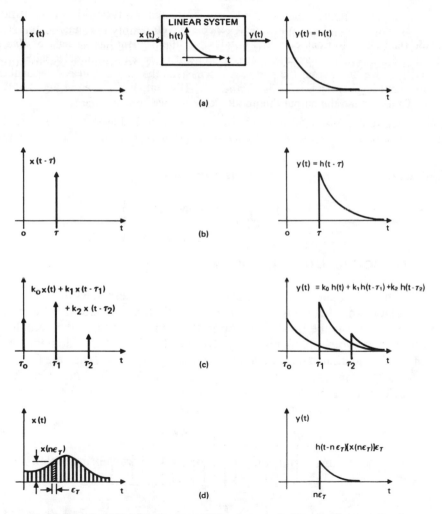

Figure 4.8 Graphical development of the characterization of a linear system by the convolution integral.

τ, then the output response is also delayed by the same duration of time. As shown in Fig. 4.8(b), the input $\delta(t - \tau)$ results in an output $h(t - \tau)$.

We also assume that the system is linear. Linear means that if the input $x_i(t)$ produces the output $y_i(t)$, then the input $k_1 x_1(t) + k_2 x_2(t)$ produces the output $k_1 y_1(t) + k_2 y_2(t)$. As illustrated in Fig. 4.8(c), a system input composed of a series of delayed impulses of varying amplitudes yields an output consisting of delayed impulse response functions whose amplitudes are determined by the amplitude of the input impulse causing the response. The sum of these individual impulse responses is the system output and is computed by the sum

$$y(t) = \sum_{i=1}^{3} k_i(t - \tau_i) \tag{4.11}$$

To extend Eq. (4.11) to include a general waveform, consider Fig. 4.8(d). We divide the input into small elements of width ϵ_τ; the element has a height of $x(n\epsilon_\tau)$ and a width of ϵ_τ. If we assume that this element represents an impulse with area $[x(n\epsilon_\tau)][\epsilon_\tau]$, we know from the previous discussions that the output corresponding to this input is given by $[h(t - n\epsilon_\tau)][x(n\epsilon_\tau)][\epsilon_\tau]$ This output is shown in Fig. 4.8(d).

To determine the output due to all elements, we compute the sum

$$y(t) = \sum_{n=-\infty}^{\infty} [h(t - n\epsilon_\tau)][x(n\epsilon_\tau)][\epsilon_\tau] \tag{4.12}$$

If we let $\epsilon_\tau \to 0$ and $n \to \infty$, such that $n\epsilon_\tau \to \tau$, we obtain

$$y(t) = \int_{-\infty}^{\infty} h(t - \tau)x(\tau)\, d\tau \tag{4.13}$$

4.5 TIME-CONVOLUTION THEOREM

Possibly the most important and powerful tool in modern scientific analysis is the relationship between Eq. (4.1) and its Fourier transform. This relationship, known as the time-convolution theorem, allows one the complete freedom to convolve mathematically (or visually) in the time domain by a simple multiplication in the frequency domain. That is, if $h(t)$ has the Fourier transform $H(f)$ and $x(t)$ has the Fourier transform $X(f)$, then $h(t) * x(t)$ has the Fourier transform $H(f)X(f)$. The convolution theorem is thus given by the Fourier transform pair:

$$h(t) * x(t) \iff H(f)X(f) \tag{4.14}$$

To establish this result, first form the Fourier transform of both sides of Eq. (4.1):

$$\int_{-\infty}^{\infty} y(t)e^{-j2\pi ft}\, dt = \int_{-\infty}^{\infty} \left[\int_{-\infty}^{\infty} x(\tau)h(t - \tau)\, d\tau \right] e^{-j2\pi ft}\, dt \tag{4.15}$$

which is equivalent to (assuming the order of integration can be changed)

$$Y(f) = \int_{-\infty}^{\infty} x(\tau) \left[\int_{-\infty}^{\infty} h(t - \tau)e^{-j2\pi ft} \, dt \right] d\tau \qquad (4.16)$$

By substituting $\sigma = t - \tau$, the term in the brackets becomes

$$\int_{-\infty}^{\infty} h(\sigma)e^{-j2\pi f(\sigma + \tau)} \, d\sigma = e^{-j2\pi f\tau} \int_{-\infty}^{\infty} h(\sigma)e^{-j2\pi f\sigma} \, d\sigma$$

$$= e^{-j2\pi f\tau} H(f) \qquad (4.17)$$

Equation (4.16) can then be rewritten as

$$Y(f) = \int_{-\infty}^{\infty} x(\tau)e^{-j2\pi f\tau} H(f) \, d\tau = H(f)X(f) \qquad (4.18)$$

The converse is proven similarly.

Example 4.5   Convolution of Pulse Waveforms

To illustrate the application of the convolution theorem, consider the convolution of the two pulse functions shown in Figs. 4.9(a) and (b). As we have seen previously, the convolution of two rectangular functions is a triangular function, as shown in Fig. 4.9(e). Recall from the Fourier transform pair of Eq. (2.21) that the Fourier transform of a rectangular function is the $[\sin(f)]/f$ function illustrated in Figs. 4.9(c) and (d). The convolution theorem states that convolution in the time domain corresponds to multiplication in the frequency domain; therefore, the triangular waveform of Fig. 4.9(e) and the $[\sin^2(f)]/f^2$ function of Fig. 4.9(f) are Fourier transform pairs. Thus, we can use the theorem as a convenient tool for developing additional Fourier transform pairs.

Example 4.6   Infinite Pulse-Train Waveform

One of the most significant contributions of distribution theory results from the fact that the product of a continuous function and an impulse function is well-defined (Appendix A); hence, if $h(t)$ is continuous at $t = t_0$, then

$$h(t)\delta(t - t_0) = h(t_0)\delta(t - t_0) \qquad (4.19)$$

This result, coupled with the convolution theorem, allows one to eliminate the tedious derivation of many Fourier transform pairs. To illustrate, consider the two time functions $h(t)$ and $x(t)$ shown in Figs. 4.10(a) and (b). As described previously, the convolution of these two functions is the infinite pulse train illustrated in Fig. 4.10(e). It is desired to determine the Fourier transform of this infinite sequence of pulses. We simply use the convolution theorem: the Fourier transform of $h(t)$ is the sequence of impulse functions, the transform pair of Eq. (2.44), as illustrated in Fig. 4.10(c); and the Fourier transform of a rectangular function is the $[\sin(f)]/f$ function shown in Fig. 4.10(d). Multiplication of these two frequency functions yields the desired Fourier transform. As illustrated in Fig. 4.10(f), the Fourier transform of a pulse train is a sequence of impulse functions whose amplitude is weighted by a

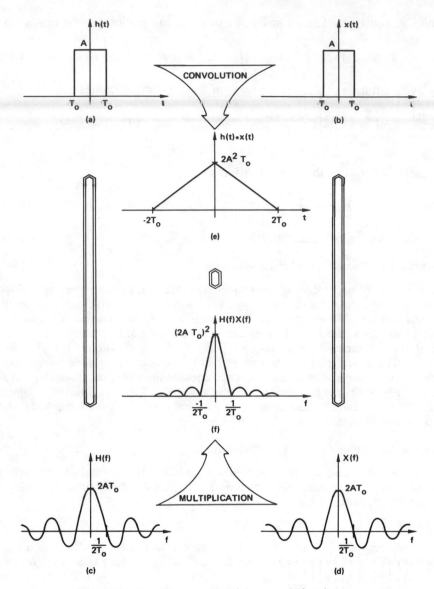

Figure 4.9 Graphical example of the convolution theorem.

$[\sin(f)]/f$ function. This is a well-known result in the field of radar systems. It is to be noted that the multiplication of the two frequency functions must be interpreted in the sense of distribution theory; otherwise, the product is meaningless. We can see that the ability to change from a convolution in the time domain to multiplication in the frequency domain often renders unwieldy problems rather straightforward.

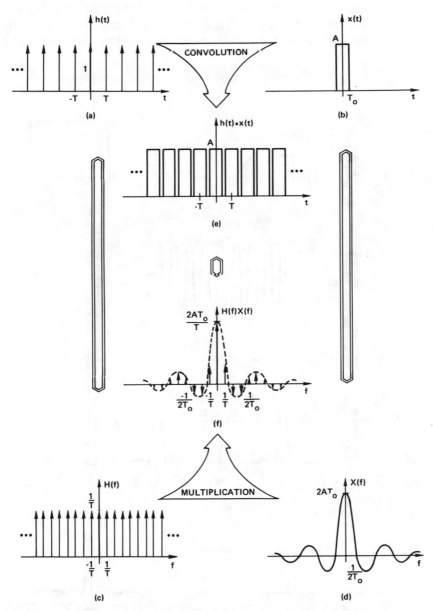

Figure 4.10   Example application of the convolution theorem.

4.6 FREQUENCY-CONVOLUTION THEOREM

We can equivalently go from convolution in the frequency domain to multiplication in the time domain by using the frequency-convolution theorem: the Fourier transform of the product $h(t)x(t)$ is equal to the convolution $H(f)$

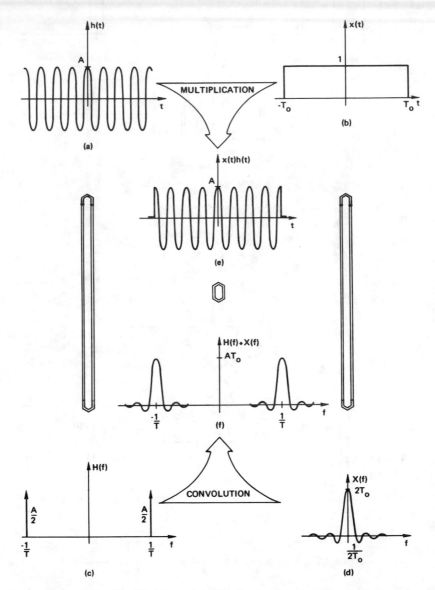

Figure 4.11   Graphical example of the frequency-convolution theorem.

$* X(f)$. The frequency-convolution theorem is

$$h(t)x(t) \quad \diamond\kern-6pt\diamond \quad H(f) * X(f) \tag{4.20}$$

This pair is established by simply substituting the Fourier transform pair of Eq. (4.14) into the symmetry Fourier transform relationship of Eq. (3.6).

Example 4.7   Modulated Pulse Waveform

To illustrate the frequency-convolution theorem, consider the cosine waveform of Fig. 4.11(a) and the rectangular waveform of Fig. 4.11(b). It is desired to determine the Fourier transform of the product of these two functions [Fig. 4.11(e)]. The Fourier transforms of the cosine and rectangular waveforms are given in Figs. 4.11(c) and (d), respectively. Convolution of these two frequency functions yields the function shown in Fig. 4.11(f); Figs. 4.11(e) and (f) are thus Fourier transform pairs. This is the well-known Fourier transform pair of a single frequency-modulated pulse.

4.7 CORRELATION THEOREM

Another integral equation of importance in both theoretical and practical application is the correlation integral:

$$z(t) = \int_{-\infty}^{\infty} x(\tau)h(t + \tau) \, d\tau \tag{4.21}$$

A comparison of the above expression and the convolution integral, Eq. (4.1), indicates that the two are closely related. The nature of this relationship is best described by the graphical illustrations of Fig. 4.12. The functions to be both *convolved* and *correlated* are shown in Fig. 4.12(a). Illustrations on the left depict the process of convolution as described in the previous section; illustrations on the right graphically portray the process of correlation. As evidenced in Fig. 4.12(b), the two integrals differ in that there is no folding of one of the integrands in correlation. The previously described rules of displacement, multiplication, and integration are performed identically for both convolution and correlation. For the special case where either $x(t)$ or $h(t)$ is an even function, convolution and correlation are equivalent; this follows because an even function and its image are identical and, thus, folding can be eliminated from the steps in computing the convolution integral.

Example 4.8   Correlation Procedure

Correlate graphically and analytically the waveforms illustrated in Fig. 4.13(a).

According to the rules for correlation, we displace $h(\tau)$ by the shift t, multiply by $x(\tau)$, and integrate the product $x(\tau)h(t + \tau)$, as illustrated in Figs. 4.13(b), (c), and (d), respectively.

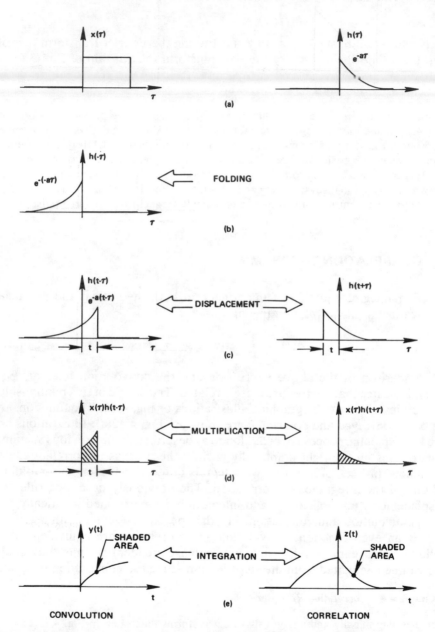

Figure 4.12 Graphical comparison of convolution and correlation.

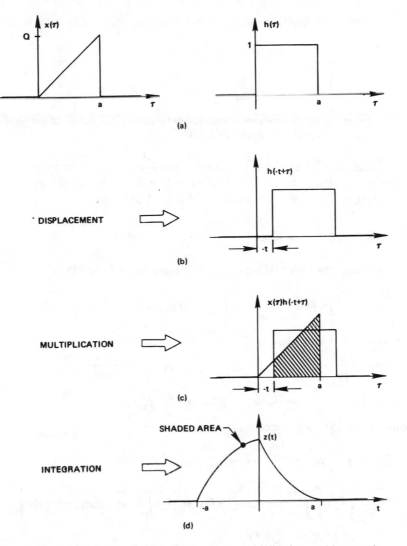

Figure 4.13 Correlation procedure: displacement, multiplication, and integeration.

From Eq. (4.21), for positive displacement t, we obtain

$$z(t) = \int_{-\infty}^{\infty} x(\tau)h(t + \tau)\, d\tau$$

$$= \int_0^{a-t} (1)\frac{Q}{a}\tau\, d\tau \qquad\qquad (4.22)$$

$$= \frac{Q}{2a}\tau^2 \bigg|_0^{a-t} = \frac{Q}{2a}(a - t)^2 \qquad 0 \le t \le a$$

For negative displacement, see Fig. 4.13(c) to justify the limits of integration.

$$z(t) = \int_t^a (1)\frac{Q}{a}\tau \, d\tau$$

$$= \frac{Q}{2a}(a^2 - t^2) \qquad -a \le t \le 0 \tag{4.23}$$

A general rule can be developed for determining the limits of integration for the correlation integral (see Problem 4.14).

Recall that convolution multiplication forms a Fourier transform pair. A similar result can be obtained for correlation. To derive this relationship, first evaluate the Fourier transform of Eq. (4.21)

$$\int_{-\infty}^{\infty} z(t)e^{-j2\pi ft} \, dt = \int_{-\infty}^{\infty} \left[\int_{-\infty}^{\infty} x(\tau)h(t + \tau) \, d\tau \right] e^{-j2\pi ft} \, dt \tag{4.24}$$

or (assuming the order of integration can be interchanged)

$$Z(f) = \int_{-\infty}^{\infty} x(\tau) \left[\int_{-\infty}^{\infty} h(t + \tau)e^{-j2\pi ft} \, dt \right] d\tau \tag{4.25}$$

Let $\sigma = t + \tau$ and rewrite the term in brackets as

$$\int_{-\infty}^{\infty} h(\sigma)e^{-j2\pi f(\sigma - \tau)} \, d\sigma = e^{j2\pi f\tau} \int_{-\infty}^{\infty} h(\sigma)e^{-j2\pi f\sigma} \, d\sigma$$

$$= e^{j2\pi f\tau} H(f) \tag{4.26}$$

Equation (4.25) then becomes

$$Z(f) = \int_{-\infty}^{\infty} x(\tau)e^{j2\pi f\tau} H(f) \, d\tau$$

$$= H(f) \left[\int_{-\infty}^{\infty} x(\tau) \cos(2\pi f\tau) \, d\tau + j \int_{-\infty}^{\infty} x(\tau) \sin(2\pi f\tau) \, d\tau \right] \tag{4.27}$$

$$= H(f)[R(f) + jI(f)]$$

Now the Fourier transform of $x(\tau)$ is given by

$$X(f) = \int_{-\infty}^{\infty} x(\tau)e^{-j2\pi f\tau} \, d\tau$$

$$= \int_{-\infty}^{\infty} x(\tau) \cos(2\pi f\tau) \, d\tau - j \int_{-\infty}^{\infty} x(\tau) \sin(2\pi f\tau) \, d\tau \tag{4.28}$$

$$= R(f) - jI(f)$$

The bracketed term of Eq. (4.27) and the expression on the right in Eq. (4.28) are called conjugates [defined in Eq. (3.25)]. Equation (4.27) can be written as

$$Z(f) = H(f)X^*(f) \tag{4.29}$$

and the Fourier transform pair for correlation is

$$\int_{-\infty}^{\infty} x(\tau)h(t + \tau) \, d\tau \iff H(f)X^*(f) \tag{4.30}$$

Note that if $x(t)$ is an even function, then $X(f)$ is purely real and $X(f) = X^*(f)$. For these conditions, the Fourier transform of the correlation integral is $H(f)X(f)$, which is identical to the Fourier transform of the convolution integral. These arguments for identity of the two integrals are simply the frequency-domain equivalents of the previously discussed time-domain requirement for equality of the two integrals.

If $x(t)$ and $h(t)$ are the same function, Eq. (4.21) is normally termed the *autocorrelation* function; if $x(t)$ and $h(t)$ differ, the term *crosscorrelation* is normally used.

Example 4.9   Autocorrelation Function

Determine the autocorrelation function of the waveform

$$
\begin{aligned}
h(t) &= e^{-at} & t &> 0 \\
&= 0 & t &< 0
\end{aligned} \tag{4.31}
$$

From Eq. (4.21),

$$
\begin{aligned}
z(t) &= \int_{-\infty}^{\infty} h(\tau)h(t + \tau) \, d\tau \\
&= \int_{0}^{\infty} e^{-a\tau} e^{-a(t+\tau)} \, d\tau & t &> 0 \\
&= \int_{t}^{\infty} e^{-a\tau} e^{-a(t+\tau)} \, d\tau & t &< 0 \\
&= \frac{e^{-a|t|}}{2a} & -\infty &< t < \infty
\end{aligned} \tag{4.32}
$$

PROBLEMS

4.1. Prove the following convolution properties:
   (a) Convolution is commutative: $[h(t) * x(t)] = [x(t) * h(t)]$
   (b) Convolution is associative: $h(t) * [g(t) * x(t)] = [h(t) * g(t)] * x(t)$
   (c) Convolution is distributive over addition: $h(t) * [g(t) + x(t)] = h(t) * g(t) + h(t) * x(t)$

4.2. Determine $h(t) * g(t)$, where
   (a) $\begin{aligned} h(t) &= e^{-at} & t &> 0 \\ &= 0 & t &< 0 \\ g(t) &= e^{-bt} & t &> 0 \\ &= 0 & t &< 0 \end{aligned}$

(b) $h(t) = te^{-t} \qquad t \geq 0$
$\quad\quad\quad\; = 0 \qquad\quad t < 0$
$\quad g(t) = e^{-t} \qquad t > 0$
$\quad\quad\quad\; = 0 \qquad\quad t < 0$

(c) $h(t) = te^{-t} \qquad t \geq 0$
$\quad\quad\quad\; = 0 \qquad\quad t < 0$
$\quad g(t) = e^{t} \qquad\; t < -1$
$\quad\quad\quad\; = 0 \qquad\quad t > -1$

(d) $h(t) = 2e^{3t} \qquad t > 1$
$\quad\quad\quad\; = 0 \qquad\quad t < 0$
$\quad g(t) = 2e^{t} \qquad t < 0$
$\quad\quad\quad\; = 0 \qquad\quad t > 0$

(e) $h(t) = \sin(2\pi t) \qquad 0 \leq t \leq \tfrac{1}{2}$
$\quad\quad\quad\; = 0 \qquad\qquad\; \text{elsewhere}$
$\quad g(t) = 1 \qquad\qquad 0 < t < \tfrac{1}{8}$
$\quad\quad\quad\; = 0 \qquad\qquad\; t < 0; t > \tfrac{1}{8}$

(f) $h(t) = 1 - t \qquad\quad 0 < t < 1$
$\quad\quad\quad\; = 0 \qquad\qquad\; t < 0; t > 1$
$\quad g(t) = h(t)$

(g) $h(t) = (a - |t|)^3 \qquad -a \leq t \leq a$
$\quad\quad\quad\; = 0 \qquad\qquad\quad \text{elsewhere}$
$\quad g(t) = h(t)$

(h) $h(t) = e^{-at} \qquad\; t > 0$
$\quad\quad\quad\; = 0 \qquad\qquad t < 0$
$\quad g(t) = 1 - t \qquad 0 < t < 1$
$\quad\quad\quad\; = 0 \qquad\qquad t < 0; t > 1$

4.3. Sketch the convolution of the functions $x(t)$ and $h(t)$ illustrated in Fig. 4.14.

4.4. Sketch the convolution of the two odd functions $x(t)$ and $h(t)$ illustrated in Fig. 4.15. Show that the convolution of two odd functions is an even function.

4.5. Use the convolution theorem to graphically determine the Fourier transform of the functions illustrated in Fig. 4.16.

4.6. Analytically, determine the Fourier transform of $e^{-\alpha t^2} * e^{-\beta t^2}$. (Hint: Use the convolution theorem.)

4.7. Use the frequency convolution theorem to graphically determine the Fourier transform of the product of the functions $x(t)$ and $h(t)$ illustrated in Fig. 4.17.

4.8. Graphically determine the correlation of the functions $x(t)$ and $h(t)$ illustrated in Fig. 4.14.

4.9. Let $h(t)$ be a time-limited function that is nonzero over the range

$$\frac{-T_0}{2} \leq t \leq \frac{T_0}{2}$$

Show that $h(t) * h(t)$ is nonzero over the range $-T_0 \leq t \leq T_0$; that is, $h(t) * h(t)$ has a "width" twice that of $h(t)$.

4.10. Show that if $h(t) = f(t) * g(t)$, then

$$\frac{dh(t)}{dt} = \frac{df(t)}{dt} * g(t) = f(t) * \frac{dg(t)}{dt}$$

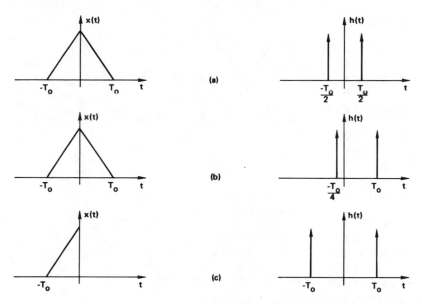

Figure 4.14   Functions $x(t)$ and $h(t)$ for Problems 4.3 and 4.8.

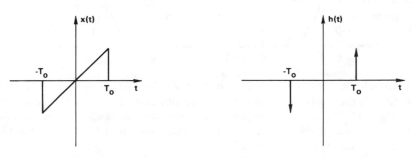

Figure 4.15   Functions $x(t)$ and $h(t)$ for Problem 4.4.

4.11. By means of the frequency-convolution theorem, graphically determine the Fourier transform of a half-wave rectified waveform. Using this result, incorporate the shifting theorem to determine the Fourier transform of a full-wave rectified waveform.

4.12. Graphically find the Fourier transform of the following functions:
   (a) $h(t) = A \cos^2(2\pi f_0 t)$
   (b) $h(t) = A \sin^2(2\pi f_0 t)$
   (c) $h(t) = A \cos^2(2\pi f_0 t) + A \cos^2(\pi f_0 t)$

4.13. Graphically find the inverse Fourier transform of the following functions:
   (a) $\left[\dfrac{\sin(2\pi f)}{2\pi f} \right]^2$

   (b) $\dfrac{1}{(1 + j2\pi f)^2}$

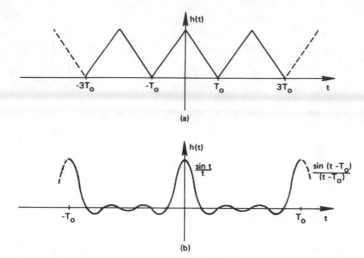

(a)

(b)

Figure 4.16   Functions for Problem 4.5.

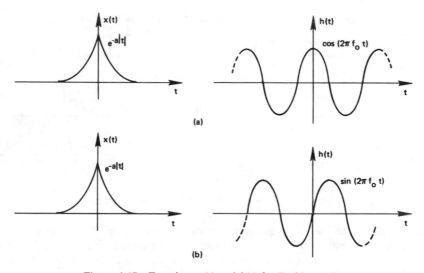

Figure 4.17   Functions $x(t)$ and $h(t)$ for Problem 4.7.

(c) $e^{-|2\pi f|}$

(d) $1 - e^{-|f|}$

4.14. Develop a set of rules for determining the limits of integration for the correlation integral.

REFERENCES

1. BRACEWELL, R. *The Fourier Transform and Its Applications*, 2d Rev. Ed. New York: McGraw-Hill, 1986.

2. GUPTA, S. *Transform and State Variable Methods in Linear Systems*. New York: Wiley, 1966.

3. HEALY, T. J. "Convolution Revisited." *IEEE Spectrum* (April 1969), Vol. 6, No. 4, pp. 87–93.

4. PAPOULIS, A. *The Fourier Integral and Its Applications*, 2d Ed. New York: McGraw-Hill, 1984.

5

FOURIER SERIES AND
SAMPLED WAVEFORMS

In the technical literature, Fourier series are normally developed independently of the Fourier integral. However, with the introduction of distribution theory, Fourier series can be theoretically derived as a special case of the Fourier integral. This approach is significant in that it is fundamental in considering the discrete Fourier transform as a special case of the Fourier integral. Also fundamental to an understanding of the discrete Fourier transform is the Fourier transform of sampled waveforms. In this chapter, we relate both of these relationships to the Fourier transform and thereby provide the framework for the development of the discrete Fourier transform in Chapter 6.

5.1 FOURIER SERIES

A periodic function $y(t)$ with period T_0 expressed as a Fourier series is given by the expression

$$y(t) = \frac{a_0}{2} + \sum_{n=1}^{\infty} [a_n \cos(2\pi n f_0 t) + b_n \sin(2\pi n f_0 t)] \qquad (5.1)$$

where f_0 is the fundamental frequency equal to $1/T_0$. The magnitude of the sinusoids or coefficients are given by the integrals

$$a_n = \frac{2}{T_0} \int_{-T_0/2}^{T_0/2} y(t) \cos(2\pi n f_0 t) \, dt \qquad n = 0, 1, 2, 3, \ldots \qquad (5.2)$$

$$b_n = \frac{2}{T_0} \int_{-T_0/2}^{T_0/2} y(t) \sin(2\pi n f_0 t) \, dt \qquad n = 1, 2, 3, \ldots \qquad (5.3)$$

By applying the identities

$$\cos(2\pi n f_0 t) = \frac{1}{2}(e^{j2\pi n f_0 t} + e^{-j2\pi n f_0 t}) \qquad (5.4)$$

and

$$\sin(2\pi n f_0 t) = \frac{1}{2j}(e^{j2\pi n f_0 t} - e^{-j2\pi n f_0 t}) \qquad (5.5)$$

the expression of Eq. (5.1) can be written as

$$y(t) = \frac{a_0}{2} + \frac{1}{2} \sum_{n=1}^{\infty} (a_n - jb_n)e^{j2\pi n f_0 t}$$
$$+ \frac{1}{2} \sum_{n=1}^{\infty} (a_n + jb_n)e^{-j2\pi n f_0 t} \qquad (5.6)$$

To simplify this expression, negative values of n are introduced in Eqs. (5.2) and (5.3).

$$a_{-n} = \frac{2}{T_0} \int_{-T_0/2}^{T_0/2} y(t) \cos(-2\pi n f_0 t) \, dt$$
$$= \frac{2}{T_0} \int_{-T_0/2}^{T_0/2} y(t) \cos(2\pi n f_0 t) \, dt \qquad (5.7)$$
$$= a_n \qquad n = 1, 2, 3, \ldots$$

$$b_{-n} = \frac{2}{T_0} \int_{-T_0/2}^{T_0/2} y(t) \sin(-2\pi n f_0 t) \, dt$$
$$= -\frac{2}{T_0} \int_{-T_0/2}^{T_0/2} y(t) \sin(2\pi n f_0 t) \, dt \qquad (5.8)$$
$$= -b_n \qquad n = 1, 2, 3, \ldots$$

Hence, we can write

$$\sum_{n=1}^{\infty} a_n e^{-j2\pi n f_0 t} = \sum_{n=-1}^{-\infty} a_n e^{j2\pi n f_0 t} \qquad (5.9)$$

and

$$\sum_{n=1}^{\infty} jb_n e^{-j2\pi n f_0 t} = - \sum_{n=-1}^{-\infty} jb_n e^{j2\pi n f_0 t} \qquad (5.10)$$

Substitution of Eqs. (5.9) and (5.10) into Eq. (5.6) yields

$$y(t) = \frac{a_0}{2} + \frac{1}{2} \sum_{n=-\infty}^{\infty} (a_n - jb_n) e^{j2\pi n f_0 t} \qquad (5.11)$$

$$= \sum_{n=-\infty}^{\infty} \alpha_n e^{j2\pi n f_0 t}$$

Equation (5.11) is the Fourier series expressed in exponential form; coefficients α_n are, in general, complex. Because

$$\alpha_n = \frac{1}{2}(a_n - jb_n) \qquad n = 0, \pm 1, \pm 2, \ldots$$

the combination of Eqs. (5.2), (5.3), (5.7), and (5.8) yields

$$\alpha_n = \frac{1}{T_0} \int_{-T_0/2}^{T_0/2} y(t) e^{-j2\pi n f_0 t} dt \qquad n = 0, \pm 1, \pm 2, \ldots \qquad (5.12)$$

The expression of the Fourier series in exponential form, Eq. (5.11), and the complex coefficients in the form of Eq. (5.12) is normally the preferred approach in analysis.

Example 5.1   Triangular-Waveform Fourier Series

Determine the Fourier series of the periodic function illustrated in Fig. 5.1.

From Eq. (5.12), because $y(t)$ is an even function, then

$$\alpha_n = \begin{cases} \dfrac{1}{T_0} \displaystyle\int_{-T_0/2}^{T_0/2} y(t) \cos(2\pi n f_0 t)\, dt \\[2mm] \dfrac{1}{T_0} \displaystyle\int_{-T_0/2}^{0} \left(\dfrac{2}{T_0} + \dfrac{4}{T_0^2} t \right) \cos(2\pi n f_0 t)\, dt \\[2mm] \quad + \dfrac{1}{T_0} \displaystyle\int_{0}^{T_0/2} \left(\dfrac{2}{T_0} - \dfrac{4}{T_0^2} t \right) \cos(2\pi n f_0 t)\, dt \qquad n = 0, 1, 3, 5, \ldots \\[2mm] 0 \qquad\qquad\qquad\qquad n = 2, 4, 6, \ldots \end{cases}$$

$$(5.13)$$

$$\alpha_n = \begin{cases} \dfrac{4}{\pi^2 T_0} \dfrac{1}{n^2} \qquad n = 1, 3, 5, \ldots \\[3mm] \dfrac{1}{T_0} \qquad\qquad n = 0 \end{cases}$$

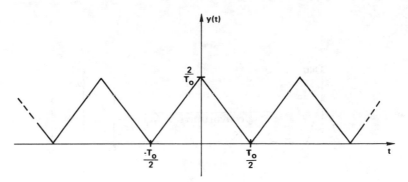

Figure 5.1   Periodic triangular waveform.

Hence,

$$y(t) = \frac{1}{T_0} + \frac{8}{\pi^2 T_0} \left[\cos(2\pi f_0 t) + \frac{1}{3^2} \cos(6\pi f_0 t) + \frac{1}{5^2} \cos(10\pi f_0 t) + \cdots \right]$$

(5.14)

where $f_0 = 1/T_0$.

5.2 FOURIER SERIES AS A SPECIAL CASE OF THE FOURIER INTEGRAL

Consider the periodic triangular function illustrated in Fig. 5.2(e). From Ex. 5.1, we know that the Fourier series of this waveform is an infinite set of sinusoids. We will now show that an identical relationship can be obtained from the Fourier integral.

To accomplish the derivation, we utilize the convolution theorem, Eq. (4.14). Note that the periodic triangular waveform (period T_0) is simply the convolution of the single triangle shown in Fig. 5.2(a) and the infinite sequence of equidistant impulses illustrated in Fig. 5.2(b). The periodic function $y(t)$ can then be expressed by

$$y(t) = h(t) * x(t)$$

(5.15)

Fourier transforms of both $h(t)$ and $x(t)$ have been determined previously and are illustrated in Figs. 5.2(c) and (d), respectively. From the convolution theorem, the desired Fourier transform is the product of these two frequency

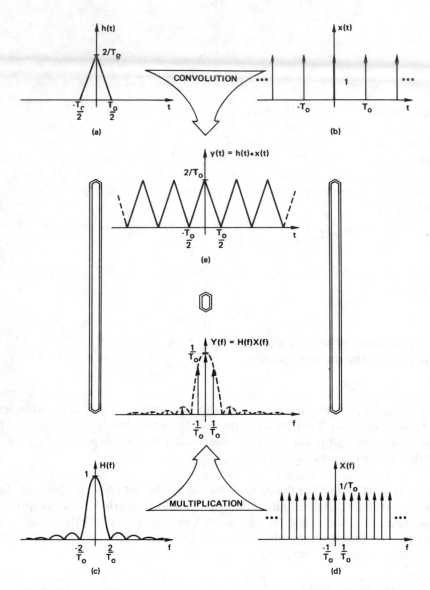

Figure 5.2 Graphical convolution theorem development of the Fourier transform of a periodic triangular waveform.

functions:

$$Y(f) = H(f)X(f)$$

$$= H(f)\frac{1}{T_0}\sum_{n=-\infty}^{\infty}\delta\left(f - \frac{n}{T_0}\right) \quad (5.16)$$

$$= \frac{1}{T_0}\sum_{n=-\infty}^{\infty}H\left(\frac{n}{T_0}\right)\delta\left(f - \frac{n}{T_0}\right)$$

Equations (2.44) and (4.19) were used to develop Eq. (5.16).

The Fourier transform of the periodic function is then an infinite set of sinusoids (i.e., an infinite sequence of equidistant impulses) with amplitudes of $H(n/T_0)$. Recall that the Fourier series of a periodic function is an infinite sum of sinusoids with amplitudes given by α_n, Eq. (5.12). But note that because the limits of integration of Eq. (5.12) are from $-T_0/2$ to $T_0/2$ and because

$$h(t) = y(t) \qquad -\frac{T_0}{2} < t < \frac{T_0}{2} \quad (5.17)$$

the function $y(t)$ can be replaced by $h(t)$ and Eq. (5.12) can be rewritten in the form:

$$\alpha_n = \frac{1}{T_0}\int_{-T_0/2}^{T_0/2} h(t)e^{-j2\pi n f_0 t}\,dt$$

$$= \frac{1}{T_0}H(nf_0) = \frac{1}{T_0}H\left(\frac{n}{T_0}\right) \quad (5.18)$$

Thus, the coefficients as derived by means of the Fourier integral and those of the conventional Fourier series are the same for a periodic function. Also, a comparison of Figs. 5.2(c) and (f) reveals that except for a factor $1/T_0$, the coefficients α_n of the Fourier series expansion of $y(t)$ equal the values of the Fourier transform $H(f)$ evaluated at n/T_0.

In summary, we point out again that the key to the preceding development is the incorporation of distribution theory into Fourier integral theory. As will be demonstrated in the discussions to follow, this unifying concept is basic to a thorough understanding of the discrete Fourier transform and hence the fast Fourier transform.

5.3 WAVEFORM SAMPLING

In the preceding chapters, we have developed a Fourier transform theory that considers both continuous and impulse functions of time. Based on these developments, it is straightforward to extend the theory to include *sampled* waveforms, which are of particular interest in this book. We have developed sufficient tools to investigate in detail the theoretical as well as the visual interpretations of sampled waveforms.

If the function $h(t)$ is continuous at $t = T$, then a sample of $h(t)$ at time equal to T is expressed as

$$\hat{h}(t) = h(t)\delta(t - T) = h(T)\delta(t - T) \qquad (5.19)$$

where the product must be interpreted in the sense of distribution theory [Eq. (A.12)]. The impulse that occurs at time T has an area equal to the function value at time T. If $h(t)$ is continuous at $t = nT$ for $n = 0, \pm 1, \pm 2, \ldots$,

$$\hat{h}(t) = \sum_{n=-\infty}^{\infty} h(nT)\delta(t - nT) \qquad (5.20)$$

$\hat{h}(t)$ is termed the sampled waveform $h(t)$ with sample interval T. Sampled $h(t)$ is then an infinite sequence of equidistant impulses, each of whose amplitude is given by the value of $h(t)$ corresponding to the time of occurrence of the impulse. Figure 5.3 illustrates graphically the sampling concept. Since Eq. (5.20) is the product of the continuous function $h(t)$ and the sequence of impulses, we can employ the frequency-convolution theorem, Eq. (4.17), to derive the Fourier transform of the sampled waveform. As illustrated in Fig. 5.3, the sampled function [Fig. 5.3(e)] is equal to the product of the waveform $h(t)$ shown in Fig. 5.3(a) and the sequence of impulses $\Delta(t)$ illustrated in Fig. 5.3(b). We call $\Delta(t)$ the sampling function; the notation $\Delta(t)$ will always imply an infinite sequence of impulses separated by T. The Fourier transforms of $h(t)$ and $\Delta(t)$ are shown in Figs. 5.3(c) and (d), respectively. Note that the Fourier transform of the sampling function $\Delta(t)$ is $\Delta(f)$; this function is termed the frequency-sampling function. From the frequency-convolution theorem, the desired Fourier transform is the convolution of the frequency functions illustrated in Figs. 5.3(c) and (d). The Fourier transform of the sampled waveform is then a periodic function, where one period is equal, within a constant, to the Fourier transform of the continuous function $h(t)$. This last statement is valid only if the sampling interval T is sufficiently small.

If T is chosen too large, the results illustrated in Fig. 5.4 are obtained. Note that as the sample interval T is increased [Figs. 5.3(b) and 5.4(b)], the equidistant impulses of $\Delta(f)$ become more closely spaced [Figs. 5.3(d) and 5.4(d)]. Because of the decreased spacing of the frequency impulses, their convolution with the frequency function $H(f)$ [Fig. 5.4(c)] results in the overlapping waveform illustrated in Fig. 5.4(f). This distortion of the desired Fourier transform of a sampled function is known as *aliasing*. As described, aliasing occurs because the time function was not sampled at a sufficiently high rate, i.e., the sample interval T is too large. It is then natural to pose the question: How does one ensure that the Fourier transform of a sampled function is not aliased? An examination of Figs. 5.4(c) and (d) points up the

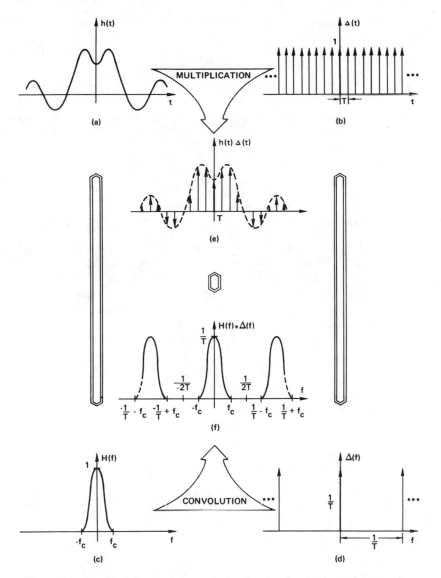

Figure 5.3 Graphical frequency-convolution theorem development of the Fourier transform of a sampled waveform.

fact that convolution overlap occurs until the separation of the impulses of $\Delta(f)$ is increased to $1/T = 2f_c$, where f_c is the highest frequency component of the Fourier transform of the continuous function $h(t)$. That is, if the sample interval T is chosen equal to one-half the reciprocal of the highest frequency component, aliasing does not occur. This is an extremely important concept in many fields of scientific application; the reason is that we need only retain

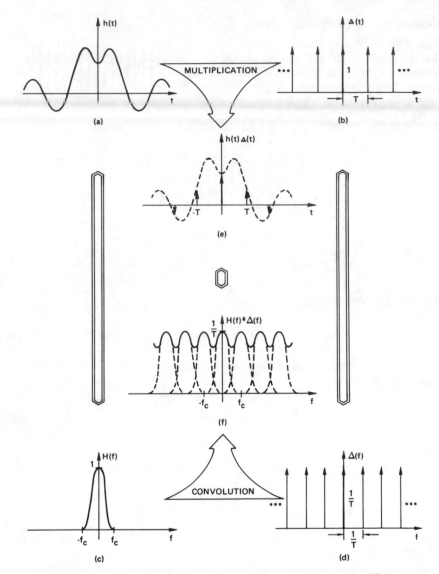

Figure 5.4   Aliased Fourier transform of a waveform sampled at an insufficient rate.

samples of the continuous waveform to determine a replica of the continuous Fourier transform. Furthermore, if a waveform is sampled such that aliasing does not occur, these samples can be appropriately combined to reconstruct identically the continuous waveform. This is merely a statement of the sampling theorem that we investigate in Sec. 5.4.

Example 5.1   Time-Domain Aliasing

Figure 5.5 illustrates the concept of aliasing from a time-domain viewpoint. Note that the sample interval T was not chosen less than one-half the period of the highest frequency component of the time waveform. As a result, the equally spaced sample values shown can represent at least two sinusoids of different frequencies. In the time domain, aliasing is characterized by the inability to distinguish the frequency of the sinusoid that the sample values represent.

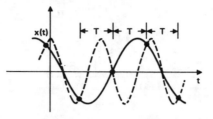

Figure 5.5   Time-domain example of aliasing.

5.4 SAMPLING THEOREMS

The sampling theorem states that if the Fourier transform of a function $h(t)$ is zero for all frequencies greater than a certain frequency f_c, then the continuous function $h(t)$ can be uniquely determined from a knowledge of its sampled values

$$\hat{h}(t) = \sum_{n=-\infty}^{\infty} h(nT)\delta(t - nT) \tag{5.21}$$

where $T = 1/2f_c$.

In particular, $h(t)$ is given by

$$h(t) = T \sum_{n=-\infty}^{\infty} h(nT) \frac{\sin 2\pi f_c(t - nT)}{\pi(t - nT)} \tag{5.22}$$

Constraints of the theorem are illustrated graphically in Fig. 5.6. First, it is necessary that the Fourier transform of $h(t)$ be zero for frequencies greater than f_c. As shown in Fig. 5.6(c), the example frequency function is *band-limited* at the frequency f_c; the term *band-limited* is a shortened way of saying that the Fourier transform is zero for $|f| > f_c$. The bandwidth of a signal is the width of the positive frequency band where the amplitude is nonzero. The bandwidth of the waveform illustrated in Fig. 5.6(c) is then f_c. The second constraint is that the sample spacing be chosen as $T = 1/2f_c$, that is, the impulse functions of Fig. 5.6(d) are required to be separated by $1/T = 2f_c$. This spacing ensures that when $\Delta(f)$ and $H(f)$ are convolved, there is no aliasing. Alternately, the functions $H(f)$ and $H(f) * \Delta(f)$, as

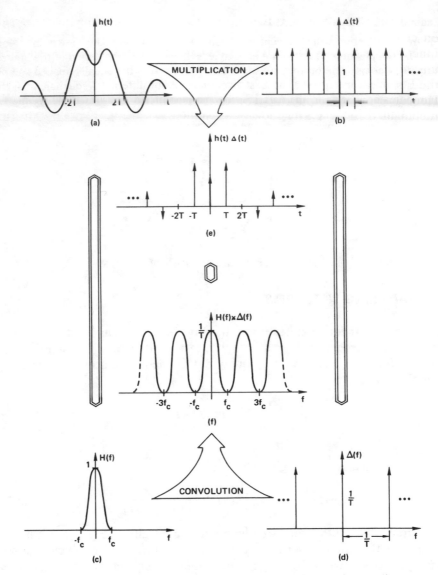

Figure 5.6   Fourier transform of a waveform sampled at the Nyquist sampling rate.

illustrated in Figs. 5.6(c) and (f), respectively, are equal in the interval $|f|$ $< f_c$, within the scaling constant T. If $T > 1/2f_c$, then aliasing will result; if $T < 1/2f_c$, the theorem still holds. The requirement that $T = 1/2f_c$ is simply the maximum spacing between samples for which the theorem holds. Frequency $1/T = 2f_c$ is known as the *Nyquist sampling rate*. Given that these two constraints are true, the theorem states that $h(t)$ [Fig. 5.6(a)] can be reconstructed from a knowledge of the impulses illustrated in Fig. 5.6(e).

To construct a proof of the sampling theorem, recall from the discussion on constraints of the theorem that the Fourier transform of the sampled function is identical, within the constant T, to the Fourier transform of the unsampled function, in the frequency range $-f_c \leq f \leq f_c$. From Fig. 5.6(f), the Fourier transform of the sampled time function is given by $H(f) * \Delta(f)$. Hence, as illustrated in Figs. 5.7(a), (b), and (e), the multiplication of a rectangular-frequency function of amplitude T with the Fourier transform

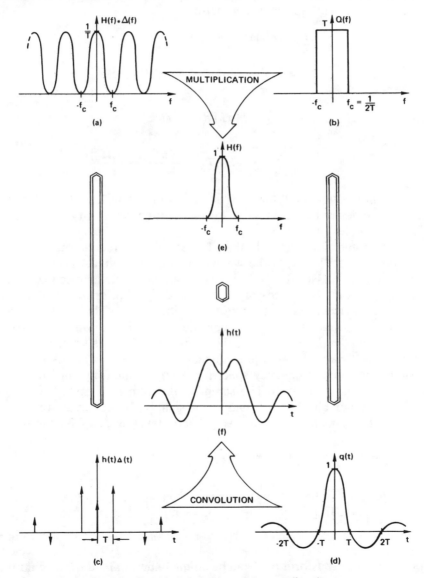

Figure 5.7 Graphical derivation of the sampling theorem.

of the sampled waveform is the Fourier transform $H(f)$:

$$H(f) = [H(f) * \Delta(f)]Q(f) \qquad (5.23)$$

The inverse Fourier transform of $H(f)$ is the original waveform $h(t)$, as shown in Fig. 5.7(f). But from the convolution theorem, $h(t)$ is equal to the convolution of the inverse Fourier transforms of $H(f) * \Delta(f)$ and of the rectangular-frequency function. Hence, $h(t)$ is given by the convolution of $h(t)\Delta(t)$ [Fig. 5.7(c)] and $q(t)$ [Fig. 5.7(d)]:

$$
\begin{aligned}
h(t) &= [h(t)\Delta(t)] * q(t) \\
&= \sum_{n=-\infty}^{\infty} [h(nT)\delta(t - nT)] * q(t) \\
&= \sum_{n=-\infty}^{\infty} h(nT)q(t - nT) \\
&= T \sum_{n=-\infty}^{\infty} h(nT) \frac{\sin[2\pi f_c(t - nT)]}{\pi(t - nT)}
\end{aligned}
\qquad (5.24)
$$

Function $q(t)$ is given by the Fourier transform pair of Eq. (2.31). Equation (5.24) is the desired expression for reconstructing $h(t)$ from a knowledge of only the samples of $h(t)$.

We should note carefully that it is possible to reconstruct a sampled waveform perfectly only if the waveform is band-limited. In practice, this condition rarely exists. The solution is to sample at such a rate that aliasing is negligible; it may be necessary to filter the signal prior to quantization to ensure that there exists, to the extent possible, a band-limited function.

The band-limited waveforms considered in this section are referred to as *baseband signals*. This nomenclature refers to signals whose frequency spectrum *generally* occupy the frequency range $0 \leq f < f_c$. A *band-pass signal* is one whose frequency spectrum occupies the frequency range $f_{low} < f < f_{high}$ and $f_{low} \gg 0$. The sampling theorem developed here can be applied to either baseband or band-pass signals. However, more efficient sampling theorems for band-pass signals are developed in Chapter 14.

Frequency-Sampling Theorem

Analogous to time-domain sampling is a sampling theorem in the frequency domain. If a function $h(t)$ is time-limited, that is,

$$h(t) = 0 \qquad |t| > T_c \qquad (5.25)$$

then its Fourier transform $H(f)$ can be uniquely determined from equidistant samples of $H(f)$. In particular, $H(f)$ is given by

$$H(f) = \frac{1}{2T_c} \sum_{n=-\infty}^{\infty} H\left(\frac{n}{2T_c}\right) \frac{\sin[2\pi T_c(f - n/2T_c)]}{\pi(f - n/2T_c)} \tag{5.26}$$

The proof is similar to the proof of the time-domain sampling theorem.

PROBLEMS

5.1. Find the Fourier series of the periodic waveforms illustrated in Fig. 5.8.

5.2. Determine the Fourier transform of the waveforms illustrated in Fig. 5.8. Compare these results with those of Problem 5.1.

5.3. By using graphical arguments similar to those of Fig. 5.4, determine the Nyquist sampling rate for the time functions whose Fourier transform magnitude functions are illustrated in Fig. 5.9.

5.4. Graphically justify the band-pass sampling theorem that states that

$$\text{Critical sampling frequency} = \frac{2f_{high}}{\text{largest integer not exceeding } \dfrac{f_{high}}{(f_{high} - f_{low})}}$$

where f_{high} and f_{low} are the upper and lower cutoff frequencies of the band-pass spectrum.

5.5. Assume that the function $h(t) = \cos(2\pi t)$ has been sampled at $t = n/4$, where $n = 0, \pm 1, \pm 2, \ldots$. Sketch $h(t)$ and indicate the sampled values. Graphically and analytically determine Eq. (5.24) for $h(t = 7/8)$, where the summation is only over $n = 2, 3, 4,$ and 5.

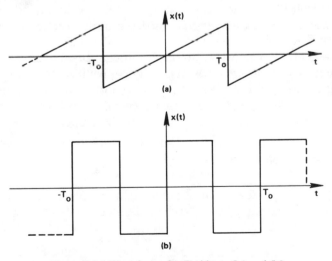

Figure 5.8 Waveforms for Problems 5.1 and 5.2.

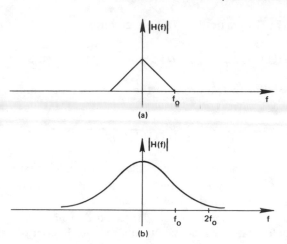

Figure 5.9 Functions for Problem 5.3.

5.6. A frequency function (say a filter frequency response) has been determined experimentally in the laboratory and is given by a graphical curve. If it is desired to sample this function for computer storage purposes, what is the minimum frequency-sampling interval if the frequency function is to be later totally reconstructed? State all assumptions.

REFERENCES

1. BRACEWELL, R. *The Fourier Transform and Its Applications*, 2d Ed. New York: McGraw-Hill, 1986.

2. PAPOULIS, A. *The Fourier Integral and Its Applications*, 2d Ed. New York: McGraw-Hill, 1984.

3. SCHWARTZ, M., AND L. SHAW. *Signal Processing: Discrete Spectral Analysis, Detection, and Estimation*. New York: McGraw-Hill, 1975.

6

THE DISCRETE FOURIER TRANSFORM

Normally, a discussion of the discrete Fourier transform is based on an initial definition of the finite-length discrete transform; from this assumed axiom, those properties of the transform implied by this definition are derived. This approach is unrewarding in that at its conclusion there is always the unanswered question: How does the discrete Fourier transform relate to the continuous Fourier transform? To answer this question, we find it preferable to derive the discrete Fourier transform as a special case of continuous Fourier transform theory. Clearly, the discrete Fourier transform can be defined independently of the Fourier transform. However, many applications involving the continuous Fourier transform rely on a digital computer for implementation, which leads to the use of the discrete Fourier transform and hence the FFT. Both approaches yield identical results; the distinction is in the interpretation of the results.

In this chapter, we develop a special case of the continuous Fourier transform that is amenable to machine computation. The approach is to develop the discrete Fourier transform from a graphical derivation based on continuous Fourier transform theory. These graphical arguments are then substantiated by a theoretical development. Both approaches emphasize the modifications of continuous Fourier transform theory that are necessary to define a computer-oriented transform pair. We also develop properties of the discrete Fourier transform.

6.1 A GRAPHICAL DEVELOPMENT

Consider the example function $h(t)$ and its Fourier transform $H(f)$, as illustrated in Fig. 6.1(a). It is desired to modify this Fourier transform pair in such a manner that the pair is amenable to digital computer computation. This modified pair, termed the *discrete Fourier transform*, is to approximate as closely as possible the continuous Fourier transform.

To determine the Fourier transform of $h(t)$ by means of digital analysis techniques, it is necessary to sample $h(t)$, as described in Chapter 5. Sampling is accomplished by multiplying $h(t)$ by the sampling function illustrated in Fig. 6.1(b). The sample interval is T. Sampled function $\hat{h}(t)$ and its Fourier transform are illustrated in Fig. 6.1(c). This Fourier transform pair represents the first modification to the original pair, which is necessary in defining a discrete transform pair. Note that to this point the modified transform pair differs from the original transform pair only by the aliasing effect that results from sampling. As discussed in Sec. 5.3, if the waveform $h(t)$ is sampled at a frequency of at least twice the largest frequency component of $h(t)$, there is no loss of information as a result of sampling. If the function $h(t)$ is not band-limited, i.e., $H(f) \neq 0$ for some $|f| > f_c$, then sampling will introduce aliasing, as illustrated in Fig. 6.1(c). To reduce this error, we have only one recourse, and that is to sample faster, that is, choose T smaller.

The Fourier transform pair in Fig. 6.1(c) is not suitable for machine computation because an infinity of samples of $h(t)$ is considered; it is necessary to truncate the sampled $\hat{h}(t)$ so that only a finite number of points, say N, are considered. The rectangular, or truncation, function and its Fourier transform are illustrated in Fig. 6.1(d). The product of the infinite sequence of impulse functions representing $h(t)$ and the truncation function yields the finite-length time function illustrated in Fig. 6.1(e). Truncation introduces the second modification of the original Fourier transform pair; this effect is to convolve the aliased frequency transform of Fig. 6.1(c) with the Fourier transform of the truncation function [Fig. 6.1(d)]. As shown in Fig. 6.1(e), the frequency transform now has a *ripple* to it; this effect has been accentuated in the illustration for emphasis. To reduce this effect, recall the inverse relation that exists between the *width* of a time function and its Fourier transform (Sec. 3.3). Hence, if the truncation (rectangular) function is increased in length, then the $[\sin(f)]/f$ function approaches an impulse; the more closely the $[\sin(f)]/f$ function approximates an impulse, the less ripple or error is introduced by the convolution that results from truncation. Therefore, it is desirable to choose the length of the truncation function as long as possible. We investigate the effect of truncation in detail in Sec. 6.4.

The modified transform pair of Fig. 6.1(e) is still not an acceptable discrete Fourier transform pair because the frequency transform is a continuous function. For machine computation, only sample values of the fre-

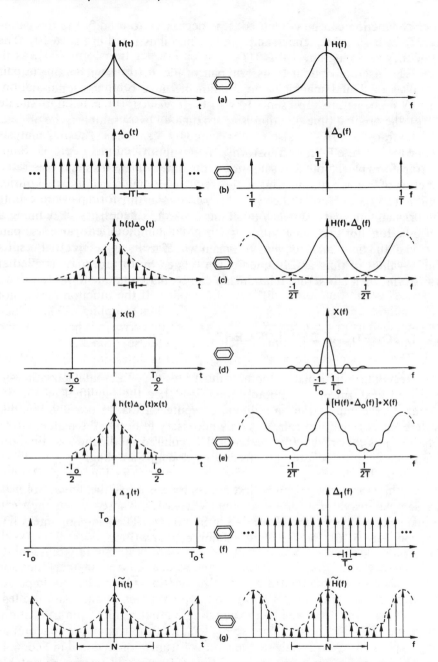

Figure 6.1 Graphical development of the discrete Fourier transform.

quency function can be computed; it is necessary to modify the frequency transform by the frequency-sampling function illustrated in Fig. 6.1(f). The frequency-sampling interval is $1/T_0$.

The discrete Fourier transform pair of Fig. 6.1(g) is acceptable for the purposes of digital machine computation because both the time and frequency domains are represented by discrete values. As illustrated in Fig. 6.1(g), the original time function is approximated by N samples; the original Fourier transform $H(f)$ is also approximated by N samples. These N samples define the discrete Fourier transform pair and approximate the original Fourier transform pair. Note that sampling in the time domain results in a periodic function of frequency; sampling in the frequency domain results in a periodic function of time. Hence, the discrete Fourier transform requires that both the original time and frequency functions be modified such that they become periodic functions. N time samples and N frequency values represent one period of the time- and frequency-domain waveforms, respectively. Because the N values of time and frequency are related by the continuous Fourier transform, then a discrete relationship can be derived.

6.2 THEORETICAL DEVELOPMENT

The preceding graphical development illustrates the point that if a continuous Fourier transform pair is suitably modified, then the modified pair is acceptable for computation on a digital computer. Thus, to develop this discrete Fourier transform pair, it is only necessary to derive the mathematical relationships that result from each of the required modifications: time-domain sampling, truncation, and frequency-domain sampling.

Consider the Fourier transform pair illustrated in Fig. 6.2(a). To discretize this transform pair, it is first necessary to sample the waveform $h(t)$; the sampled waveform can be written as $h(t)\Delta_0(t)$, where $\Delta_0(t)$ is the time-domain sampling function illustrated in Fig. 6.2(b). The sampling interval is T. From Eq. (5.20), the sampled function can be written as

$$h(t)\Delta_0(t) = h(t) \sum_{k=-\infty}^{\infty} \delta(t - kT)$$

$$= \sum_{k=-\infty}^{\infty} h(kT)\delta(t - kT)$$

(6.1)

The result of this multiplication is illustrated in Fig. 6.2(c). Note the aliasing effect that results from the choice of T.

Next, the sampled function is truncated by multiplication with the rec-

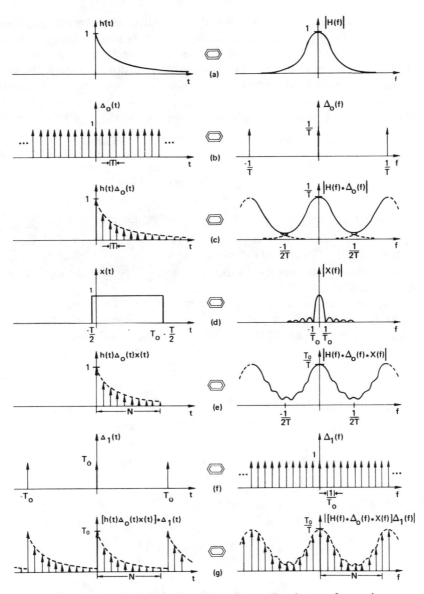

Figure 6.2 Graphical derivation of the discrete Fourier transform pair.

tangular function $x(t)$, as illustrated in Fig. 6.2(d):

$$x(t) = 1 \qquad -\frac{T}{2} < t < T_0 - \frac{T}{2}$$

$$= 0 \qquad \text{otherwise}$$

$$(6.2)$$

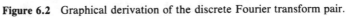

where T_0 is the duration of the truncation function. An obvious question at this point is: Why is the rectangular function $x(t)$ not centered at zero or $T_0/2$? Centering of $x(t)$ at zero is avoided to alleviate notation problems. The reason for not centering the rectangular function at $T_0/2$ will become obvious later in the development.

Truncation yields

$$h(t)\Delta_0(t)x(t) = \left[\sum_{k=-\infty}^{\infty} h(kT)\delta(t - kT) \right] x(t)$$

$$= \sum_{k=0}^{N-1} h(kT)\delta(t - kT)$$

(6.3)

where it has been assumed that there are N equidistant impulse functions lying within the truncation interval, that is, $N = T_0/T$. The sampled truncated waveform and its Fourier transform are illustrated in Fig. 6.2(e). As in the previous example, truncation in the time domain results in *rippling* in the frequency domain.

The final step in modifying the original Fourier transform pair to a discrete Fourier transform pair is to sample the Fourier transform of Eq. (6.3). In the time domain, this product is equivalent to convolving the sampled truncated waveform of Eq. (6.3) and the time function $\Delta_1(t)$, as illustrated in Fig. 6.2(f). Function $\Delta_1(t)$ is given by Fourier transform pair of Eq. (2.44) as

$$\Delta_1(t) = T_0 \sum_{r=-\infty}^{\infty} \delta(t - rT_0)$$

(6.4)

The desired relationship is $[h(t)\Delta_0(t)x(t)] * \Delta_1(t)$; hence,

$$[h(t)\Delta_0(t)x(t)] * \Delta_1(t) = \left[\sum_{k=0}^{N-1} h(kT)\delta(t - kT) \right]$$

$$* \left[T_0 \sum_{r=-\infty}^{\infty} \delta(t - rT_0) \right]$$

$$= \ldots + T_0 \sum_{k=0}^{N-1} h(kT)\delta(t + T_0 - kT)$$

(6.5)

$$+ T_0 \sum_{k=0}^{N-1} h(kT)\delta(t - kT)$$

$$+ T_0 \sum_{k=0}^{N-1} h(kT)\delta(t - T_0 - kT) + \ldots$$

Note that Eq. (6.5) is periodic with period T_0; in compact notation form, the equation can be written as

$$\bar{h}(t) = T_0 \sum_{r=-\infty}^{\infty} \left[\sum_{k=0}^{N-1} h(kT)\delta(t - kT - rT_0) \right] \tag{6.6}$$

We choose the notation $\bar{h}(t)$ to imply that $\bar{h}(t)$ is an approximation to the function $h(t)$.

Choice of the rectangular function $x(t)$, as described by Eq. (6.2), can now be explained. Note that the convolution result of Eq. (6.6) is a periodic function with period T_0 that consists of N samples. If the rectangular function had been chosen such that a sample value coincided with each end point of the rectangular function, the convolution of the rectangular function with impulses spaced at intervals of T_0 would result in time-domain aliasing. That is, the Nth point of one period would coincide with (and add to) the first point of the next period. To ensure that time-domain aliasing does not occur, it is necessary to choose the truncation interval as illustrated in Fig. 6.2(d). (The truncation function can also be chosen as illustrated in Fig. 6.1(d), but note that the end points of the truncation function lie at the midpoint of two adjacent sample values to avoid time-domain aliasing.)

To develop the Fourier transform of Eq. (6.6), recall from the discussion on Fourier series, Sec. 5.1, that the Fourier transform of a periodic function $h(t)$ is a sequence of equidistant impulses:

$$\tilde{H}\left(\frac{n}{T_0}\right) = \sum_{n=-\infty}^{\infty} \alpha_n \delta(f - nf_0) \qquad f_0 = \frac{1}{T_0} \tag{6.7}$$

where

$$\alpha_n = \frac{1}{T_0} \int_{-T/2}^{T_0-T/2} \bar{h}(t) e^{-j2\pi nt/T_0}\, dt \qquad n = 0, \pm 1, \pm 2, \ldots \tag{6.8}$$

Substituting Eq. (6.6) in (6.8) yields

$$\alpha_n = \frac{1}{T_0} \int_{-T/2}^{T_0-T/2} T_0 \sum_{r=-\infty}^{\infty} \sum_{k=0}^{N-1} h(kT)\delta(t - kT - rT_0) e^{-j2\pi nt/T_0}\, dt$$

Integration is only over one period; hence,

$$\alpha_n = \int_{-T/2}^{T_0-T/2} \sum_{k=0}^{N-1} h(kT)\delta(t - kT) e^{-j2\pi nt/T_0}\, dt$$

$$= \sum_{k=0}^{N-1} h(kT) \int_{-T/2}^{T_0-T/2} e^{-j2\pi nt/T_0} \delta(t - kT)\, dt \tag{6.9}$$

$$= \sum_{k=0}^{N-1} h(kT) e^{-j2\pi knT/T_0}$$

Because $T_0 = NT$, Eq. (6.9) can be rewritten as

$$\alpha_n = \sum_{k=0}^{N-1} h(kT)e^{-j2\pi kn/N} \qquad n = 0, \pm 1, \pm 2, \ldots \qquad (6.10)$$

and the Fourier transform of Eq. (6.6) is

$$\tilde{H}\left(\frac{n}{NT}\right) = \sum_{n=-\infty}^{\infty} \sum_{k=0}^{N-1} h(kT)e^{j2\pi kn/N} \qquad (6.11)$$

For a cursory evaluation of Eq. (6.11), it is not obvious that the Fourier transform $\tilde{H}(n/NT)$ is periodic, as illustrated in Fig. 6.2(g). However, there are only N distinct complex values computable from Eq. (6.11). To establish this fact, let $n = r$, where r is an arbitrary integer; Eq. (6.11) becomes

$$\tilde{H}\left(\frac{r}{NT}\right) = \sum_{k=0}^{N-1} h(kT)e^{-j2\pi kr/N} \qquad (6.12)$$

Now let $n = r + N$; note that

$$e^{-j2\pi k(r+N)/N} = e^{-j2\pi kr/N}e^{-j2\pi k} \qquad (6.13)$$
$$= e^{-j2\pi kr/N}$$

because $e^{-j2\pi k} = \cos(2\pi k) - j\sin(2\pi k) = 1$ for k integer valued. Thus, for $n = r + N$,

$$\tilde{H}\left(\frac{r+N}{NT}\right) = \sum_{k=0}^{N-1} h(kT)e^{-j2\pi k(r+N)/N}$$
$$= \sum_{k=0}^{N-1} h(kT)e^{-j2\pi kr/N} \qquad (6.14)$$
$$= \tilde{H}\left(\frac{r}{NT}\right)$$

Therefore, there are only N distinct values for which Eq. (6.11) can be evaluated; $\tilde{H}(n/NT)$ is periodic with a period of N samples. The Fourier transform, Eq. (6.11), can be expressed equivalently as

$$\tilde{H}\left(\frac{n}{NT}\right) = \sum_{k=0}^{N-1} h(kT)e^{-j2\pi nk/N} \qquad n = 0, 1, \ldots, N-1 \qquad (6.15)$$

Equation (6.15) is the desired discrete Fourier transform; the expression relates N samples of time and N samples of frequency by means of the continuous Fourier transform. The discrete Fourier transform is then a special case of the continuous Fourier transform. If it is assumed that the N samples of the original function $h(t)$ are one period of a periodic waveform, the Fourier transform of this periodic function is given by the N samples, as computed by Eq. (6.15). Notation $\tilde{H}(n/NT)$ indicates that the discrete

Fourier transform is an approximation to the continuous Fourier transform. Normally, Eq. (6.15) is written as

$$G\left(\frac{n}{NT}\right) = \sum_{k=0}^{N-1} g(kT)e^{-j2\pi nk/N} \qquad n = 0, 1, \ldots, N - 1 \qquad (6.16)$$

because the Fourier transform of the sampled periodic function $g(kT)$ is identically $G(n/NT)$.

6.3 DISCRETE INVERSE FOURIER TRANSFORM

The discrete inverse Fourier transform is given by

$$g(kT) = \frac{1}{N} \sum_{n=0}^{N-1} G\left(\frac{n}{NT}\right) e^{j2\pi nk/N} \qquad k = 0, 1, \ldots, N - 1 \qquad (6.17)$$

To prove that Eq. (6.17) and the transform relation, Eq. (6.16), form a discrete Fourier transform pair, substitute Eq. (6.17) into Eq. (6.16).

$$G\left(\frac{n}{NT}\right) = \sum_{k=0}^{N-1} \left[\frac{1}{N} \sum_{r=0}^{N-1} G\left(\frac{r}{NT}\right) e^{j2\pi rk/N}\right] e^{-j2\pi nk/N}$$

$$= \frac{1}{N} \sum_{r=0}^{N-1} G\left(\frac{r}{NT}\right) \left[\sum_{k=0}^{N-1} e^{j2\pi rk/N} e^{-j2\pi nk/N}\right] \qquad (6.18)$$

$$= G\left(\frac{n}{NT}\right)$$

The identity of Eq. (6.18) follows from the orthogonality relationship:

$$\sum_{k=0}^{N-1} e^{j2\pi rk/N} e^{-j2\pi nk/N} = \begin{cases} N & r = n \\ 0 & \text{otherwise} \end{cases} \qquad (6.19)$$

The discrete inversion formula, Eq. (6.17), exhibits periodicity in the same manner as the discrete transform; the period is defined by N samples of $g(kT)$. This property results from the periodic nature of $e^{j2\pi nk/N}$. Hence, $g(kT)$ is actually defined on the complete set of integers $k = 0, \pm 1, \pm 2, \ldots$ and is constrained by the identity

$$g(kT) = g[(rN + k)T] \qquad r = 0, \pm 1, \pm 2, \ldots \qquad (6.20)$$

In summary, the discrete Fourier transform pair is given by

$$g(kT) = \frac{1}{N} \sum_{n=0}^{N-1} G\left(\frac{n}{NT}\right) e^{j2\pi nk/N} \quad \Longleftrightarrow \quad G\left(\frac{n}{NT}\right) = \sum_{k=0}^{N-1} g(kT)e^{-j2\pi nk/N}$$

$$(6.21)$$

It is important to remember that the transform pair of Eq. (6.21) requires both the time- and frequency-domain functions to be periodic:

$$G\left(\frac{n}{NT}\right) = G\left[\frac{(rN + n)}{NT}\right] \qquad r = 0, \pm 1, \pm 2, \ldots \qquad (6.22)$$

$$g(kT) = g[(rN + k)T] \qquad r = 0, \pm 1, \pm 2, \ldots \qquad (6.23)$$

6.4 RELATIONSHIP BETWEEN THE DISCRETE AND CONTINUOUS FOURIER TRANSFORM

The discrete Fourier transform is of interest primarily because it approximates the continuous Fourier transform. Validity of this approximation is strictly a function of the waveform being analyzed. In this section, we use graphical analysis to indicate for general classes of functions the degree of equivalence between the discrete and continuous transform. As will be stressed, differences in the two transforms arise because of the discrete transform requirement for sampling and truncation.

Band-Limited Periodic Waveforms: Truncation Internal Equal to Period

Consider the function $h(t)$ and its Fourier transform, as illustrated in Fig. 6.3(a). We wish to sample $h(t)$, truncate the sampled function to N samples, and apply the discrete Fourier transform of Eq. (6.16). Rather than applying this equation directly, we develop its application graphically. Waveform $h(t)$ is sampled by multiplication with the sampling function, as illustrated in Fig. 6.3(b). Sampled waveform $h(kT)$ and its Fourier transform are illustrated in Fig. 6.3(c). Note that for this example there is no aliasing. Also observe that as a result of time-domain sampling, the frequency domain has been scaled by the factor $1/T$; the Fourier transform impulse now has an area of $A/2T$ rather than the original area of $A/2$. The sampled waveform is truncated by multiplication with the rectangular function, as illustrated in Fig. 6.3(d); Fig. 6.3(e) illustrates the sampled and truncated waveform. As shown, we chose the rectangular function, so that the N sample values remaining after truncation equate to one period of the original waveform $h(t)$.

The Fourier transform of the finite-length sampled waveform [Fig. 6.3(e)] is obtained by convolving the frequency-domain impulse functions of Fig. 6.3(c) and the $[\sin(f)]/f$ frequency function of Fig. 6.3(d). Figure 6.3(e) illustrates the convolution results; an expanded view of this convolution is shown in Fig. 6.4(b). A $[\sin(f)]/f$ function (dashed line) is centered on each impulse of Fig. 6.4(a) and the resultant waveforms are additively combined (solid line) to form the convolution result.

With respect to the original transform $H(f)$, the convolved frequency function [Fig. 6.4(b)] is significantly distorted. However, when this function

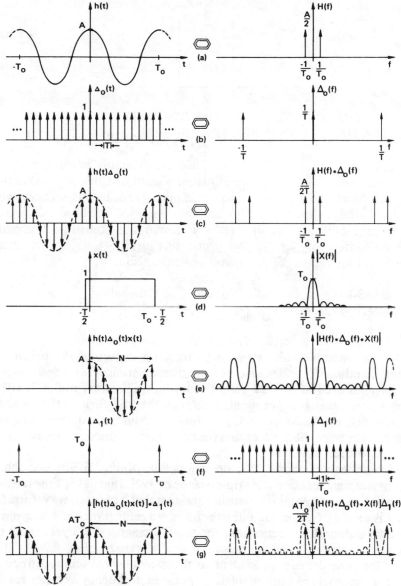

Figure 6.3 Discrete Fourier transform of a band-limited periodic waveform: the truncation interval is equal to one period.

is sampled by the frequency-sampling function illustrated in Fig. 6.3(f), the distortion is eliminated. This follows because the equidistant impulses of the frequency-sampling function are separated by $1/T_0$; at these frequencies, the solid line of Fig. 6.4(b) is zero except at frequency $\pm 1/T_0$. Frequency

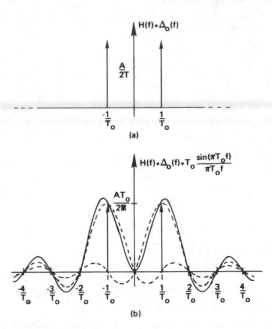

Figure 6.4 Expanded illustration of the convolution of Fig. 6.3(e).

$\pm 1/T_0$ corresponds to the frequency-domain impulses of the original frequency function $H(f)$. Because of time-domain truncation, these impulses
now have an area of $AT_0/2T$ rather than the original area of $A/2$. (Figure
6.4(b) does not take into account that the Fourier transform of the truncation
function $x(t)$, as illustrated in Fig. 6.4(d), is actually a complex frequency
function; however, had we considered a complex function, similar results
would have been obtained.)

Multiplication of the frequency function of Fig. 6.3(e) and the frequency-sampling function $\Delta_1(f)$ implies the convolution of the time functions
shown in Figs. 6.3(e) and (f). Because the sampled truncated waveform [Fig.
6.3(e)] is exactly one period of the original waveform $h(t)$ and because the
time-domain impulse functions of Fig. 6.3(f) are separated by T_0, then their
convolution yields a periodic function, as illustrated in Fig. 6.3(g). This is
simply the time-domain equivalent to the previously discussed frequency
sampling that yielded only a single impulse or frequency component. The
time function of Fig. 6.3(g) has a maximum amplitude of AT_0 as compared
to the original maximum value of A as a result of frequency-domain sampling.

Examination of Fig. 6.3(g) indicates that we have taken our original
time function, sampled it, and then multiplied each sample by T_0. The Fourier transform of this function is related to the original frequency function by
the factor $AT_0/2T$. Factor T_0 is common and can be eliminated. If we desire
to compute the Fourier transform by means of the discrete Fourier trans-

form, it is necessary to multiply the discrete time function by the factor T, which yields the desired $A/2$ area for the frequency function; Eq. (6.16) thus becomes

$$H\left(\frac{n}{NT}\right) = T \sum_{k=0}^{N-1} h(kT)e^{-j2\pi nk/N} \qquad (6.24)$$

We expect this result because the relationship of Eq. (6.24) is simply the rectangular rule for integration of the continuous Fourier transform.

This example represents the only class of waveforms for which the discrete and continuous Fourier transforms are exactly the same within a scaling constant. Equivalence of the two transforms requires (1) the time function $h(t)$ must be periodic, (2) $h(t)$ must be band-limited, (3) the sampling rate must be at least two times the largest frequency component of $h(t)$, and (4) the truncation function $x(t)$ must be nonzero over exactly one period (or integer multiple period) of $h(t)$.

Band-Limited Periodic Waveforms: Truncation Interval Not Equal to Period

If a band-limited periodic function is sampled and truncated to consist of other than an integer multiple of the period, the resulting discrete and continuous Fourier transform differs considerably. To examine this effect, consider the illustrations of Fig. 6.5. This example differs from the preceding only in the frequency of the sinusoidal waveform $h(t)$. As before, function $h(t)$ is sampled [Fig. 6.5(c)] and truncated [Fig. 6.5(e)]. Note that the sampled truncated function is not an integer multiple of the period of $h(t)$; therefore, when the time functions of Figs. 6.5(c) and (f) are convolved, the periodic waveform of Fig. 6.5(g) results. Although this function is periodic, it is not a replica of the original periodic function $h(t)$. We would not expect the Fourier transform of the time waveforms of Fig. 6.5(a) and (g) to be equivalent. It is of value to examine these same relationships in the frequency domain.

Fourier transform of the sampled truncated waveform of Fig. 6.5(e) is obtained by convolving the frequency-domain impulse functions of Fig. 6.5(c) and the $[\sin(f)]/f$ function illustrated in Fig. 6.5(d). This convolution is graphically illustrated in an expanded view in Fig. 6.6. Sampling of the resulting convolution at frequency intervals of $1/T_0$ yields the impulses as illustrated in Fig. 6.6 and, equivalently, Fig. 6.5(g). These sample values represent the Fourier transform of the periodic time waveform of Fig. 6.5(g). Note that there is an impulse at zero frequency. This component represents the average value of the truncated waveform; because the truncated waveform is not an even number of cycles, the average value is not expected to be zero. The remaining frequency-domain impulses occur because the zeros

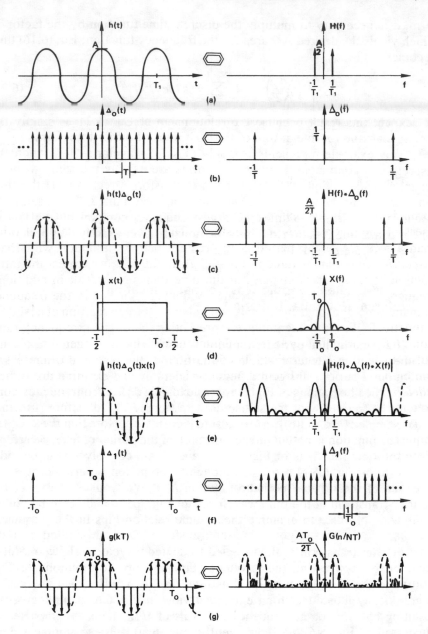

Figure 6.5 Discrete Fourier transform of a band-limited periodic waveform: the truncation interval is not equal to one period.

of the $[\sin(f)]/f$ function are not coincident with each sample value as was the case in the previous example.

This discrepancy between the continuous and discrete Fourier trans-

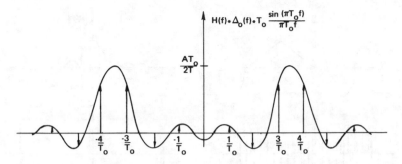

Figure 6.6   Expanded illustration of the convolution of Fig. 6.5(e); $T_0 = 3.5T_1$.

forms is probably the one most often encountered and least understood by users of the discrete Fourier transform. The effect of truncation at other than a multiple of the period is to create a periodic function with sharp *discontinuities*, as illustrated in Fig. 6.5(g). Intuitively, we expect the introduction of these sharp changes in the time domain to result in additional frequency components in the frequency domain. Viewed in the frequency domain, time-domain truncation is equivalent to the convolution of a $[\sin(f)]/f$ function with the single impulse representing the original frequency function $H(f)$. Consequently, the frequency function is no longer a single impulse, but rather a continuous function of frequency with a local maximum centered at the original impulse and a series of other peaks that are termed *sidelobes*. These sidelobes are responsible for the additional frequency components that occur after frequency-domain sampling. This effect is termed *leakage* and is inherent in the discrete Fourier transform because of the required time-domain truncation. Techniques for reducing leakage are explored in Sec. 9.2.

Finite-Duration Waveforms

The preceding two examples have explored the relationship between the discrete and continuous Fourier transforms for band-limited periodic functions. Another class of functions of interest is that of finite duration, such as the function $h(t)$ illustrated in Fig. 6.7. If $h(t)$ is time-limited, its Fourier transform cannot be band-limited; sampling must result in aliasing. It is necessary to choose the sample interval T such that aliasing is reduced to an acceptable range. As illustrated in Fig. 6.7(c), the sample interval T was chosen too large and as a result there is significant aliasing.

If the finite-length waveform is sampled and if N is chosen equal to the number of samples of the time-limited waveform, then it is not necessary to truncate in the time domain. Truncation is omitted and the Fourier transform of the time-sampled function [Fig. 6.7(c)] is multiplied by $\Delta_1(f)$, the frequency-domain sampling function. The time-domain equivalent to this

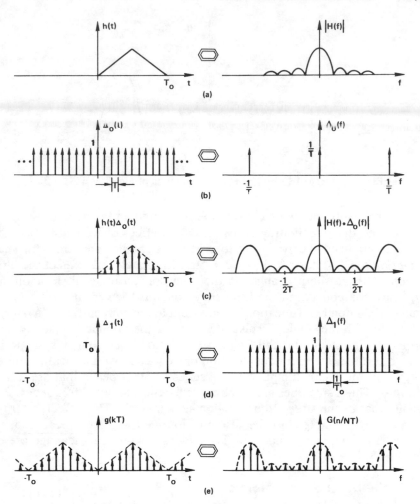

Figure 6.7 Discrete Fourier transform of a time-limited waveform.

product is the convolution of the time functions shown in Figs. 6.7(c) and (d). The resulting waveform is periodic, where a period is defined by the N samples of the original function, and thus is a replica of the original function. The Fourier transform of this periodic function is the sampled function illustrated in Fig. 6.7(e).

For this class of functions, if N is chosen equal to the number of samples of the finite-length function, then the only error is that introduced by aliasing. Errors introduced by aliasing are reduced by choosing the sample interval T sufficiently small. For this case, the discrete Fourier transform sample values agree (within a constant) reasonably well with samples of the con-

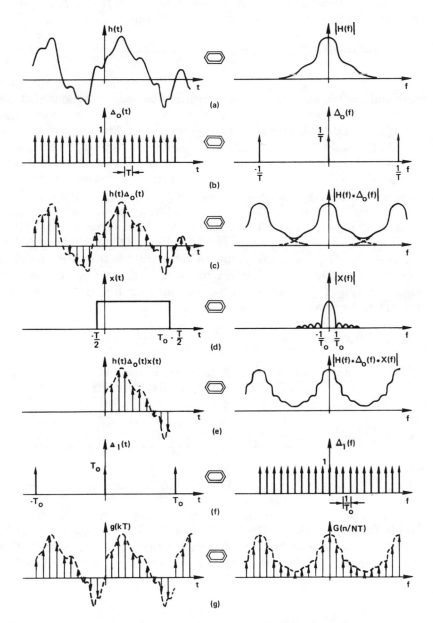

Figure 6.8 Discrete Fourier transform of a general waveform.

tinuous Fourier transform. Unfortunately, there exist few applications of discrete Fourier transform for this class of functions.

General Periodic Waveforms

Figure 6.7 can also be used to illustrate the relationship between the discrete and continuous Fourier transform for periodic functions that are not band-limited. Assume that $h(t)$, as illustrated in Fig. 6.7(a), is only one period of a periodic waveform. If this periodic waveform is sampled and truncated at exactly the period, then the resulting waveform is identical to the time waveform of Fig. 6.7(c). Instead of the continuous frequency function, as illustrated in Fig. 6.7(c), the frequency transform is an infinite series of equidistant impulses separated by $1/T_0$, whose areas are given exactly by the continuous frequency function. Because the frequency-sampling function $\Delta_1(f)$, as illustrated in Fig. 6.7(d), is an infinite series of equidistant impulses separated by $1/T_0$, then the result is identical to those of Fig. 6.7(e). As before, the only error source is that of aliasing if the truncation function is chosen exactly equal to an integer multiple of the period. If the time-domain truncation is not equal to a period, then results as described previously are to be expected.

General Waveforms

The most important class of functions are those that are neither time-limited nor band-limited. An example of this class of functions is illustrated in Fig. 6.8(a). Sampling results in the aliased frequency function illustrated in Fig. 6.8(c). Time-domain truncation introduces rippling in the frequency domain of Fig. 6.8(e). Frequency sampling results in the Fourier transform pair illustrated in Fig. 6.8(g). The time-domain function of this pair is a periodic function, where the period is defined by the N points of the original function after sampling and truncation. The frequency-domain function of the pair is also a periodic function, where a period is defined by N points, whose values differ from the original frequency function by the errors introduced in aliasing and time-domain truncation. The aliasing error can be reduced to an acceptable level by decreasing the sample interval T. Procedures for reducing time-domain truncation errors are addressed in Sec. 9.2.

Summary

We have shown that if care is exercised, then there exist many applications where the discrete Fourier transform can be employed to derive results essentially equivalent to the continuous Fourier transform. The most important concept to keep in mind is that the discrete Fourier transform implies periodicity in both the time and frequency domains. If one remembers that the N sample values of the time-domain function represent one sample of a periodic function, then application of the discrete Fourier transform should result in few surprises.

6.5 DISCRETE FOURIER TRANSFORM PROPERTIES

Properties established for the Fourier transform in Chapter 3 can be extended to the discrete Fourier transform because we have shown that the discrete transform is a special case of the continous transform. Although we often use the continuous equivalents in our problem-solving thought process, it is the discrete properties that form the theoretical basis for applications of the FFT. We replace kT by k and n/NT by n for convenience of notation.

Linearity

If $x(k)$ and $y(k)$ have discrete Fourier transforms $X(n)$ and $Y(n)$, respectively, then

$$x(k) + y(k) \iff X(n) + Y(n) \tag{6.25}$$

The discrete Fourier transform pair of Eq. (6.25) follows directly from the discrete Fourier transform pair of Eq. (6.21).

Symmetry

If $h(k)$ and $H(n)$ are a discrete Fourier transform pair, then

$$\frac{1}{N} H(k) \iff h(-n) \tag{6.26}$$

The discrete Fourier transform pair of Eq. (6.26) is established by rewriting Eq. (6.17):

$$h(-k) = \frac{1}{N} \sum_{k=0}^{N-1} H(n) e^{j2\pi n(-k)/N} \tag{6.27}$$

and by interchanging the parameters k and n:

$$h(-n) = \frac{1}{N} \sum_{n=0}^{N-1} H(k) e^{-j2\pi nk/N} \tag{6.28}$$

Time Shifting

If $h(k)$ is shifted by the integer i, then

$$h(k - i) \iff H(n) e^{-j2\pi ni/N} \tag{6.29}$$

To verify Eq. (6.29), substitute $r = k - i$ into the inverse discrete Fourier transform:

$$h(r) = \frac{1}{N} \sum_{n=0}^{N-1} H(n)e^{j2\pi nr/N}$$

$$h(k - i) = \frac{1}{N} \sum_{n=0}^{N-1} H(n)e^{j2\pi n(k-i)/N} \tag{6.30}$$

$$= \frac{1}{N} \sum_{n=0}^{N-1} [H(n)e^{-j2\pi ni/N}]e^{j2\pi nk/N}$$

Frequency Shifting

If $H(n)$ is shifted by the integer i, then its inverse discrete Fourier transform is multiplied by $e^{j2\pi ik/N}$

$$h(k)e^{j2\pi ik/N} \Longleftrightarrow H(n - i) \tag{6.31}$$

This discrete Fourier transform pair is established by substituting $r = n - i$ into the discrete Fourier transform

$$H(r) = \sum_{k=0}^{N-1} h(k)e^{-j2\pi rk/N}$$

$$H(n - i) = \sum_{k=0}^{N-1} h(k)e^{-j2\pi(n-i)k/N} \tag{6.32}$$

$$= \sum_{k=0}^{N-1} [h(k)e^{j2\pi ik/N}]e^{-j2\pi nk/N}$$

Alternate Inversion Formula

The discrete inversion formula, Eq. (6.17), can also be written as

$$h(k) = \frac{1}{N} \left[\sum_{k=0}^{N-1} H*(n)e^{-j2\pi nk/N} \right]^* \tag{6.33}$$

where $*$ implies conjugation. To prove Eq. (6.33), we simply perform the indicated conjugation. Let $H(n) = R(n) + jI(n)$; hence, $H*(n) = R(n) - jI(n)$ and Eq. (6.33) becomes

$$h(k) = \frac{1}{N} \left[\sum_{n=0}^{N-1} [R(n) - jI(n)]e^{-j2\pi nk/N} \right]^*$$

$$= \frac{1}{N} \left[\sum_{n=0}^{N-1} [R(n) - jI(n)] \left[\cos\left(\frac{2\pi nk}{N}\right) - j\sin\left(\frac{2\pi nk}{N}\right) \right] \right]^*$$

$$= \frac{1}{N}\left[\sum_{n=0}^{N-1} R(n) \cos\left(\frac{2\pi nk}{N}\right) - I(n) \sin\left(\frac{2\pi nk}{N}\right) \right.$$

$$\left. - j \sum_{n=0}^{N-1} R(n) \sin\left(\frac{2\pi nk}{N}\right) + I(n) \cos\left(\frac{2\pi nk}{N}\right) \right]^{*}$$

$$= \frac{1}{N}\left[\sum_{n=0}^{N-1} R(n) \cos\left(\frac{2\pi nk}{N}\right) - I(n) \sin\left(\frac{2\pi nk}{N}\right) \right.$$

$$\left. + j \sum_{n=0}^{N-1} R(n) \sin\left(\frac{2\pi nk}{N}\right) + I(n) \cos\left(\frac{2\pi nk}{N}\right) \right]$$

$$= \frac{1}{N} \sum_{n=0}^{N-1} [R(n) + jI(n)]\left[\cos\left(\frac{2\pi nk}{N}\right) + j \sin\left(\frac{2\pi nk}{N}\right) \right]$$

$$= \frac{1}{N} \sum_{n=0}^{N-1} H(n) e^{j2\pi nk/N}$$

(6.34)

The significance of the alternate inversion formula is that the discrete transform, Eq. (6.34), can be used to compute both the Fourier transform and its inversion. Hence, one only needs to develop a single FFT computer program.

Even Functions

If $h_e(k)$ is an even function, then $h_e(k) = h_e(-k)$ and the discrete Fourier transform of $h_e(k)$ is an even function and is real:

$$h_e(k) \iff R_e(n) = \sum_{n=0}^{N-1} h_e(k) \cos\left(\frac{2\pi nk}{N}\right) \qquad (6.35)$$

To verify Eq. (6.35), we simply manipulate the defining relationships:

$$H_e(n) = \sum_{k=0}^{N-1} h_e(k) e^{-j2\pi nk/N}$$

$$= \sum_{k=0}^{N-1} h_e(k) \cos\left(\frac{2\pi nk}{N}\right) + j \sum_{k=0}^{N-1} h_e(k) \sin\left(\frac{2\pi nk}{N}\right)$$

$$= \sum_{k=0}^{N-1} h_e(k) \cos\left(\frac{2\pi nk}{N}\right)$$

$$= R_e(n)$$

(6.36)

The imaginary summation is zero because the summation is over an even number of cycles of an odd function. Because $h_e(k) \cos(2\pi nk/N) = h_e(k)$ $\{\cos[2\pi(-n)k/N]\}$, then $H_e(n) = H_e(-n)$ and the frequency function is even. The inversion formula is proven similarly. Hence, if $H(n)$ is given as a real and even function, then its inverse discrete Fourier transform is an even function.

Odd Functions

If $h_0(k) = -h_0(-k)$, then $h_0(k)$ is an odd function and its discrete Fourier transform is an odd and imaginary function.

$$H_0(n) = \sum_{k=0}^{N-1} h_0(k)e^{-j2\pi nk/N}$$

$$= \sum_{k=0}^{N-1} h_0(k) \cos\left(\frac{2\pi nk}{N}\right) - j\sum_{k=0}^{N-1} h_0(k) \sin\left(\frac{2\pi nk}{N}\right) \qquad (6.37)$$

$$= -j\sum_{k=0}^{N-1} h_0(k) \sin\left(\frac{2\pi nk}{N}\right)$$

$$= jI_0(n)$$

The real summation is zero because summation is over an even number of cycles of an odd function. For $H(n)$ given as an odd and imaginary function, the proof that $h_0(k)$ is an odd function is established similarly; therefore,

$$h_0(k) \Longleftrightarrow jI_0(n) = -j\sum_{k=0}^{N-1} h_0(k) \sin\left(\frac{2\pi nk}{N}\right) \qquad (6.38)$$

Waveform Decomposition

To decompose an arbitrary function $h(k)$ into an even and an odd function, we simply add and subtract the common function $h(-k)/2$.

$$h(k) = \frac{h(k)}{2} + \frac{h(k)}{2}$$

$$= \left[\frac{h(k)}{2} + \frac{h(-k)}{2}\right] + \left[\frac{h(k)}{2} - \frac{h(-k)}{2}\right] \qquad (6.39)$$

$$= h_e(k) + h_0(k)$$

The terms in brackets satisfy the definition of an even and an odd function, respectively. Because $h(k)$ is periodic with period N, then

$$h(-k) = h(N - k) \qquad (6.40)$$

and

$$h_e(k) = \frac{h(k)}{2} + \frac{h(N - k)}{2}$$

$$\qquad\qquad\qquad\qquad\qquad\qquad\qquad (6.41)$$

$$h_0(k) = \frac{h(k)}{2} - \frac{h(N - k)}{2}$$

For discrete periodic functions, Eq. (6.41) is the desired relationship for decomposition. From Eqs. (6.35) and (6.38), the discrete Fourier transform of Eq. (6.39) is

$$H(n) = R(n) + jI(n) = H_e(n) + H_0(n) \qquad (6.42)$$

where

$$H_e(n) = R(n) \qquad \text{and} \qquad H_0(n) = jI(n) \qquad (6.43)$$

Complex Time Functions

If $h(k) = h_r(k) + jh_i(k)$, where $h_r(k)$ and $h_i(k)$ are, respectively, the real and imaginary parts of the complex time function $h(k)$, then the discrete Fourier transform becomes

$$
\begin{aligned}
H(n) &= \sum_{k=0}^{N-1} [h_r(k) + jh_i(k)]e^{-j2\pi nk/N} \\
&= \sum_{k=0}^{N-1} h_r(k) \cos\left(\frac{2\pi nk}{N}\right) + h_i(k) \sin\left(\frac{2\pi nk}{N}\right) \qquad (6.44) \\
&\quad - j\left[\sum_{k=0}^{N-1} h_r(k) \sin\left(\frac{2\pi nk}{N}\right) - h_i(k) \cos\left(\frac{2\pi nk}{N}\right)\right]
\end{aligned}
$$

The first expression of Eq. (6.44) is $R(n)$, the real part of the discrete transform, and the latter expression is $I(n)$, the imaginary part of the discrete transform. If $h(k)$ is real, then $h(k) = h_r(k)$, and from Eq. (6.44),

$$R_e(n) = \sum_{k=0}^{N-1} h_r(k) \cos\left(\frac{2\pi nk}{N}\right) \qquad (6.45)$$

$$I_0(n) = -j\sum_{k=0}^{N-1} h_r(k) \sin\left(\frac{2\pi nk}{N}\right) \qquad (6.46)$$

Note that $\cos(2\pi nk/N) = \cos(-2\pi nk/N)$; thus, $R_e(n) = R_e(-n)$, and $R_e(n)$ is an even function. Similarly, $I_0(n) = -I_0(-n)$ and $I_0(n)$ is an odd function.

For $h(k)$ purely imaginary, $h(k) = jh_i(k)$ and from Eq. (6.44),

$$R_0(n) = \sum_{k=0}^{N-1} h_i(k) \sin\left(\frac{2\pi nk}{N}\right) \qquad (6.47)$$

$$I_e(n) = \sum_{k=0}^{N-1} h_i(k) \cos\left(\frac{2\pi nk}{N}\right) \qquad (6.48)$$

For $h(k)$ imaginary, the real part of its transform is odd and the imaginary part of its transform is even.

Time-Convolution Theorem

Discrete convolution is defined by the summation (see Chapter 7):

$$y(k) = \sum_{i=0}^{N-1} x(i)h(k - i) \qquad (6.49)$$

where $x(k)$, $h(k)$, and $y(k)$ are periodic functions with period N.

Analogous to Fourier transform theory, one of the most important properties of the discrete Fourier transform is exhibited by the discrete Fourier transform of Eq. (6.49). Discrete Fourier transformation yields the discrete convolution theorem that is expressed as

$$\sum_{i=0}^{N-1} x(i)h(k - i) \iff X(n)H(n) \qquad (6.50)$$

To establish this result, substitute Eq. (6.17) into the left-hand side of Eq. (6.50):

$$
\begin{aligned}
\sum_{i=0}^{N-1} x(i)h(k - i) &= \sum_{i=0}^{N-1} \frac{1}{N} \sum_{n=0}^{N-1} X(n)e^{j2\pi ni/N} \\
&\quad \times \frac{1}{N} \sum_{m=0}^{N-1} H(m)e^{j2\pi m(k-i)/N} \\
&= \frac{1}{N} \sum_{n=0}^{N-1} \sum_{m=0}^{N-1} X(n)H(m)e^{j2\pi mk/N} \\
&\quad \times \frac{1}{N} \left[\sum_{i=0}^{N-1} e^{j2\pi in/N} e^{-j2\pi im/N} \right]
\end{aligned}
\qquad (6.51)
$$

The bracketed term of Eq. (6.51) is simply the orthogonality relationship of Eq. (6.19) and is equal to N if $m = n$; therefore,

$$\sum_{i=0}^{N-1} x(i)h(k - i) = \frac{1}{N} \sum_{n=0}^{N-1} X(n)H(n)e^{j2\pi nk/N} \qquad (6.52)$$

Thus, the discrete Fourier transform of the convolution of two periodic sampled functions with period N is equal to the product of the discrete Fourier transform of the periodic functions.

Frequency-Convolution Theorem

Consider the frequency convolution:

$$Y(n) = \sum_{i=0}^{N-1} X(i)H(n - i) \qquad (6.53)$$

We can establish the frequency-convolution theorem by substitution into Eq. (6.53):

$$\sum_{i=0}^{N-1} X(i)H(n-i) = \sum_{i=0}^{N-1} \left[\sum_{m=0}^{N-1} x(m)e^{-j2\pi mi/N} \right]$$

$$\times \left[\sum_{k=0}^{N-1} h(k)e^{-j2\pi k(n-i)/N} \right] \qquad (6.54)$$

$$= \sum_{m=0}^{N-1} \sum_{k=0}^{N-1} x(m)h(k)e^{-j2\pi kn/N}$$

$$\times \left[\sum_{i=0}^{N-1} e^{-j2\pi mi/N} e^{j2\pi ki/N} \right]$$

The bracketed term of Eq. (6.54) is the orthogonality relationship of Eq. (6.19) and is equal to N if $m = k$; therefore,

$$\sum_{i=0}^{N-1} X(i)H(n-i) = N \sum_{k=0}^{N-1} x(k)h(k)e^{-j2\pi nk/N} \qquad (6.55)$$

and the discrete transform pair is established:

$$x(k)h(k) \iff \frac{1}{N} \sum_{i=0}^{N-1} X(i)H(n-i) \qquad (6.56)$$

Correlation Theorem

Discrete correlation is defined as

$$z(k) = \sum_{i=0}^{N-1} x(i)h(k+i) \qquad (6.57)$$

where $x(k)$, $h(k)$, and $z(k)$ are periodic functions with period N.

The transform pair

$$\sum_{i=0}^{N-1} x(i)h(k+i) \iff X*(n)H(n) \qquad (6.58)$$

is termed the *discrete correlation theorem*. By means of the correlation theorem, correlation can be determined equivalently in the transform domain. To verify this relationship, substitute the discrete Fourier transform

TABLE 6.1 Continuous and Discrete Fourier Transform Properties

Fourier Transform	Property	Discrete Fourier Transform
$x(t) + y(t) \Leftrightarrow X(f) + Y(f)$	(3.2) Linearity	(6.25) $x(k) + y(k) \Leftrightarrow X(n) + Y(n)$
$H(t) \Leftrightarrow h(-f)$	(3.6) Symmetry	(6.26) $\dfrac{1}{N}H(k) \Leftrightarrow h(-n)$
$h(t - t_0) \Leftrightarrow H(f)e^{-j2\pi f t_0}$	(3.21) Time shifting	(6.29) $h(k - i) \Leftrightarrow H(n)e^{-j2\pi n i/N}$
$h(t)e^{j2\pi f t_0} \Leftrightarrow H(f - f_0)$	(3.23) Frequency shifting	(6.31) $h(k)e^{j2\pi k i/N} \Leftrightarrow H(n - i)$
$\left[\displaystyle\int_{-\infty}^{\infty} H^*(f)e^{-j2\pi f t}\,df \right]^*$	(3.25) Alternate inversion formula	(6.33) $\left[\dfrac{1}{N}\displaystyle\sum_{n=0}^{N-1} H^*(n)e^{-j2\pi k n/N} \right]^*$
$h_e(t) \Leftrightarrow R_e(f)$	(3.27) Even functions	(6.35) $h_e(k) \Leftrightarrow R_e(n)$
$h_o(t) \Leftrightarrow jI_0(f)$	(3.32) Odd functions	(6.38) $h_o(k) \Leftrightarrow jI_0(n)$
$h(t) = h_e(t) + h_o(t)$ $= \left[\dfrac{h(t)}{2} + \dfrac{h(-t)}{2} \right]$ $+ \left[\dfrac{h(t)}{2} - \dfrac{h(-t)}{2} \right]$	(3.33) Decomposition	(6.39) $h(k) = h_e(k) + h_o(k)$ $= \left[\dfrac{h(k)}{2} + \dfrac{h(N - k)}{2} \right]$ $+ \left[\dfrac{h(k)}{2} - \dfrac{h(N - k)}{2} \right]$
$y(t) = \displaystyle\int_{-\infty}^{\infty} x(\tau)h(t - \tau)\,d\tau$ $= x(t) * h(t)$	(4.1) Convolution	(6.49) $y(k) = \displaystyle\sum_{i=0}^{N-1} x(i)h(k - i) = x(k) * h(k)$
$y(t) * h(t) \Leftrightarrow Y(f)H(f)$	(4.14) Time convolution theorem	(6.50) $y(k) * h(k) \Leftrightarrow Y(n)H(n)$
$z(t) = \displaystyle\int_{-\infty}^{\infty} x(\tau)h(t + \tau)\,d\tau$	(4.21) Correlation	(6.57) $y(k) = \displaystyle\sum_{i=0}^{N-1} x(i)h(k + i)$
$y(t)h(t) \Leftrightarrow Y(f) * H(f)$	(4.20) Frequency convolution theorem	(6.56) $y(k)h(k) \Leftrightarrow \dfrac{1}{N}Y(n) * H(n)$
$\displaystyle\int_{-\infty}^{\infty} h^2(t)\,dt = \int_{-\infty}^{\infty} |H(f)|^2\,df$	Parseval's Theorem	$\displaystyle\sum_{k=0}^{N-1} h^2(k) = \dfrac{1}{N}\sum_{n=0}^{N-1} |H(n)|^2$

into the left-hand side of Eq. (6.58):

$$\sum_{i=0}^{N-1} x(i)h(k + i) = \sum_{i=0}^{N-1} \left[\frac{1}{N} \sum_{n=0}^{N-1} X(n)e^{j2\pi in/N} \right]$$

$$\times \frac{1}{N} \sum_{m=0}^{N-1} H(m)e^{j2\pi m(k+i)/N}$$

$$= \sum_{i=0}^{N-1} \left[\frac{1}{N} \sum_{n=0}^{N-1} X * (n)e^{-j2\pi in/N} \right]^{*} \qquad (6.59)$$

$$\times \left[\frac{1}{N} \sum_{m=0}^{N-1} H(m)e^{j2\pi m(k+i)/N} \right]$$

where the alternate inversion formula, Eq. (6.33), has been utilized to introduce the conjugate of $X(n)$. Note that the second conjugation indicated in Eq. (6.33) can be omitted if only real functions are considered. For this case, Eq. (6.59) can be rewritten as:

$$\sum_{i=0}^{N-1} x(i)h(k + i)$$

$$= \frac{1}{N} \sum_{n=0}^{N-1} \sum_{m=0}^{N-1} X * (n)H(m)e^{j2\pi mk/N} \left[\frac{1}{N} \sum_{i=0}^{N-1} e^{-j2\pi in/N} e^{j2\pi im/N} \right] \qquad (6.60)$$

From the orthogonality relationship of Eq. (6.19), the bracketed term is equal to N if $n = m$. Hence, Eq. (6.60) becomes

$$\sum_{i=0}^{N-1} x(i)h(k + i) = \frac{1}{N} \sum_{n=0}^{N-1} X * (n)H(n)e^{j2\pi nk/N} \qquad (6.61)$$

Summary Table of Discrete Fourier Transform Properties

For future reference, the discrete Fourier transform properties are summarized in Table 6.1. The continuous Fourier transform properties are also tabled for purposes of comparison. Appropriate equation numbers are listed in order that one can easily locate the continuous or discrete development for each property.

PROBLEMS

6.1. Repeat the graphical development of Fig. 6.1 for the following functions:

(a) $h(t) = | t | e^{-a|t|}$

(b) $h(t) = 1 - | t | \qquad | t | \leq 1$

$\qquad = 0 \qquad\qquad | t | > 1$

(c) $h(t) = \cos t$

6.2. Retrace the development of the discrete Fourier transform [Eqs. (6.1) through (6.16)]. Write all steps of the derivation in detail.

6.3. Repeat the graphical derivation of Fig. 6.3 for $h(t) = \sin(2\pi f_0 t)$. Show the effect of setting the truncation interval unequal to the period. What is the result of setting the truncation interval equal to two periods?

6.4. Consider Fig. 6.7. Assume that $h(t)\Delta_0(t)$ is represented by N nonzero samples. What is the effect of truncating $h(t)\Delta_0(t)$ so that only $3N/4$ nonzero samples are considered? What is the effect of truncating $h(t)\Delta_0(t)$ so that the N nonzero samples and $N/4$ zero samples are considered?

6.5. Repeat the graphical derivation of Fig. 6.7 for $h(t) = \sum_{n=-\infty}^{\infty} e^{-a|t-nT_0|}$. What are the error sources?

6.6. To establish the concept of rippling, perform the following graphical convolutions:
   (a) An impulse with $(\sin t)/t$.
   (b) A narrow pulse with $(\sin t)/t$.
   (c) A wide pulse with $(\sin t)/t$.
   (d) A single triangular waveform with $(\sin t)/t$.

6.7. Write several terms of Eq. (6.19) to establish the orthogonality relationship.

6.8. The truncation interval is termed the *record length*. In terms of the record length, write an equation defining the *resolution* or frequency spacing of the frequency-domain samples of the discrete Fourier transform.

6.9. Comment on the following: The discrete Fourier transform is analogous to a bank of band-pass filters.

Let $x(k)$ and $y(k)$ be discrete periodic functions:

$$x(k) = \begin{cases} \frac{1}{2} & k = 0, 4 \\ 1 & k = 1, 2, 3 \\ 0 & k = 5, 6, 7 \end{cases}$$
$$x(k + 8r) = x(k) \qquad r = 0, \pm 1, \pm 2, \ldots$$
$$y(k) = x(k)$$
$$y(k + 8r) = y(k) \qquad r = 0, \pm 1, \pm 2, \ldots$$

6.10. Compute $X(n)$ and $Y(n)$. Add these results to determine $[X(n) + Y(n)]$. Determine $z(k) = x(k) + y(k)$. Compute $Z(n)$. Discuss your results in terms of the linearity property.

6.11. Demonstrate the symmetry property of Eq. (6.26) for $x(k)$.

6.12. Compute the discrete Fourier transform of $x(k - 3)$. Compare results with those obtained from the time-shifting relationship of Eq. (6.29).

6.13. Compute the inverse discrete Fourier transform of $X(n - 1)$. Repeat this computation by applying the frequency-shifting theorem of Eq. (6.31) and compare results.

6.14. Compute the inverse discrete Fourier transform of $X(n)$ using the alternate inversion formula of Eq. (6.33).

6.15. Compute the discrete Fourier transform of $x(k - 2)$. Investigate the even-odd relationship of $x(k - 2)$ and the real-imaginary relationship of its discrete transform.

6.16. Let $z(k) = x(k) - y(k - 4)$. Compute the discrete Fourier transform of $z(k)$.

6.17. Let $z(k) = y(k) + y(k - 2) - x(k - 4)$. Decompose $z(k)$ into even and odd functions both analytically and graphically. Demonstrate Eq. (6.42) with $z(k)$.

6.18. Demonstrate the frequency convolution theorem using $x(k)$ and $y(k)$.

6.19. Demonstrate the discrete correlation theorem using $x(k)$ and $y(k)$.

REFERENCES

1. COOLEY, J. W., P. A. W. LEWIS, AND P. D. WELCH. "The Finite Fourier Transform." *IEEE Trans. on Audio and Electroacoustics* (June 1969), Vol. AU-17, No. 2, pp. 77–85.

2. BERGLAND, G. D. "A Guided Tour of the Fast Fourier Transform." *IEEE Spectrum* (July 1969), Vol. 6, No. 7, pp. 41–52.

7

DISCRETE CONVOLUTION
AND CORRELATION

Possibly the most important discrete Fourier transform properties are those of convolution and correlation. This follows because the importance of the fast Fourier transform is primarily a result of its efficiency in computing discrete convolution or correlation. In this chapter, we examine, analytically and graphically, the discrete convolution and correlation equations. The relationship between discrete and continuous convolution is also explored in detail.

7.1 DISCRETE CONVOLUTION

Discrete convolution is defined by the summation:

$$y(kT) = \sum_{i=0}^{N-1} x(iT)h[(k-i)T] \qquad (7.1)$$

where both $x(kT)$ and $h(kT)$ are periodic functions with period N,

$$x(kT) = x[(k+rN)T] \qquad r = 0, \pm 1, \pm 2, \ldots \qquad (7.2)$$
$$h(kT) = h[(k+rN)T] \qquad r = 0, \pm 1, \pm 2, \ldots$$

For convenience of notation, discrete convolution is normally written as

$$y(kT) = x(kT) * h(kT) \qquad (7.3)$$

To examine the discrete convolution equation, consider the illustrations of Fig. 7.1. Both functions $x(kT)$ and $h(kT)$ are periodic with period

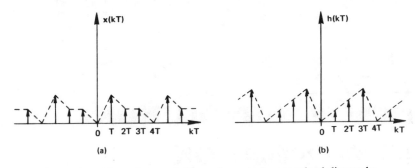

Figure 7.1   Example sampled waveforms to be convolved discretely.

$N = 4$. From Eq. (7.1), functions $x(iT)$ and $h[(k - i)T]$ are required. Function $h(-iT)$ is the image of $h(iT)$ about the ordinate axis, as illustrated in Fig. 7.2(a); function $h[(k - i)T]$ is simply the function $h(-iT)$ shifted by the amount kT. Figure 7.2(b) illustrates $h[(k - i)T]$ for the shift $2T$. Equation (7.1) is evaluated for each kT shift by performing the required multiplications and additions.

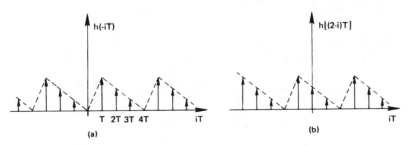

Figure 7.2   Graphical description of discrete convolution shifting operation.

7.2 GRAPHICAL INTERPRETATION OF DISCRETE CONVOLUTION

The discrete convolution process is illustrated graphically in Fig. 7.3. Sample values of $x(kT)$ and $h(kT)$ are denoted by *dots* and *crosses*, respectively. Figure 7.3(a) illustrates the desired computation for $k = 0$. The value of each *dot* is multiplied by the value of the *cross* that occurs at the same abscissa value; these products are summed over the $N = 4$ discrete values indicated. Computation of Eq. (7.1) is graphically evaluated for $k = 1$ in Fig. 7.3(b); multiplication and summation is over the N points indicated. Figures 7.3(c) and (d) illustrate the convolution computation for $k = 2$ and $k = 3$, respectively. Note that for $k = 4$ [Fig. 7.3(e)], the terms multiplied and summed are identical to those of Fig. 7.3(a). This is expected because

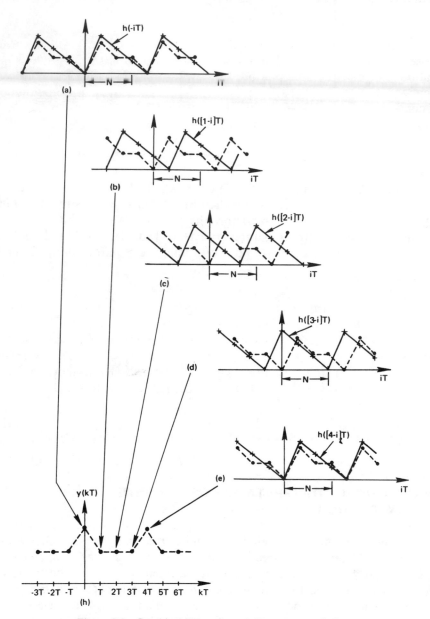

Figure 7.3 Graphical illustration of discrete convolution.

both $x(kT)$ and $h(kT)$ are periodic with a period of four terms. Therefore,

$$y(kT) = y[(k + rN)T] \qquad r = 0, \pm 1, \pm 2, \ldots \qquad (7.4)$$

Steps for graphically computing the discrete convolution differ from those of continuous convolution only in that integration is replaced by summation. For discrete convolution, these steps are (1) folding, (2) displacement or shifting, (3) multiplication, and (4) summation. As in the convolution of continuous functions, either the sequence $x(kT)$ or $h(kT)$ can be selected for displacement. Equation (7.1) can be written equivalently as

$$y(kT) = \sum_{i=0}^{N-1} x[(k - i)T]h(iT) \qquad (7.5)$$

7.3 RELATIONSHIP BETWEEN DISCRETE AND CONTINUOUS CONVOLUTION

If we only consider periodic functions represented by equally spaced impulse functions, discrete convolution relates identically to its continuous equivalent. This follows because, as we show in Appendix A (Eq. A.14), continuous convolution is well-defined for impulse functions.

The most important application of discrete convolution is not to sampled periodic functions but rather to approximate the continuous convolutions of general waveforms. For this reason, we will now explore in detail the relationship between discrete and continuous convolution.

Discrete Convolution of Finite-Duration Waveforms

Consider the functions $x(t)$ and $h(t)$, as illustrated in Fig. 7.4(a). We wish to convolve these two functions both continuously and discretely and to compare these results. Continuous convolution $y(t)$ of the two functions is also shown in Fig. 7.4(a). To evaluate the discrete convolution, we sample both $x(t)$ and $h(t)$ with sample interval T and we assume that both sample functions are periodic with period N. As illustrated in Fig. 7.4(b), the period has been chosen as $N = 9$ and both $x(kT)$ and $h(kT)$ are represented by $P = Q = 6$ samples; the remaining samples defining a period are set to zero. Figure 7.4(b) also illustrates the discrete convolution $y(kT)$ for the period $N = 9$; for this choice of N, the discrete convolution is a very poor approximation of the continuous case because the periodicity constraint results in an overlap of the desired periodic output. That is, we did not choose the period sufficiently large so that the convolution result of one period would not *interfere* or *overlap* the convolution result of the succeeding period. It is obvious that if we wish the discrete convolution to approximate continuous

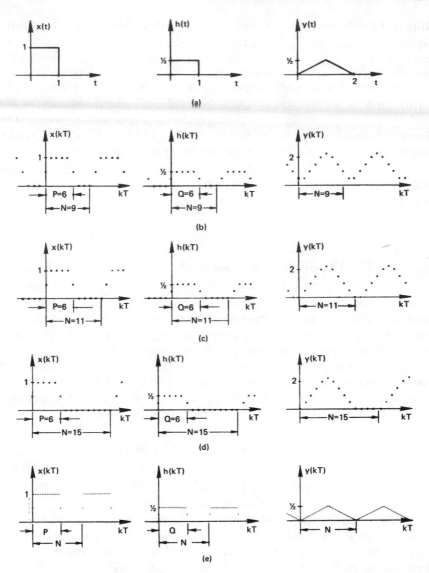

Figure 7.4 Relationship between discrete and continuous convolution: finite-duration waveforms.

convolution, then it is necessary that the period be chosen so that there is no overlap.

Choose the period according to the relationship

$$N = P + Q - 1 \tag{7.6}$$

This situation is illustrated in Fig. 7.4(c), where $N = P + Q - 1 = 11$.

Note that for this choice of N there is no overlap in the resulting convolution. Equation (7.6) is based on the fact that the convolution of a function represented by P samples and a function represented by Q samples is a function described by $P + Q - 1$ samples.

There is no advantage in choosing $N > P + Q - 1$; as shown in Fig. 7.4(d), for $N = 15$, the nonzero values of the discrete convolution are identical to those of Fig. 7.4(c). As long as N is chosen according to Eq. (7.6), discrete convolution results in a periodic function, where each period approximates the continuous convolution results.

Figure 7.4(c) illustrates the fact that discrete convolution results are scaled differently than that of continuous convolution. This scaling constant is T; modifying the discrete convolution Eq. (7.1), we obtain

$$y(kT) = T \sum_{i=0}^{N-1} x(iT)h[(k - i)T] \tag{7.7}$$

The relationship of Eq. (7.7) is simply the continuous convolution integral for time-limited functions evaluated by rectangular integration. Thus, for finite-length time functions, discrete convolution approximates continuous convolution within the error introduced by rectangular integration. As illustrated in Fig. 7.4(e), if the sample interval T is made sufficiently small, then the error introduced by the discrete convolution Eq. (7.7) is negligible.

Example 7.1   Circular Convolution

Discrete convolution yields periodic results because of the periodicity of the functions being convolved. This periodicity gives rise to what is commonly called *circular convolution*. Figure 7.5 illustrates this concept.

In Figure 7.5(a), we show two example discrete periodic waveforms to be convolved. For the shift $k = 2$, Fig. 7.5(b) illustrates the appropriate folding and shifting operations. Multiplication and addition over the $N = 8$ points of the period yield the convolution results for $k = 2$. An alternate way of displaying the discrete convolution of Fig. 7.5(b) for shift $k = 2$ is shown in Fig. 7.5(c). The rings represent one period of the two periodic functions; the inner ring is $h(iT)$ and is the function being shifted. As illustrated, the function is set for a shift of $k = 2$. The outer ring corresponds to the function $x(iT)$. Appropriate values to be multiplied are adjacent to each other. These multiplied results are then summed around the circle (i.e., over one period).

The inner ring is turned for each shift of k. As the ring is turned, it returns to its original position every eight shifts. Hence, the same values will be computed. This corresponds to the periodic convolution results discussed previously. Figure 7.5(c) can also be used to illustrate the problem of overlap. As the inner ring turns, there must be a sufficient number of zero values in the outer ring so that a convolution value is not computed, which is a function of both ends of the data used to form the outer ring. If sufficient zeros are appended to the nonzero sample values of each function to be convolved, then the finite-duration convolution result does not overlap with the following period.

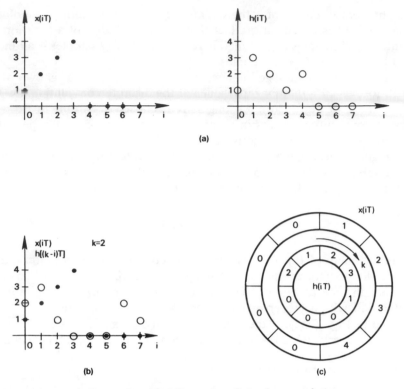

Figure 7.5   Graphical illustration of *circular convolution.*

Discrete Convolution of an Infinite- and a Finite-Duration Waveform

The previous example considered the case for which both $x(kT)$ and $h(kT)$ were of finite duration. Another case of interest is that where only one of the time functions to be convolved is finite. To explore the relationship of the discrete and continuous convolution for this case, consider the illustrations of Fig. 7.6. As illustrated in Fig. 7.6(a), function $h(t)$ is assumed to be of finite duration and $x(t)$ of infinite duration; convolution of these two functions is shown in Fig. 7.6(b). Because the discrete convolution requires that both the sampled functions $x(kT)$ and $h(kT)$ be periodic, we obtain the illustrations of Fig. 7.6(c); period N has been chosen [Figs. 7.6(a) and (c)]. For $x(kT)$ infinite in duration, the imposed periodicity introduces what is known as an *end effect*.

Compare the discrete convolution of Fig. 7.6(d) and the continuous convolution [Fig. 7.6(b)]. As illustrated, the two results agree reasonably well, with the exception of the first $Q - 1$ samples of the discrete convolution. To establish this fact more clearly, consider the illustrations of Fig.

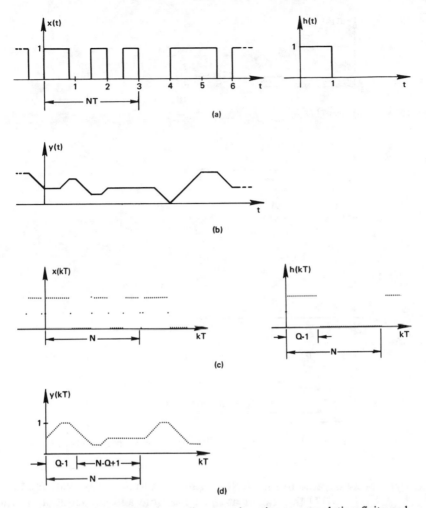

Figure 7.6 Relationship between discrete and continuous convolution: finite- and infinite-duration waveforms.

7.7. We show only one period of $x(iT)$ and $h[(5 - i)T]$. To compute the discrete convolution, Eq. (7.1), for this shift, we multiply those samples of $x(iT)$ and $h[(5 - i)T]$ that occur at the same time [Fig. 7.7(a)] and add. The convolution result is a function of $x(iT)$ at both ends of the period. Such a condition obviously has no meaningful interpretation in terms of the desired continuous convolution. Similar results are obtained for each shift value until the Q points of $h(iT)$ are shifted by $Q - 1$, that is, the end effect exists until shift $k = Q - 1$.

Note that the end effect does not occur at the right end of the N sample values; functions $h(iT)$ for the shift $k = N - 1$ (therefore maximum shift)

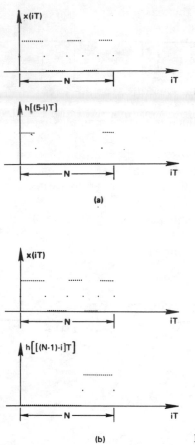

(a)

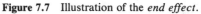

(b)

Figure 7.7 Illustration of the *end effect*.

and $x(iT)$ are illustrated in Fig. 7.7(b). Multiplication of those values of $x(iT)$ and $h[(N - 1 - i)T]$ that occur at the same time and subsequent addition yield the desired convolution; the result is only a function of the correct values of $x(iT)$.

If the sample interval T is chosen sufficiently small, then the discrete convolution for this example class of functions closely approximate the continuous convolution except for the end effect.

Summary

We have emphasized the point that discrete convolution is defined only for periodic functions. However, as illustrated graphically, the implications of this requirement are negligible if at least one of the functions to be convolved is of finite duration. For this case, discrete convolution is approximately equivalent to continuous convolution where the differences in the two methods are due to rectangular integration and to the end effect.

In general, it is impossible to discretely convolve two functions of infinite duration.

The convolution waveform illustrated could have been computed equivalently by means of the convolution theorem. Recall that the discrete convolution of Eq. (7.1) was defined in such a manner that those functions being convolved were assumed to be periodic. The underlying reason for this assumption is to enable the discrete convolution theorem, Eq. (6.50), to hold. If we compute the discrete Fourier transform of each of the periodic sequences $x(kT)$ and $h(kT)$, multiply the resulting transforms, and then compute the inverse discrete Fourier transform of this product, we obtain identical results to those illustrated. As is discussed in Chapter 10, it is normally faster computationally to use the discrete Fourier transform to compute the discrete convolution if the FFT is employed.

7.4 GRAPHICAL INTERPRETATION OF DISCRETE CORRELATION

Discrete correlation is defined as

$$z(kT) = \sum_{i=0}^{N-1} x(iT)h[(k + i)T] \tag{7.8}$$

where $x(kT)$, $h(kT)$, and $z(kT)$ are periodic functions.

$$z(kT) = z[(k + rN)T] \qquad r = 0, \pm 1, \pm 2, \ldots$$
$$x(kT) = x[(k + rN)T] \qquad r = 0, \pm 1, \pm 2, \ldots \tag{7.9}$$
$$h(kT) = h[(k + rN)T] \qquad r = 0, \pm 1, \pm 2, \ldots$$

As in the continuous case, discrete correlation differs from convolution in that there is no folding operation. Hence, the remaining rules for displacement, multiplication, and summation are performed exactly as for the case of discrete convolution.

To illustrate the process of discrete correlation or *lagged products*, as it sometimes is referred, consider Fig. 7.8. The discrete functions to be correlated are shown in Fig. 7.8(a). According to the rules for correlation, we shift, multiply, and sum, as illustrated in Figs. 7.8(b), (c), and (d), respectively. Compare with the results of Ex. 4.8. In Chapter 10, we discuss the application of the FFT for efficient computation of Eq. (7.8).

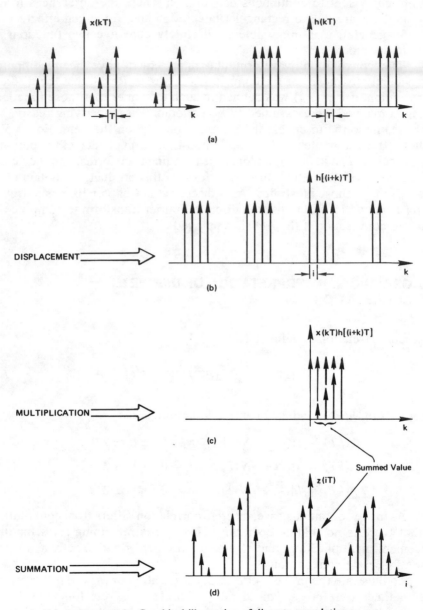

Figure 7.8 Graphical illustration of discrete correlation.

PROBLEMS

7.1. Let

$$x(kT) = e^{-kT} \qquad k = 0, 1, 2, 3$$
$$= 0 \qquad k = 4, 5, \ldots, N$$
$$= x[(k + rN)T] \qquad r = 0, \pm 1, \pm 2, \ldots$$

and

$$h(kT) = 1 \qquad k = 0, 1, 2$$
$$= 0 \qquad k = 3, 4, \ldots, N$$
$$= h[(k + rN)T] \qquad r = 0, \pm 1, \pm 2, \ldots$$

With $T = 1$, graphically and analytically determine $x(kT) * h(kT)$. Choose N less than, equal to, and greater than Eq. (7.6).

7.2. Consider the continuous functions $x(t)$ and $h(t)$, as illustrated in Fig. 4.14(a). Sample both functions with sample interval $T = T_0/4$ and assume both sample functions are periodic with period N. Choose N according to relationship of Eq. (7.6). Determine $x(kT) * h(kT)$ both analytically and graphically. Investigate the results of an incorrect choice of N. Compare results with continuous convolution results.

7.3. Repeat Problem 7.2 for Figs. 4.14(b) and (c).

7.4. Refer to Fig. 7.6. Let $x(t)$ be defined as illustrated in Fig. 7.6(a). Function $h(t)$ is given as

(a) $h(t) = \delta(t)$

(b) $h(t) = \delta(t) + \delta\left(t - \dfrac{3}{2}\right)$

(c) $h(t) = 0 \qquad t < 0$

$$= 1 \qquad 0 < t < \frac{1}{2}$$

$$= 0 \qquad \frac{1}{2} < t < 1$$

$$= 1 \qquad 1 < t < \frac{3}{2}$$

$$= 0 \qquad t > \frac{3}{2}$$

Following Fig. 7.6, graphically determine the discrete convolution in each case. Compare the discrete and continuous convolution in each case. Investigate the end effect in each case.

7.5. It is desired to discretely convolve a finite-duration and an infinite-duration waveform. Assume that a hardware device is to be used that is limited in capacity to N sample values of each function. Describe a procedure that allows one to perform successive N-point discrete convolutions and combine the two to eliminate the end effect. Demonstrate your concept by repeating the illustrations of Fig. 7.6 for the case $NT = 1.5$. Successively apply the developed technique to determine the discrete convolution $y(kT)$ for $0 \leq kT \leq 3$.

7.6. Derive the discrete convolution theorem for the following:

(a) $\sum\limits_{i=0}^{N-1} h(iT)x[(k - i)T]$

(b) $\sum\limits_{i=0}^{N-1} h(iT)h[(k - i)T]$

7.7. Let $x(kT)$ and $h(kT)$ be defined by Problem 7.1. Determine the discrete correlation of Eq. (7.11) both analytically and graphically. What are the contraints on N?

7.8. Repeat Problem 7.2 for discrete correlation.

7.9. Repeat Problem 7.3 for discrete correlation.

7.10. Repeat Problem 7.4 for discrete correlation.

REFERENCES

1. COOLEY, J. W., P. A. W. LEWIS, AND P. D. WELCH. "The Finite Fourier Transform." *IEEE Trans. on Audio and Electroacoustics* (June 1969), Vol. AU17, No. 2, pp. 77–85.

2. COOLEY, J. W., P. A. W. LEWIS, AND P. D. WELCH. "The Fast Fourier Transform and its Applications." *IEEE Trans. on Education* (March 1969), Vol. 12, pp. 27–34.

3. BERGLAND, G. D. "A Guided Tour of the Fast Fourier Transform." *IEEE Spectrum* (1969), Vol. 6, No. 7, pp. 41–52.

THE FAST FOURIER
TRANSFORM (FFT)

Interpretation of fast Fourier transform results does not require a well-grounded education in the algorithm itself, but rather a thorough understanding of the discrete Fourier transform. This follows from the fact that the FFT is simply an algorithm (i.e., a particular method of performing a series of computations) that can compute the discrete Fourier transform much more rapidly than other available algorithms. For this reason, our discussion of the FFT addresses only the computational aspect of the algorithm.

A simple matrix-factoring example is used to intuitively justify the FFT algorithm. The factored matrices are alternatively represented by signal flow graphs. From these graphs, we construct the logic of an FFT computer program. Theoretical developments of various forms of the FFT algorithm are then presented.

8.1 MATRIX FORMULATION

Consider the discrete Fourier transform, Eq. (6.16):

$$X(n) = \sum_{k=0}^{N-1} x_0(k)e^{-j2\pi nk/N} \qquad n = 0, 1, \ldots, N - 1 \qquad (8.1)$$

where we have replaced kT by k and n/NT by n for convenience of notation. We note that Eq. (8.1) describes the computation of N equations. For example, if $N = 4$ and if we let

$$W = e^{-j2\pi/N} \tag{8.2}$$

then Eq. (8.1) can be written as

$$X(0) = x_0(0)W^0 + x_0(1)W^0 + x_0(2)W^0 + x_0(3)W^0$$

$$X(1) = x_0(0)W^0 + x_0(1)W^1 + x_0(2)W^2 + x_0(3)W^3 \tag{8.3}$$

$$X(2) = x_0(0)W^0 + x_0(1)W^2 + x_0(2)W^4 + x_0(3)W^6$$

$$X(3) = x_0(0)W^0 + x_0(1)W^3 + x_0(2)W^6 + x_0(3)W^9$$

Equations (8.3) can be more easily represented in matrix form:

$$\begin{bmatrix} X(0) \\ X(1) \\ X(2) \\ X(3) \end{bmatrix} = \begin{bmatrix} W^0 & W^0 & W^0 & W^0 \\ W^0 & W^1 & W^2 & W^3 \\ W^0 & W^2 & W^4 & W^6 \\ W^0 & W^3 & W^6 & W^9 \end{bmatrix} \begin{bmatrix} x_0(0) \\ x_0(1) \\ x_0(2) \\ x_0(3) \end{bmatrix} \tag{8.4}$$

or more compactly as

$$X(n) = W^{nk}x_0(k) \tag{8.5}$$

We will denote a matrix by **boldface italic** type.

Examination of Eq. (8.4) reveals that since W and possibly $x_0(k)$ are complex, then N^2 complex multiplications and $(N)(N - 1)$ complex additions are necessary to perform the required matrix computation. The FFT owes its success to the fact that the algorithm reduces the number of multiplications and additions required in the computation of Eq. (8.4). We will now discuss, on an intuitive level, how this reduction is accomplished. A proof of the FFT algorithm is delayed until Sec. 8.9.

8.2 INTUITIVE DEVELOPMENT

To illustrate the FFT algorithm, it is convenient to choose the number of sample points of $x_0(k)$ according to the relation $N = 2^\gamma$, where γ is an integer. Later developments remove this restriction. Recall that Eq. (8.4) results from the choice of $N = 4 = 2^\gamma = 2^2$; therefore, we can apply the FFT to the computation of Eq. (8.4).

The first step in developing the FFT algorithm for this example is to rewrite Eq. (8.4) as

$$\begin{bmatrix} X(0) \\ X(1) \\ X(2) \\ X(3) \end{bmatrix} = \begin{bmatrix} 1 & 1 & 1 & 1 \\ 1 & W^1 & W^2 & W^3 \\ 1 & W^2 & W^0 & W^2 \\ 1 & W^3 & W^2 & W^1 \end{bmatrix} \begin{bmatrix} x_0(0) \\ x_0(1) \\ x_0(2) \\ x_0(3) \end{bmatrix} \tag{8.6}$$

Matrix Eq. (8.6) was derived from Eq. (8.4) by using the relationship W^{nk} = $W^{nk \bmod(N)}$. Recall that $[nk \bmod(N)]$ is the remainder upon division of nk by N; hence if $N = 4$, $n = 2$, and $k = 3$, then

$$W^6 = W^2 \tag{8.7}$$

because

$$W^{nk} = W^6 = \exp\left[\left(\frac{-j2\pi}{4}\right)(6)\right] = \exp[-j3\pi]$$
$$= \exp[-j\pi] = \exp\left[\left(\frac{-j2\pi}{4}\right)(2)\right] = W^2 = W^{nk \bmod(N)} \tag{8.8}$$

The second step in the development is to factor the square matrix in Eq. (8.6) as follows:

$$
\begin{bmatrix} X(0) \\ X(2) \\ X(1) \\ X(3) \end{bmatrix}
=
\begin{bmatrix}
1 & W^0 & 0 & 0 \\
1 & W^2 & 0 & 0 \\
0 & 0 & 1 & W^1 \\
0 & 0 & 1 & W^3
\end{bmatrix}
\begin{bmatrix}
1 & 0 & W^0 & 0 \\
0 & 1 & 0 & W^0 \\
1 & 0 & W^2 & 0 \\
0 & 1 & 0 & W^2
\end{bmatrix}
\begin{bmatrix} x_0(0) \\ x_0(1) \\ x_0(2) \\ x_0(3) \end{bmatrix}
\tag{8.9}
$$

The method of factorization is based on the theory of the FFT algorithm developed in Sec. 8.9. For the present, it suffices to show that multiplication of the two square matrices of Eq. (8.9) yields the square matrix of Eq. (8.6) with the exception that rows 1 and 2 have been interchanged (the rows are numbered 0, 1, 2, and 3). Note that this interchange has been taken into account in Eq. (8.9) by rewriting the column vector $X(n)$; let the row-interchanged vector be denoted by

$$
\overline{X(n)} =
\begin{bmatrix} X(0) \\ X(2) \\ X(1) \\ X(3) \end{bmatrix}
\tag{8.10}
$$

Repeating, the reader should verify that Eq. (8.9) yields Eq. (8.6) with the interchanged rows as noted. This factorization is the key to the efficiency of the FFT algorithm.

Having accepted the fact that Eq. (8.9) is correct, although the results are *scrambled*, one should then examine the number of multiplications required to compute the equation. First, let

$$
\begin{bmatrix} x_1(0) \\ x_1(1) \\ x_1(2) \\ x_1(3) \end{bmatrix}
=
\begin{bmatrix}
1 & 0 & W^0 & 0 \\
0 & 1 & 0 & W^0 \\
1 & 0 & W^2 & 0 \\
0 & 1 & 0 & W^2
\end{bmatrix}
\begin{bmatrix} x_0(0) \\ x_0(1) \\ x_0(2) \\ x_0(3) \end{bmatrix}
\tag{8.11}
$$

That is, column vector $x_1(k)$ is equal to the product of the two matrices on the right in Eq. (8.9).

Element $x_1(0)$ is computed by one complex multiplication and one complex addition (W^0 is not reduced to unity in order to develop a generalized result).

$$x_1(0) = x_0(0) + W^0 x_0(2) \tag{8.12}$$

Element $x_1(1)$ is also determined by one complex multiplication and addition. Only one complex addition is required to compute $x_1(2)$. This follows because $W^0 = -W^2$; hence,

$$\begin{aligned} x_1(2) &= x_0(0) + W^2 x_0(2) \\ &= x_0(0) - W^0 x_0(2) \end{aligned} \tag{8.13}$$

where the complex multiplication $W^0 x_0(2)$ has already been computed in the determination of $x_1(0)$ [Eq. (8.12)]. By the same reasoning, $x_1(3)$ is computed by only one complex addition and no multiplications. The intermediate vector $x_1(k)$ is then determined by four complex additions and two complex multiplications.

Let us continue by completing the computation of Eq. (8.9)

$$\begin{bmatrix} X(0) \\ X(2) \\ X(1) \\ X(3) \end{bmatrix} = \begin{vmatrix} x_2(0) \\ x_2(1) \\ x_2(2) \\ x_2(3) \end{vmatrix} = \begin{vmatrix} 1 & W^0 & 0 & 0 \\ 1 & W^2 & 0 & 0 \\ 0 & 0 & 1 & W^1 \\ 0 & 0 & 1 & W^3 \end{vmatrix} \begin{vmatrix} x_1(0) \\ x_1(1) \\ x_1(2) \\ x_1(3) \end{vmatrix} \tag{8.14}$$

Term $x_2(0)$ is determined by one complex multiplication and addition:

$$x_2(0) = x_1(0) + W^0 x_1(1) \tag{8.15}$$

Element $x_2(1)$ is computed by one addition because $W^0 = -W^2$. By similar reasoning, $x_2(2)$ is determined by one complex multiplication and addition, and $x_2(3)$ by only one addition.

Computation of $\overline{X(n)}$ by means of Eq. (8.9) requires a total of four complex multiplications and eight complex additions. Computation of $X(n)$ by (8.4) requires 16 complex multiplications and 12 complex additions. Note that the matrix-factorization process introduces zeros into the factored matrices and, as a result, reduces the required number of multiplications. For this example, the matrix-factorization process reduces the number of multiplications by a factor of two. Because computation time is largely governed by the required number of multiplications, we see the reason for the efficiency of the FFT algorithm.

For $N = 2^\gamma$, the FFT algorithm is then simply a procedure for factoring an $N \times N$ matrix into γ matrices (each $N \times N$), such that each of the factored matrices has the special property of minimizing the number of complex multiplications and additions. If we extend the results of the previous example, we note that the FFT requires $N\gamma/2 = 4$ *complex* multiplications and $N\gamma = 8$ *complex* additions, whereas the direct method [Eq. (8.4)] requires N^2 *complex* multiplications and $N(N - 1)$ *complex* additions. If we assume that computing time is proportional to the number of multiplications, then the approximate ratio of direct to FFT computing time is given by

$$\frac{N^2}{N\gamma/2} = \frac{2N}{\gamma} \qquad (8.16)$$

which for $N = 1024 = 2^{10}$ is a computational reduction of more than 200 to 1. Figure 8.1 illustrates the relationship between the number of multiplications required using the FFT algorithm compared with the number of multiplications using the direct method.

The matrix-factoring procedure does introduce one discrepancy. Recall that the computation of Eq. (8.9) yields $\overline{X(n)}$ instead of $X(n)$; that is,

$$\overline{X(n)} = \begin{bmatrix} X(0) \\ X(2) \\ X(1) \\ X(3) \end{bmatrix} \text{ instead of } X(n) = \begin{bmatrix} X(0) \\ X(1) \\ X(2) \\ X(3) \end{bmatrix} \qquad (8.17)$$

This rearrangement is inherent in the matrix-factoring process and is a minor problem because it is straightforward to generalize a technique for *unscrambling* $\overline{X(n)}$ to obtain $X(n)$.

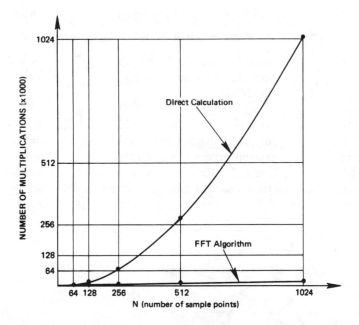

Figure 8.1   Comparison of multiplications required by direct calculation and FFT algorithm.

Rewrite $\overline{X(n)}$ by replacing argument n with its binary equivalent:

$$\begin{bmatrix} X(0) \\ X(2) \\ X(1) \\ X(3) \end{bmatrix} \quad \text{becomes} \quad \begin{bmatrix} X(00) \\ X(10) \\ X(01) \\ X(11) \end{bmatrix} \tag{8.18}$$

Observe that if the binary arguments of Eq. (8.18) are *flipped* or *bit-reversed* (i.e., 01 becomes 10, 10 becomes 01, etc.), then

$$\overline{X(n)} = \begin{bmatrix} X(00) \\ X(10) \\ X(01) \\ X(11) \end{bmatrix} \quad \text{flips to} \quad \begin{bmatrix} X(00) \\ X(01) \\ X(10) \\ X(11) \end{bmatrix} = X(n) \tag{8.19}$$

It is straightforward to develop a generalized result for unscrambling the FFT.

For N greater than 4, it is cumbersome to describe the matrix-factorization process analogous to Eq. (8.9). For this reason, we interpret (8.9) in a graphical manner. Using this graphical formulation, we can describe sufficient generalities to develop a flow graph for a computer program.

8.3 SIGNAL FLOW GRAPH

We convert Eq. (8.9) into the signal flow graph illustrated in Fig. 8.2. As shown, we represent the data vector or array $x_0(k)$ by a vertical column of nodes on the left of the graph. The second vertical array of nodes is the vector $x_1(k)$ computed in Eq. (8.11), and the next vertical array corresponds to the vector $x_2(k) = \overline{X(n)}$, Eq. (8.14). In general, there will be γ computational arrays where $N = 2^\gamma$.

The signal flow graph is interpreted as follows. Each node is entered

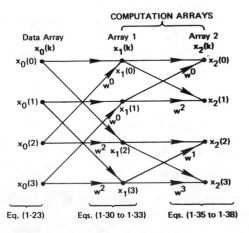

Figure 8.2 FFT signal flow graph, $N = 4$.

by two solid lines representing *transmission* paths from previous nodes. A path transmits or brings a quantity from a node in one array, multiplies the quantity by W^p, and inputs the result into the node in the next array. Factor W^p appears near the arrowhead of the transmission path; absence of this factor implies that $W^p = 1$. Results entering a node from the two transmission paths are combined additively.

To illustrate the interpretation of the signal flow graph, consider node

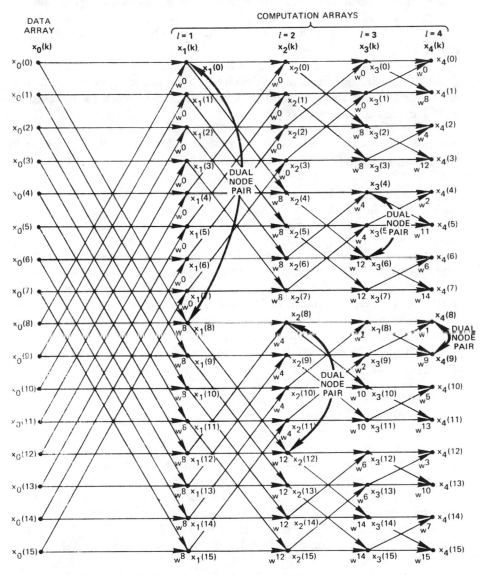

Figure 8.3   Example of dual nodes.

$x_1(2)$ in Fig. 8.2. According to the rules for interpreting the signal flow graph,

$$x_1(2) = x_0(0) + W^2 x_0(2) \tag{8.20}$$

which is simply Eq. (8.13). Each node of the signal flow graph is expressed similarly.

The signal flow graph is then a concise method for representing the computations required in the factored matrix FFT algorithm of Eq. (8.9). Each computational column of the graph corresponds to a factored matrix; γ vertical arrays of N points each ($N = 2^\gamma$) are required. Utilization of this graphical presentation allows us to easily describe the matrix-factoring process for large N.

We show in Fig. 8.3 the signal flow graph for $N = 16$. With a flow graph of this size, it is possible to develop general properties concerning the matrix-factorization process and thus provide a framework for developing a FFT computer program flowchart.

8.4 DUAL NODES

Inspection of Fig. 8.3 reveals that in every array we can always find two nodes whose input transmission paths stem from the same pair of nodes in the previous array. For example, nodes $x_1(0)$ and $x_1(8)$ are computed in terms of nodes $x_0(0)$ and $x_0(8)$. Note that nodes $x_0(0)$ and $x_0(8)$ do not enter into the computation of any other node. We define two such nodes as a *dual-node pair*.

Because the computation of a dual-node pair is independent of other nodes, it is possible to perform *in-place computation*. To illustrate, note from Fig. 8.3 that we can simultaneously compute $x_1(0)$ and $x_1(8)$ in terms of $x_0(0)$ and $x_0(8)$ and return the results to the storage locations previously occupied by $x_0(0)$ and $x_0(8)$. Storage requirements are then limited to the data array $x_0(k)$ only. As each array is computed, the results are returned to this array.

Dual-Node Spacing

Let us now investigate the spacing (measured vertically in terms of the index k) between a dual-node pair. The following discussion will refer to Fig. 8.3. First, in array $l = 1$, a dual-node pair, say $\{x_1(0), x_1(8)\}$, is separated by $k = 8 = N/2^l = N/2^1$. In array $l = 2$, a dual-node pair, say $\{x_2(8), x_2(12)\}$, is separated by $k = 4 = N/2^l = N/2^2$. Similarly, a dual-node pair, $\{x_3(4), x_3(6)\}$, in array $l = 3$ is separated by $k = 2 = N/2^l = N/2^3$; and in array $l = 4$, a dual-node pair, $\{x_4(8), x_4(9)\}$, is separated by $k = 1 = N/2^l = N/2^4$.

Generalizing these results, we observe that the spacing between dual nodes in array l is given by $N/2^l$. Thus, if we consider a particular node

$x_l(k)$, then its dual node is $x_l(k + N/2^l)$. This property allows us to easily identify a dual-node pair.

Dual-Node Computation

The computation of a dual-node pair requires only one complex multiplication. To clarify this point, consider node $x_2(8)$ and its dual $x_2(12)$, as

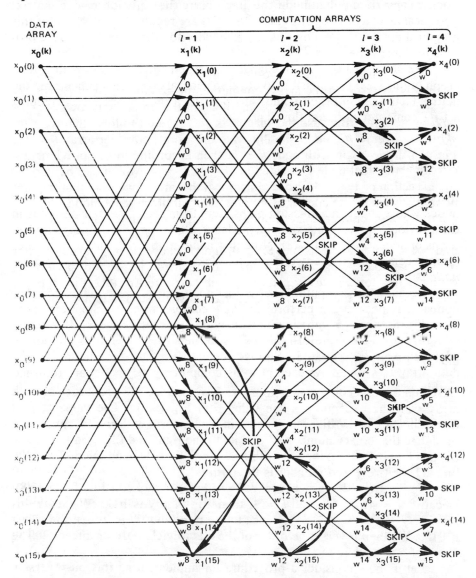

Figure 8.4 Example of nodes to be skipped when computing a signal flow graph.

illustrated in Fig. 8.3. The transmission paths stemming from node $x_1(12)$ are multiplied by W^4 and W^{12} prior to input at nodes $x_2(8)$ and $x_2(12)$, respectively. It is important to note that $W^4 = -W^{12}$ and that only one multiplication is required because the same data $x_1(12)$ is to be multiplied by these terms. In general, if the weighting factor at one node is W^p, then the weighting factor at the dual node is $W^{p+N/2}$. Because $W^p = -W^{p+N/2}$, only one multiplication is required in the computation of a dual-node pair. The computation of any dual-node pair is given by the equation pair:

$$x_l(k) = x_{l-1}(k) + W^p x_{l-1}(k + N/2^l)$$
$$x_l(k + N/2^l) = x_{l-1}(k) - W^p x_{l-1}(k + N/2^l)$$

$$(8.21)$$

In computing an array, we normally begin with node $k = 0$ and sequentially work down the array, computing the equation pair of Eq. (8.21). As stated previously, the dual of any node in the lth array is always down $N/2^l$ in the array. Because the spacing is $N/2^l$, then it follows that we must *skip* after every $N/2^l$ node. To illustrate this point, consider array $l = 2$ in Fig. 8.4. If we begin with node $k = 0$, then according to our previous discussions, the dual node is located at $k = N/2^2 = 4$, which can be verified by inspection of Fig. 8.4. Proceeding down this array, we note that the dual node is always located down by 4 in the array until we reach node 4. At this point, we have entered a set of nodes previously encountered, that is, these nodes are the duals for nodes $k = 0, 1, 2$, and 3. It is necessary to *skip over* nodes $k = 4, 5, 6$, and 7. Nodes 8, 9, 10, and 11 follow the original convention of the dual node being located 4 down in the array. In general, if we work from the top down in array l, then we will compute Eq. (8.21) for the first $N/2^l$ nodes, skip the next $N/2^l$, etc. We know to stop skipping when we reach a node index greater than $N - 1$.

8.5 W^p DETERMINATION

Based on the preceding discussions, we have defined the properties of each array with the exception of the value p in Eq. (8.21). The value of p is determined by (a) writing the index k in binary form with γ bits, (b) scaling or sliding this binary number $\gamma - l$ bits to the right and filling in the newly opened bit position on the left with zeros, and (c) reversing the order of the bits. This bit-reversed number is the term p.

To illustrate this procedure, refer to Fig. 8.4 and consider node $x_3(8)$. Because $\gamma = 4$, $k = 8$, and $l = 3$, then k in binary is 1000. We scale this number $\gamma - l = 4 - 3 = 1$ places to the right and fill in zeros; the result is 0100. We then reverse the order of the bits to yield 0010 or integer 2. The value of p is then 2.

Let us now consider a procedure for implementing this bit-reversing operation. We know that a binary number, say $a_4 a_3 a_2 a_1 \cdot$, can be written in base 10 as $a_4 \times 2^3 + a_3 \times 2^2 + a_2 \times 2^1 + a_1 \times 2^0$. The bit-reversed number that we are trying to describe is given by $a_1 \times 2^3 + a_2 \times 2^2 + a_3 \times 2^1 + a_4 \times 2^0$. If we describe a technique for determining the binary bits a_4, a_3, a_2, and a_1, then we have defined a bit-reversing operation.

Now assume that M is a binary number equal to $a_4a_3a_2a_1\cdot$. Divide M by 2, truncate, and multiply the truncated results by 2. Then compute $a_4a_3a_2a_1\cdot - 2(a_4a_3a_2\cdot)$. If the bit a_1 is 0, then this difference is zero because division by 2, truncation, and subsequent multiplication by 2 does not alter M. However, if the bit a_1 is 1, truncation changes the value of M and the above difference expression is nonzero. We observe that by this technique, we can determine if the bit a_1 is 0 or 1.

We can identify the bit a_2 in a similar manner. The appropriate difference expression is $a_4a_3a_2\cdot - 2(a_4a_3\cdot)$. If this difference is zero, then a_2 is zero. Bits a_3 and a_4 are determined similarly. This procedure forms the basis for developing a bit-reversing computer routine in Sec. 8.7.

8.6 UNSCRAMBLING THE FFT

The final step in computing the FFT is to *unscramble* the results analogous to Eq. (8.19). Recall that the procedure for unscrambling the vector $\overline{X(n)}$ is to write n in binary and reverse or flip the binary number. We show in Fig. 8.5 the results of this bit-reversing operation: terms $x_4(k)$ and $x_4(i)$ have simply been interchanged, where i is the integer obtained by bit-reversing the integer k.

Note that a situation similar to the dual-node concept exists when we unscramble the output array. If we proceed down the array, interchanging $x(k)$ with the appropriate $x(i)$, we eventually encounter a node that has previously been interchanged. For example, in Fig. 8.5, node $k = 0$ remains in its location, nodes $k = 1$, 2, and 3 are interchanged with nodes 8, 4, and 12, respectively. The next node to be considered is node 4, but this node was previously interchanged with node 2. To eliminate the possibility of considering a node that has previously been interchanged, we simply check to see if i (the integer obtained by bit-reversing k) is less than k. If so, this implies that the node has been interchanged by a previous operation. With this check, we can ensure a straightforward unscrambling procedure.

8.7 FFT COMPUTATION FLOWCHART

Using the discussed properties of the FFT signal flow graph, we can easily develop a flowchart for programming the algorithm on a digital computer. We know from the previous discussions that we first compute array $l = 1$ by starting at node $k = 0$ and working down the array. At each node k, we

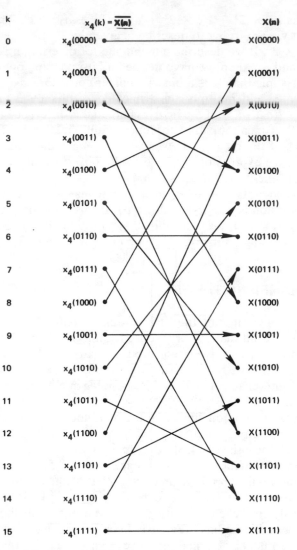

Figure 8.5 Example of the bit-reversing operation for $N = 16$.

compute the equation pair of Eq. (8.21), where p is determined by the described procedure. We continue down the array computing the equation pair of Eq. (8.21) until we reach a region of nodes that must be skipped over. We skip over the appropriate nodes and continue until we have computed the entire array. We then proceed to compute the remaining arrays using the same procedures. Finally, we unscramble the final array to obtain the desired results. Figure 8.6 illustrates a flowchart for computer programming the FFT algorithm.

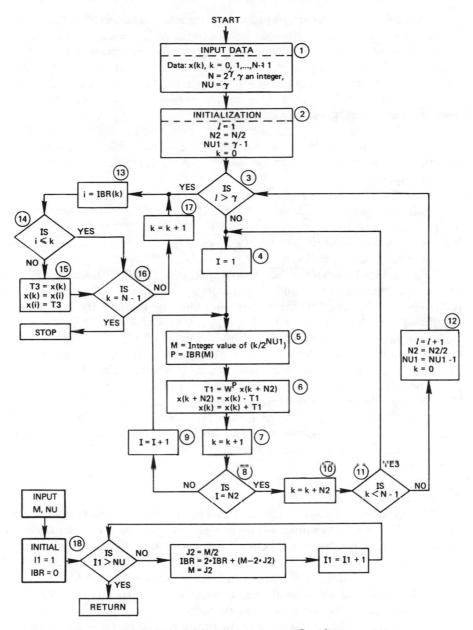

Figure 8.6 FFT computer program flowchart.

Box 1 describes the necessary input data. Data vector $x_0(k)$ is assumed to be complex and is indexed as $k = 0, 1, \ldots, N - 1$. If $x_0(k)$ is real, then the imaginary part should be set to zero. The number of sample points N must satisfy the relationship $N = 2^\gamma$, where γ is integer valued.

Initialization of the various program parameters is accomplished in Box 2. Parameter l is the array number being considered. We start with array $l = 1$. The spacing between dual nodes is given by the parameter $N2$; for array $l = 1$, $N2 = N/2$ and is initialized as such. Parameter $NU1$ is the right shift required when determining the value of p in Eq. (8.21); $NU1$ is initialized to $\gamma - 1$. The index k of the array is initialized to $k = 0$; thus, we will work from the top and progress down the array.

Box 3 checks to see if the array l to be computed is greater than γ. If yes, then the program branches to Box 13 to unscramble the computed results by bit inversion. If all arrays have not been computed, then we proceed to Box 4.

Box 4 sets a counter $I = 1$. This counter monitors the number of dual-node pairs that have been considered. Recall from Sec. 8.4 that it is necessary to skip certain nodes in order to ensure that previously considered nodes are not encountered a second time. Counter I is the control for determining when the program must skip.

Boxes 5 and 6 perform the computation of Eq. (8.21). Because k and l have been initialized to 0 and 1, respectively, the initial node considered is the first node of the first array. To determine the factor p for this node, recall that we must first scale the binary number k to the right $\gamma - l$ bits. To accomplish this, we compute the integer value of $k/2^{\gamma - l} = k/2^{NU1}$ and set the result to M as shown in Box 5. According to the procedure for determining p, we must bit reverse M, where M is represented by $\gamma = NU$ bits. The function $IBR(M)$ denoted in Box 5 is a special function routine for bit inversion; this routine is described later.

Box 6 is the computation of Eq. (8.21). We compute the product W^p $x(k + N2)$ and assign the result to a temporary storage location. Next, we add and subtract this term according to Eq. (8.21). The result is the dual-node output.

We then proceed down the array to the next node. As shown in Box 7, k is incremented by 1.

To avoid recomputing a dual node that has been considered previously, we check Box 8 to determine if the counter I is equal to $N2$. For array 1, the number of nodes that can be considered consecutively without skipping is equal to $N/2 = N2$. Box 8 determines this condition. If I is not equal to $N2$, then we proceed down the array and increment the counter I, as shown in Box 9. Recall that we have already incremented k in Box 7. Boxes 5 and 6 are then repeated for the new value of k.

If $I = N2$ in Box 8, then we know that we have reached a node previously considered. We then skip $N2$ nodes by setting $k = k + N2$. Because k has already been incremented by 1 in Box 7, it is sufficient to skip the previously considered nodes by incrementing k by $N2$.

Before we perform the required computations indicated by Boxes 5 and 6 for the new node $k = k + N2$, we must first check to see that we have not exceeded the array size. As shown in Box 11, if k is less than N − 1 (recall k is indexed from 0 to N − 1), then we reset the counter I to 1 in Box 4 and repeat Boxes 5 and 6.

If $k > N$ − 1 in Box 11, we know that we must proceed to the next array. Hence, as shown in Box 12, l is indexed by 1. The new spacing $N2$ is simply $N2/2$ (recall the spacing is $N/2^l$). $NU1$ is decremented by 1 ($NU1$ is equal to γ − l), and k is reset to zero. We then check Box 3 to see if all arrays have been computed. If so, then we proceed to unscramble the final results. This operation is performed by Boxes 13 through 17.

Box 13 bit-reverses the integer k to obtain the integer i. Again we use the bit-reversing function $IBR(k)$, which is explained later. Recall that to unscramble the FFT, we simply interchange $x(k)$ and $x(i)$. This manipulation is performed by the operations indicated in Box 15. However, before Box 15 is entered, it is necessary to determine, as shown in Box 14, if i is less than or equal to k. This step is necessary to prohibit the altering of previously unscrambled nodes.

Box 16 determines when all nodes have been unscrambled and Box 17 is simply an index for k.

In Box 18, we describe the logic of the bit-reversing function $IBR(k)$. We have implemented the bit-reversing procedure discussed in Sec. 8.5.

When one proceeds to implement the flow graph of Fig. 8.6 into a computer program, it is necessary to consider the variables $x(k)$ and W^p as complex numbers and they must be handled accordingly.

8.8 FFT BASIC AND PASCAL COMPUTER PROGRAMS

A listing of a BASIC program based on the FFT algorithm flowchart in Fig. 8.6 is shown in Fig. 8.7. The program does not attempt to accomplish the utlimate in efficiency but rather is designed to acquaint the reader with the computer programming procedure of the FFT algorithm. Efficient programming results in a slight increase in computing speed.

The inputs to the FFT program are XREAL(N%), the real part of the function to be discrete Fourier transformed; XIMAG(N%), the imaginary part; N%, the number of sample points; and NU%, where N% = $2^{NU\%}$.

```
10000 REM:    FFT SUBROUTINE- THE CALLING PROGRAM SHOULD
10002 REM:    DIMENSION XREAL(I%) AND XIMAG(I%).
10004 REM:    N% AND NU% MUST BE INITIALIZED.
10010 N2% = N%/2
10020 NU1% = NU% - 1
10030 K% = 0
10040 FOR L% = 1 TO NU% STEP 1
```

Figure 8.7   FFT BASIC computer subroutine.

```
10050            FOR I% = 1 TO N2% STEP 1
10060               J% = K%\2^NU1%
10070               GOSUB 10410
10080                ARG = 6.283185# * IBITR%/N%
10090                C = COS(ARG)
10100                S = SIN(ARG)
10110                K1% = K% + 1
10120                K1N2% = K1% + N2%
10130                TREAL =   XREAL(K1N2%) * C + XIMAG(K1N2%)*S
10140                TIMAG = XIMAG(K1N2%) * C - XREAL(K1N2%) * S
10150                XREAL(K1N2%) = XREAL(K1%) - TREAL
10160                XIMAG(K1N2%) = XIMAG(K1%) - TIMAG
10170                XREAL(K1%) = XREAL(K1%) + TREAL
10180                XIMAG(K1%) = XIMAG(K1%) + TIMAG
10190                K% = K% + 1
10200            NEXT I%
10210      K% = K% + N2%
10220      IF K%<N%  GOTO 10050
10230      K% = 0
10240      NU1% = NU1% -1
10250      N2% = N2% / 2
10260 NEXT L%
10270 FOR K% = 1 TO N% STEP 1
10280         J% = K% - 1
10290         GOSUB 10410
10300         I%= IBITR% + 1
10310         IF(I%<=K%) GOTO 10380
10320         TREAL = XREAL(K%)
10330         TIMAG = XIMAG(K%)
10340            XREAL(K%) = XREAL(I%)
10350            XIMAG(K%) = XIMAG(I%)
10360            XREAL(I%) = TREAL
10370            XIMAG(I%) = TIMAG
10380 NEXT K%
10390 RETURN
10400 END
10410 REM:  BIT REVERSAL SUB-ROUTINE
10420 J1% = J%
10430 IBITR% = 0
10440 FOR I1% = 1 TO NU% STEP 1
10450    J2% = J1%\2
10460    IBITR% = IBITR%*2 + (J1% - 2*J2%)
10470    J1% = J2%
10480 NEXT I1%
10490 RETURN
10500 END
```

Figure 8.7 *(continued)*

Upon completion, XREAL(N%) is the real part of the transform and XIMAG(N%) is the imaginary part of the transform. Input data is destroyed. Note that the backslash (\) implies integer division in BASIC. Integer division is required for the subroutine to operate correctly. Integer variables are specified by the percent symbol (%) and the pound symbol (#) denotes double precision. The program uses the convention that array indexing begins with 1. (Computing recipes throughout this book are more readily implemented if the complier permits the use of an array index of zero.) Because FFT computations are recursive, double-precision arithmetic may be required.

In Fig. 8.8, we show a PASCAL program based on the flowchart of Fig. 8.6. Input and output variables are identical to those described for the BASIC program.

```
TYPE REALARRAY=ARRAY[0..31] OF REAL;

FUNCTION IBITR (J,NU:INTEGER): INTEGER;
VAR    I,J1,J2,K:    INTEGER; .
BEGIN
       J1 := J;
       K  := 0;
       FOR I := 1 TO NU DO
       BEGIN
              J2 := J1 DIV 2;
              K  := K*2+(J1-2*J2);
              J1 := J2
       END;
       IBITR := K
END;  (IBITR)


PROCEDURE FFT (VAR XREAL,XIMAG: REALARRAY; N,NU: INTEGER);
VAR    N2,NU1,I,L,K,M: INTEGER;
       TREAL,TIMAG,P,ARG,C,S: REAL;
LABEL LBL;
BEGIN
       N2 := N DIV 2;
       NU1 := NU-1;
       K := 0;
       FOR L := 1 TO NU DO
       BEGIN
           LBL:
              FOR I := 1 TO N2 DO
              BEGIN
                     M := K DIV ROUND(EXP (NU1 * LN (2)));
                     P := IBITR (M,NU);
```

Figure 8.8   FFT PASCAL computer subroutine.

```
                    ARG  := 6.283185*P/N;
                    C  := COS (ARG);
                    S  := SIN (ARG);
                    TREAL  := XREAL[K+N2]*C+XIMAG[K+N2]*S;
                    TIMAG  := XIMAG[K+N2]*C-XREAL[K+N2]*S;
                    XREAL[K+N2]  := XREAL[K]-TREAL;
                    XIMAG[K+N2]  := XIMAG[K]-TIMAG;
                    XREAL[K]  := XREAL[K]+TREAL;
                    XIMAG[K]  := XIMAG[K]+TIMAG;
                    K  := K+1
              END;
              K  := K+N2;
              IF K<N THEN GOTO LBL;
              K  := 0;
              NU1  := NU1-1;
              N2  := N2 DIV 2
        END;
        FOR K  := 0 TO N-1 DO
        BEGIN
              I  := IBITR (K,NU);
              IF I>K THEN
              BEGIN
                    TREAL  := XREAL[K];
                    TIMAG  := XIMAG[K];
                    XREAL[K]  := XREAL[I];
                    XIMAG[K]  := XIMAG[I];
                    XREAL[I]  := TREAL;
                    XIMAG[I]  := TIMAG
              END
        END
END; {FFT}
```

Figure 8.8   (*continued*)

8.9 THEORETICAL DEVELOPMENT OF THE BASE-2 FFT ALGORITHM

In Sec. 8.2, we used a matrix argument to develop an understanding of why the FFT is an efficient algorithm. We then constructed a signal flow graph that described the algorithm for any $N = 2^\gamma$. In this section, we relate each of these developments to a theoretical basis. First, we will develop a theoretical proof of the algorithm for the case $N = 4$. We then extend these arguments to the case $N = 8$. The reason for these developments for specific cases is to establish the notation that we use in the final derivation of the algorithm for the case $N = 2^\gamma$, where γ is integer valued.

Definition of Notation

Consider the discrete Fourier transform relationship of Eq. (8.1)

$$X(n) = \sum_{k=0}^{N-1} x_0(k)W^{nk} \qquad n = 0, 1, \ldots, N - 1 \qquad (8.22)$$

where we have set $W = e^{-j2\pi/N}$. It is desirable to represent the integers n and k as binary numbers; that is, if we assume $N = 4$, then $\gamma = 2$ and we can represent k and n as two-bit binary numbers,

$$k = 0, 1, 2, 3 \qquad \text{or} \qquad k = (k_1, k_0) = 00, 01, 10, 11$$

$$n = 0, 1, 2, 3 \qquad \text{or} \qquad n = (n_1, n_0) = 00, 01, 10, 11$$

A compact method of writing k and n is

$$k = 2k_1 + k_0 \qquad n = 2n_1 + n_0 \qquad (8.23)$$

where k_0, k_1, n_0, and n_1 can take the values of 0 and 1 only. Equation (8.23) is simply the method of writing a binary number as its base-10 equivalent.

Using the representation of Eq. (8.23), we can rewrite Eq. (8.22) for the case $N = 4$ as

$$X(n_1, n_0) = \sum_{k_0=0}^{1} \sum_{k_1=0}^{1} x_0(k_1, k_0)W^{(2n_1 + n_0)(2k_1 + k_0)} \qquad (8.24)$$

Note that the single summation in Eq. (8.22) must now be replaced by γ summations in order to enumerate all the bits of the binary representation of k.

Factorization of W^P

Now consider the W^P term. Because $W^{a+b} = W^a W^b$, then

$$W^{(2n_1 + n_0)(2k_1 + k_0)} = W^{(2n_1 + n_0)2k_1} W^{(2n_1 + n_0)k_0}$$

$$= [W^{4n_1 k_1}] W^{2n_0 k_1} W^{(2n_1 + n_0)k_0} \qquad (8.25)$$

$$= W^{2n_0 k_1} W^{(2n_1 + n_0)k_0}$$

Note that the term in brackets is equal to unity because

$$W^{4n_1 k_1} = [W^4]^{n_1 k_1} = [e^{-j2\pi 4/4}]^{n_1 k_1} = [1]^{n_1 k_1} = 1 \qquad (8.26)$$

Thus, Eq. (8.24) can be written in the form:

$$X(n_1, n_0) = \sum_{k_0=0}^{1} \left[\sum_{k_1=0}^{1} x_0(k_1, k_0)W^{2n_0 k_1} \right] W^{(2n_1 + n_0)k_0} \qquad (8.27)$$

This equation represents the foundation of the FFT algorithm. To demonstrate this point, let us consider each of the summations of Eq. (8.27) individually. First, rewrite the summation in brackets as

$$x_1(n_0,k_0) = \sum_{k_1=0}^{1} x_0(k_1,k_0)W^{2n_0k_1} \tag{8.28}$$

Enumerating the equations represented by Eq. (8.28), we obtain

$$x_1(0,0) = x_0(0,0) + x_0(1,0)W^0$$

$$x_1(0,1) = x_0(0,1) + x_0(1,1)W^0$$

$$x_1(1,0) = x_0(0,0) + x_0(1,0)W^2 \tag{8.29}$$

$$x_1(1,1) = x_0(0,1) + x_0(1,1)W^2$$

If we rewrite Eq. (8.29) in matrix notation, we have

$$\begin{bmatrix} x_1(0,0) \\ x_1(0,1) \\ x_1(1,0) \\ x_1(1,1) \end{bmatrix} = \begin{bmatrix} 1 & 0 & W^0 & 0 \\ 0 & 1 & 0 & W^0 \\ 1 & 0 & W^2 & 0 \\ 0 & 1 & 0 & W^2 \end{bmatrix} \begin{bmatrix} x_0(0,0) \\ x_0(0,1) \\ x_0(1,0) \\ x_0(1,1) \end{bmatrix} \tag{8.30}$$

Note that Eq. (8.30) is exactly the factored matrix equation of Eq. (8.11), developed in Sec. 8.2, with the index k written in binary notation. Thus, the inner summation of Eq. (8.27) specifies the first of the factored matrices for the example developed in Sec. 8.2 or, equivalently, the array $l = 1$ of the signal flow graph illustrated in Fig. 8.2.

Similarly, if we write the outer summation of Eq. (8.27) as

$$x_2(n_0,n_1) = \sum_{k_0=0}^{1} x_1(n_0,k_0)W^{(2n_1+n_0)k_0} \tag{8.31}$$

and enumerate the results in matrix form, we obtain

$$\begin{bmatrix} x_2(0,0) \\ x_2(0,1) \\ x_2(1,0) \\ x_2(1,1) \end{bmatrix} = \begin{bmatrix} 1 & W^0 & 0 & 0 \\ 1 & W^2 & 0 & 0 \\ 0 & 0 & 1 & W^1 \\ 0 & 0 & 1 & W^3 \end{bmatrix} \begin{bmatrix} x_1(0,0) \\ x_1(0,1) \\ x_1(1,0) \\ x_1(1,1) \end{bmatrix} \tag{8.32}$$

which is Eq. (8.14). Thus, the outer summation of Eq. (8.27) determines the second of the factored matrices of the example in Sec. 8.2.

From Eqs. (8.27) and (8.31) we have

$$X(n_1,n_0) = x_2(n_0,n_1) \tag{8.33}$$

That is, the final results $x_2(n_0,n_1)$ as obtained from the outer sum are in bit-reversed order with respect to the desired values $X(n_1,n_0)$. This is simply the scrambling that results from the FFT algorithm.

If we combine Eqs. (8.28), (8.31), and (8.33),

$$x_1(n_0,k_0) = \sum_{k_1=0}^{1} x_0(k_1,k_0)W^{2n_0k_1}$$

$$x_2(n_0,n_1) = \sum_{k_0=0}^{1} x_1(n_0,k_0)W^{(2n_1+n_0)k_0} \tag{8.34}$$

$$X(n_1,n_0) = x_2(n_0,n_1)$$

then the set of Eq. (8.34) represents the original Cooley-Tukey [3] formulation of the FFT algorithm for $N = 4$. We term these equations *recursive* in that the second is computed in terms of the first.

Example 8.1   Cooley-Tukey Algorithm: $N = 8$

To illustrate further the notation associated with the Cooley-Tukey formulation of the FFT, consider Eq. (8.22) for the case $N = 2^3 = 8$. For this case,

$$n = 4n_2 + 2n_1 + n_0 \qquad n_i = 0 \text{ or } 1$$
$$k = 4k_2 + 2k_1 + k_0 \qquad k_i = 0 \text{ or } 1 \tag{8.35}$$

and Eq. (8.22) becomes

$$X(n_2,n_1,n_0) = \sum_{k_0=0}^{1} \sum_{k_1=0}^{1} \sum_{k_2=0}^{1} x_0(k_2,k_1,k_0)W^{(4n_2+2n_1+n_0)(4k_2+2k_1+k_0)} \tag{8.36}$$

Rewriting W^p, we obtain

$$W^{(4n_2+2n_1+n_0)(4k_2+2k_1+k_0)} = W^{(4n_2+2n_1+n_0)(4k_2)}W^{(4n_2+2n_1+n_0)(2k_1)}$$
$$\times W^{(4n_2+2n_1+n_0)(k_0)} \tag{8.37}$$

We note that because $W^8 = [e^{j2\pi/8}]^8 = 1$, then

$$W^{(4n_2+2n_1+n_0)(4k_2)} = [W^{8(2n_2k_2)}][W^{8(n_1k_2)}]W^{4n_0k_2} = W^{4n_0k_2}$$
$$W^{(4n_2+2n_1+n_0)(2k_1)} = [W^{8(n_2k_1)}]W^{(2n_1+n_0)(2k_1)} = W^{(2n_1+n_0)(2k_1)} \tag{8.38}$$

Hence, Eq. (8.36) can be rewritten as

$$X(n_2,n_1,n_0) = \sum_{k_0=0}^{1} \sum_{k_1=0}^{1} \sum_{k_2=0}^{1} x_0(k_2,k_1,k_0)W^{4n_0k_2}$$
$$\times W^{(2n_1+n_0)(2k_1)}W^{(4n_2+2n_1+n_0)(k_0)} \tag{8.39}$$

If we let

$$x_1(n_0,k_1,k_0) = \sum_{k_2=0}^{1} x_0(k_2,k_1,k_0)W^{4n_0k_2} \tag{8.40}$$

$$x_2(n_0,n_1,k_0) = \sum_{k_1=0}^{1} x_1(n_0,k_1,k_0)W^{(2n_1+n_0)(2k_1)} \tag{8.41}$$

$$x_3(n_0,n_1,n_2) = \sum_{k_0=0}^{1} x_2(n_0,n_1,k_0)W^{(4n_2+2n_1+n_0)(k_0)} \tag{8.42}$$

$$X(n_2,n_1,n_0) = x_3(n_0,n_1,n_2) \tag{8.43}$$

then we have determined the required matrix factorization or, equivalently, the signal flow graph for $N = 8$. The signal flow graph is shown in Fig. 8.9.

Derivation of the Cooley-Tukey Algorithm for $N = 2^\gamma$

When $N = 2^\gamma$, n and k can be represented in binary form as

$$\begin{aligned} n &= 2^{\gamma-1}n_{\gamma-1} + 2^{\gamma-2}n_{\gamma-2} + \cdots + n_0 \\ k &= 2^{\gamma-1}k_{\gamma-1} + 2^{\gamma-2}k_{\gamma-2} + \cdots + k_0 \end{aligned} \tag{8.44}$$

Using this representation, we can rewrite Eq. (8.22) as

$$X(n_{\gamma-1},n_{\gamma-2},\ldots,n_0) = \sum_{k_0=0}^{1} \sum_{k_1=0}^{1} \cdots \sum_{k_{\gamma-1}=0}^{1} x(k_{\gamma-1},k_{\gamma-2},\ldots,k_0)W^P \tag{8.45}$$

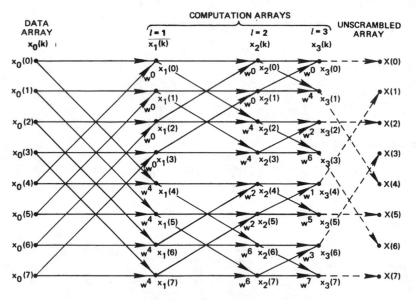

Figure 8.9 FFT signal flow graph for $N = 8$.

where

$$p = (2^{\gamma-1}n_{\gamma-1} + 2^{\gamma-2}n_{\gamma-2} + \cdots + n_0)$$
$$\times\ (2^{\gamma-1}k_{\gamma-1} + 2^{\gamma-2}k_{\gamma-2} + \cdots + k_0) \qquad (8.46)$$

Because $W^{a+b} = W^a W^b$, we rewrite W^p as

$$W^p = W^{(2^{\gamma-1}n_{\gamma-1}+2^{\gamma-2}n_{\gamma-2}+\cdots+n_0)(2^{\gamma-1}k_{\gamma-1})}$$
$$\times\ W^{(2^{\gamma-1}n_{\gamma-1}+2^{\gamma-2}n_{\gamma-2}+\cdots+n_0)(2^{\gamma-2}k_{\gamma-2})} \qquad (8.47)$$
$$\times\ \cdots\ W^{(2^{\gamma-1}n_{\gamma-1}+2^{\gamma-2}n_{\gamma-2}+\cdots+n_0)k_0}$$

Now consider the first term of Eq. (8.47):

$$W^{(2^{\gamma-1}n_{\gamma-1}+2^{\gamma-2}n_{\gamma-2}+\cdots+n_0)(2^{\gamma-1}k_{\gamma-1})} = [W^{2^\gamma(2^{\gamma-2}n_{\gamma-1}k_{\gamma-1})}]$$
$$\times\ [W^{2^\gamma(2^{\gamma-3}n_{\gamma-2}k_{\gamma-1})}]$$
$$\times\ \cdots\ [W^{2^\gamma(n_1 k_{\gamma-1})}] \qquad (8.48)$$
$$\times\ W^{2^{\gamma-1}(n_0 \quad k_{\gamma-1})}$$
$$= W^{2^{\gamma-1}(n_0 \quad k_{\gamma-1})}$$

because

$$W^{2^\gamma} = W^N = [e^{-j2\pi/N}]^N = 1 \qquad (8.49)$$

Similarly, the second term of Eq. (8.47) yields

$$W^{(2^{\gamma-1}n_{\gamma-1}+2^{\gamma-2}n_{\gamma-2}+\cdots+n_0)(2^{\gamma-2}k_{\gamma-2})} = [W^{2^\gamma(2^{\gamma-3}n_{\gamma-1}k_{\gamma-2})}]$$
$$\times\ [W^{2^\gamma(2^{\gamma-4}n_{\gamma-2}k_{\gamma-2})}]$$
$$\times\ \cdots\ W^{2^{\gamma-1}(n_1 \quad k_{\gamma-2})} \qquad (8.50)$$
$$\times\ W^{2^{\gamma-2}(n_0 \quad k_{\gamma-2})}$$
$$= W^{(2n_1+n_0)2^{\gamma-2}k_{\gamma-2}}$$

Note that as we progress through the terms of Eq. (8.47), we add another factor that does not cancel by the condition $W^{2^\gamma} = 1$. This process continues until we reach the last term in which there is no cancellation.

Using these relationships, Eq. (8.45) can be rewritten as

$$X(n_{\gamma-1},n_{\gamma-2},\ldots,n_0) = \sum_{k_0=0}^{1} \sum_{k_1=0}^{1} \cdots \sum_{k_{\gamma-1}=0}^{1} x_0(k_{\gamma-1},k_{\gamma-2},\ldots,k_0)$$
$$\times\ W^{2^{\gamma-1}(n_0 \quad k_{\gamma-1})} W^{(2n_1+n_0)2^{\gamma-2}k_{\gamma-2}} \cdots \qquad (8.51)$$
$$\times\ W^{(2^{\gamma-1}n_{\gamma-1}+2^{\gamma-2}n_{\gamma-2}+\cdots+n_0)k_0}$$

Performing each of the summations separately and labeling the intermediate results, we obtain

$$x_1(n_0,k_{\gamma-2},...,k_0) = \sum_{k_{\gamma-1}=0}^{1} x_0(k_{\gamma-1},k_{\gamma-2},...,k_0)W^{(2^{\gamma-1})(n_0 \quad k_{\gamma-1})}$$

$$x_2(n_0,n_1,k_{\gamma-3},...,k_0) = \sum_{k_{\gamma-2}=0}^{1} x_1(n_0,k_{\gamma-2},...,k_0)W^{(2n_1+n_0)(2^{\gamma-2}k_{\gamma-2})}$$

$$\vdots$$

(8.52)

$$x_\gamma(n_0,n_1,...,n_{\gamma-1}) = \sum_{k_0=0}^{1} x_{\gamma-1}(n_0,n_1,...,k_0)$$

$$\times W^{(2^{\gamma-1}n_{\gamma-1}+2^{\gamma-2}n_{\gamma-2}+\cdots+n_0)k_0}$$

$$X(n_{\gamma-1},n_{\gamma-2},...,n_0) = x_\gamma(n_0,n_1,...,n_{\gamma-1})$$

This set of recursive equations represents the original Cooley-Tukey formulation of the FFT, $N = 2^\gamma$. Recall that the direct evaluation of an N-point transform requires approximately N^2 complex multiplications. Now consider the number of multiplications required to compute the relationships of Eq. (8.52). There are γ summation equations that each represent N equations. Each of the latter equations contains two *complex* multiplications; however, the first multiplication of each equation is actually a multiplication by unity. This follows because the first multiplication is always of the form $W^{ak_{\gamma-i}}$, where $k_{\gamma-i} = 0$. Thus, only $N\gamma$ complex multiplications are required. It can be shown that in the computation of an array, there occurs the relationship $W^p = -W^{p+N/2}$; the number of multiplications can be reduced by another factor of 2. The number of complex multiplications for $N = 2^\gamma$ is then $N\gamma/2$. Similarly, one can reason that there are $N\gamma$ complex additions.

Canonic Forms of the FFT

There exist many variations of the FFT algorithm that are *canonic*. Each particular algorithm variation is formulated to exploit either a particular property of the data being transformed or the computer architecture. The Cooley-Tukey algorithm is illustrated by the signal flow graph of Fig. 8.9. We observe from the signal flow graph that this form of the algorithm can be computed in place, that is, a dual-node pair can be computed and the results stored in the original data storage locations. Further, we observe that with this form of the algorithm, the input data is in natural order and the output data is in scrambled order. In addition, the powers of W are in bit-reversed order.

If one desires, it is possible to rearrange the signal flow graph shown in Fig. (8.9) in order that the input data is in *scrambled* order and the output data is in natural order. The resulting signal flow can be computed in place and the powers of W necessary to perform the computation occur in natural order.

These two algorithms are often referred to in the literature by the term *decimation in time*. This terminology arises because alternate derivations of the algorithm [8] are structured to appeal to the concept of sample-rate reduction or throwing away samples.

Another distinct form of the FFT is due to Sande [9]. To develop this form, let $N = 4$ and write

$$X(n_1, n_0) = \sum_{k_0=0}^{1} \sum_{k_1=0}^{1} x_0(k_1, k_0) W^{(2n_1 + n_0)(2k_1 + k_0)} \tag{8.53}$$

In contrast to the Cooley-Tukey approach, we separate the components of n instead of the components of k.

$$
\begin{aligned}
W^{(2n_1 + n_0)} W^{(2k_1 + k_0)} &= W^{(2n_1)(2k_1 + k_0)} W^{n_0(2k_1 + k_0)} \\
&= [W^{4n_1 k_1}] W^{2n_1 k_0} W^{n_0(2k_1 + k_0)} \\
&= W^{2n_1 k_0} W^{n_0(2k_1 + k_0)}
\end{aligned}
\tag{8.54}
$$

where $W^4 = 1$.

Thus, Eq. (8.53) can be written as

$$X(n_1, n_0) = \sum_{k_0=0}^{1} \left[\sum_{k_1=0}^{1} x_0(k_1, k_0) W^{2n_0 k_1} W^{n_0 k_0} \right] W^{2n_1 k_0} \tag{8.55}$$

If we define the intermediate computational steps, then

$$x_1(n_0, k_0) = \sum_{k_1=0}^{1} x_0(k_1, k_0) W^{2n_0 k_1} W^{n_0 k_0}$$

$$x_2(n_0, n_1) = \sum_{k_0=0}^{1} x_1(n_0, k_0) W^{2n_1 k_0} \tag{8.56}$$

$$X(n_1, n_0) = x_2(n_0, n_1)$$

The signal flow graph describing the Sande-Tukey algorithm is shown in Fig. 8.10 for the case $N = 8$. We note that the input data is in natural order, the output data is in scrambled order, and the powers of W occur in natural order. A signal flow graph that yields results in natural order can be developed by proceeding as in the Cooley-Tukey case and interchanging nodes. The input data is now in bit-reversed order and the powers of W occur in bit-reversed order.

These two forms of the FFT algorithm are known by the term *decimation in frequency*, where the reasoning for the terminology is analogous to that for *decimation in time*.

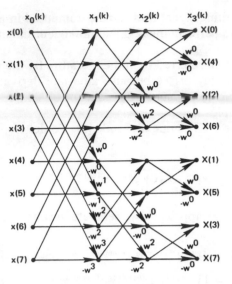

Figure 8.10 Sande-Tukey FFT algorithm signal flow graph.

8.10 FFT ALGORITHMS FOR ARBITRARY FACTORS

In the discussions to this point, we have assumed that the number of points N to be Fourier transformed satisfies the relationship $N = 2^\gamma$, where γ is integer valued. As we saw, this base-2 algorithm resulted in a tremendous savings in computation time; however, the constraint $N = 2^\gamma$ can be rather restrictive. In this section, we develop FFT algorithms that remove this assumption. We will show that significant time savings can be obtained as long as N is highly composite, that is, $N = r_1 r_2 \ldots r_m$, where r_i is an integer.

To develop the FFT algorithm for arbitrary factors, we first consider the case of $N = r_1 r_2$. This approach allows us to develop the notation required in the proof for the general case. Examples for the base-4 and base-"4 + 2" algorithms are used to further develop the case $N = r_1 r_2$. The Cooley-Tukey algorithm for the case $N = r_1 r_2 \ldots r_m$ is then developed as well as *twiddle factor* algorithms.

FFT Algorithm for $N = r_1 r_2$

Assume that the number of points N satisfies the relationship $N = r_1 r_2$, where r_1 and r_2 are integer valued. To derive the FFT algorithm for this case, we first express the n and k indices in Eq. (8.22) as

$$n = n_1 r_1 + n_0 \qquad n_0 = 0, 1, \ldots, r_1 - 1 \qquad n_1 = 0, 1, \ldots, r_2 - 1$$

$$k = k_1 k_2 + k_0 \qquad k_0 = 0, 1, \ldots, r_2 - 1 \qquad k_1 = 0, 1, \ldots, r_1 - 1$$

$$(8.57)$$

We observe that this method of writing the indices allows us to give a unique representation of each decimal integer. Using Eq. (8.57), we can rewrite Eq. (8.22) as

$$X(n_1,n_0) = \sum_{k_0=0}^{r_2-1} \left[\sum_{k_1=0}^{r_1-1} x_0(k_1,k_0)W^{nk_1r_2} \right] W^{nk_0} \qquad (8.58)$$

Rewriting $W^{nk_1r_2}$, we obtain

$$\begin{aligned} W^{nk_1r_2} &= W^{(n_1r_1+n_0)k_1r_2} \\ &= W^{r_1r_2n_1k_1}W^{n_0k_1r_2} \\ &= [W^{r_1r_2}]^{n_1k_1}W^{n_0k_1r_2} \\ &= W^{n_0k_1r_2} \end{aligned} \qquad (8.59)$$

where we have used the fact that $W^{r_1r_2} = W^N = 1$.

From Eq. (8.59), we rewrite the inner sum of Eq. (8.58) as a new array:

$$x_1(n_0,k_0) = \sum_{k_1=0}^{r_1-1} x_0(k_1,k_0)W^{n_0k_1r_2} \qquad (8.60)$$

If we expand the terms W^{nk_0}, the outer loop can be written as

$$x_2(n_0,n_1) = \sum_{k_0=0}^{r_2-1} x_1(n_0,k_0)W^{(n_1r_1+n_0)k_0} \qquad (8.61)$$

The final result can be written as

$$X(n_1,n_0) = x_2(n_0,n_1) \qquad (8.62)$$

Thus, as in the base-2 algorithm, the results are in reverse order.

Equations (8.60), (8.61), and (8.62) are the defining FFT algorithm relationships for the case $N = r_1r_2$. To further illustrate this particular algorithm, consider the following examples.

Example 8.1   Base-4 Algorithm for $N = 16$

Let us consider the case $N = r_1r_2 = 4 \times 4 = 16$, that is, we will develop the base-4 algorithm for the case $N = 16$. Using Eq. (8.57), we represent the variables n and k in Eq. (8.22) in a base-4 or quaternary number system:

$$\begin{aligned} n &= 4n_1 + n_0 & n_1,n_0 &= 0, 1, 2, 3 \\ k &= 4k_1 + k_0 & k_1,k_0 &= 0, 1, 2, 3 \end{aligned} \qquad (8.63)$$

Equation (8.58) then becomes

$$X(n_1,n_0) = \sum_{k_0=0}^{3} \left[\sum_{k_1=0}^{3} x_0(k_1,k_0)W^{4nk_1} \right] W^{nk_0} \qquad (8.64)$$

Rewriting W^{4nk_1}, we obtain

$$
\begin{aligned}
W^{4nk_1} &= W^{4(4n_1 + n_0)k_1} \\
&= W^{16n_1k_1} W^{4n_0k_1} \\
&= [W^{16}]^{n_1k_1} W^{4n_0k_1} \\
&= W^{4n_0k_1}
\end{aligned}
\tag{8.65}
$$

The term in brackets is equal to unity because $W^{16} = 1$.

Substitution of Eq. (8.65) into Eq. (8.60) yields the inner sum of the base-4 algorithm:

$$
x_1(n_0,k_0) = \sum_{k_1=0}^{3} x_0(k_1,k_0)W^{4n_0k_1}
\tag{8.66}
$$

From Eq. (8.61) the outer sum is

$$
x_2(n_0,n_1) = \sum_{k_0=0}^{3} x_1(n_0,k_0)W^{(4n_1 + n_0)k_0}
\tag{8.67}
$$

and from Eq. (8.62), the base-4 algorithm results are given by

$$
X(n_1,n_0) = x_2(n_0,n_1)
\tag{8.68}
$$

Equations (8.66), (8.67), and (8.68) define the base-4 algorithm for the case $N = 16$. Based on these equations, we can develop a base-4 signal flow graph.

Example 8.2   Base-4 Signal Flow Graph for $N = 16$

From the defining relationships of Eqs. (8.66) and (8.67), we observe that there are $\gamma = 2$ computational arrays and there are four inputs to each node. The inputs to node $x_1(n_0,k_0)$ are $x_0(0,k_0)$, $x_0(1,k_0)$, $x_0(2,k_0)$, and $x_0(3,k_0)$. That is, the four inputs to a node i in array l are those nodes in array $l - 1$ whose subscripts differ from i only in the $(\gamma - l)$th quaternary digit.

We show in Fig. 8.11 an abbreviated signal flow graph for the base-4 $N = 16$ algorithm. To alleviate confusion, only representative transmission paths are shown and all W^p factors have been omitted. W^p factors can be determined from Eqs. (8.66) and (8.67). Each pattern of transmission paths shown is applied sequentially to successive nodes until all nodes have been considered. Figure 8.11 also illustrates the unscrambling procedure for the base-4 algorithm. Enumeration of Eqs. (8.66) and (8.67) reveals that the base-4 algorithm requires approximately 30 percent fewer multiplications than the base-2 algorithm.

Example 8.3   Base-"4 + 2" Algorithm for $N = 8$

Let us now consider the case $N = r_1r_2 = 4 \times 2 = 8$. This case represents the simplest form of the base-"4 + 2" algorithm. Base "4 + 2" implies that we compute as many arrays as possible with a base-4 algorithm and then compute a base-2 array.

To develop the "4 + 2" algorithm, we first substitute $r_1 = 4$ and $r_2 = 2$ into Eq. (8.57):

$$
\begin{aligned}
n &= 4n_1 + n_0 \qquad n_0 = 0, 1, 2, 3 \qquad n_1 = 0, 1 \\
k &= 2k_1 + k_0 \qquad k_0 = 0, 1 \qquad k_1 = 0, 1, 2, 3
\end{aligned}
\tag{8.69}
$$

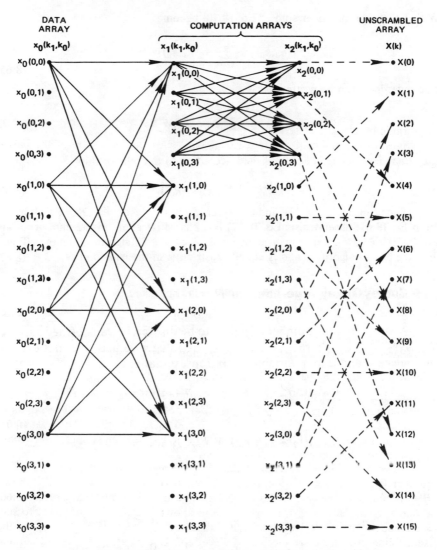

Figure 8.11   Abbreviated signal flow graph: base 4, $N = 16$.

Equation (8.58) then becomes

$$X(n_1, n_0) = \sum_{k_0 = 0}^{1} \left[\sum_{k_1 = 0}^{3} x_0(k_1, k_0) W^{2nk_1} \right] W^{nk_0} \tag{8.70}$$

Expanding W^{2nk_1}, we obtain

$$W^{2nk_1} = W^{2(4n_1 + n_0)k_1}$$
$$= [W^8]^{n_1 k_1} W^{2n_0 k_1} \tag{8.71}$$
$$= W^{2n_0 k_1}$$

With Eq. (8.71), the inner sum of Eq. (8.70) becomes

$$x_1(n_0,k_0) = \sum_{k_1=0}^{3} x_0(k_1,k_0)W^{2n_0k_1} \tag{8.72}$$

The outer loop can be written as

$$x_2(n_0,n_1) = \sum_{k_0=0}^{1} x_1(n_0,k_0)W^{(4n_1+n_0)k_0} \tag{8.73}$$

and the unscrambling is accomplished according to the relationship

$$X(n_1,n_0) = x_2(n_0,n_1) \tag{8.74}$$

Equations (8.72), (8.73), and (8.74) represent the base-"4 + 2" FFT algorithm for $N = 8$. We observe that Eq. (8.72) is a base-4 iteration on the data array and Eq. (8.73) is a base-2 iteration on array $l = 1$. The "4 + 2" algorithm is more efficient than the base-2 algorithm and is equally restrictive in the choice of N.

Cooley-Tukey Algorithm for $N = r_1 r_2 \ldots r_m$

Assume that the number of points to be discretely transformed satisfies $N = r_1 r_2 \ldots r_m$, where $r_1, r_2, \ldots, r_m$ are integer valued. We first express the indices n and k in a variable radix representation:

$$n = n_{m-1}(r_1 r_2 \ldots r_{m-1}) + n_{m-2}(r_1 r_2 \ldots r_{m-2})$$
$$+ \cdots + n_1 r_1 + n_0$$
$$k = k_{m-1}(r_2 r_3 \ldots r_m) + k_{m-2}(r_3 r_4 \ldots r_m) \tag{8.75}$$
$$+ \cdots + k_1 r_m + k_0$$

where

$$n_{i-1} = 0, 1, 2, \ldots, r_i - 1 \qquad 1 \le i \le m$$
$$k_i = 0, 1, 2, \ldots, r_{m-i} - 1 \qquad 0 \le i \le m - 1$$

We can now rewrite Eq. (8.22) as

$$X(n_{m-1}, n_{m-2}, \ldots, n_1, n_0)$$
$$= \sum_{k_0} \sum_{k_1} \cdots \sum_{k_{m-1}} x_0(k_{m-1}, k_{m-2}, \ldots, k_0)W^{nk} \tag{8.76}$$

where $\sum_{k_i}$ represents a summation over all $k_i = 0, 1, 2, \ldots, r_{m-i} - 1; 0 \le i \le m - 1$. Note that

$$W^{nk} = W^{n[k_{m-1}(r_2 r_3 \ldots r_m) + \cdots + k_0]} \tag{8.77}$$

and the first term of the summation expands to

$$W^{nk_{m-1}(r_2 r_3 \ldots r_m)} = W^{[n_{m-1}(r_1 r_2 \ldots r_{m-1}) + \cdots + n_0][k_{m-1}(r_2 r_3 \ldots r_m)]}$$

$$= [W^{r_1 r_2 \ldots r_m}]^{[n_{m-1}(r_2 r_3 \ldots r_{m-1}) + \cdots + n_1]k_{m-1}} W^{n_0 k_{m-1}(r_2 \ldots r_m)} \tag{8.78}$$

Because $W^{r_1 r_2 \ldots r_m} = W^N = 1$, then Eq. (8.78) can be written as

$$W^{n k_{m-1}(r_2 r_3 \ldots r_m)} = W^{n_0 k_{m-1}(r_2 \ldots r_m)} \tag{8.79}$$

and hence Eq. (8.77) becomes

$$W^{nk} = W^{n_0 k_{m-1}(r_2 \ldots r_m)} W^{n[k_{m-2}(r_3 \ldots r_m) + \cdots + k_0]} \tag{8.80}$$

Equation (8.76) can now be rewritten as

$$X(n_{m-1}, n_{m-2}, \ldots, n_1, n_0) = \sum_{k_0} \sum_{k_1} \cdots [\sum_{k_{m-1}} x_0(k_{m-1}, k_{m-2}, \ldots, k_0)$$

$$\times W^{n_0 k_{m-1}(r_2 \ldots r_m)}] W^{n[k_{m-2}(r_3 \ldots r_m) + \cdots + k_0]} \tag{8.81}$$

Note that the inner sum is over k_{m-1} and is only a function of the variables n_0 and $k_{m-2}, \ldots, k_0$. Thus, we define a new array as

$$x_1(n_0, k_{m-2}, \ldots, k_0) = \sum_{k_{m-1}} x_0(k_{m-1}, \ldots, k_0) W^{n_0 k_{m-1}(r_2 \ldots r_m)} \tag{8.82}$$

Equation (8.81) can now be rewritten as

$$X(n_{m-1}, n_{m-2}, \ldots, n_1, n_0) = \sum_{k_0} \sum_{k_1} \cdots \sum_{k_{m-2}} x_1(n_0, k_{m-2}, \ldots, k_0)$$

$$\times W^{n[k_{m-2}(r_3 \ldots r_m) + \cdots + k_0]} \tag{8.83}$$

By arguments analogous to those leading to Eq. (8.79), we obtain

$$W^{n k_{m-2}(r_3 r_4 \ldots r_m)} = W^{(n_1 r_1 + n_0) k_{m-2}(r_3 r_4 \ldots r_m)} \tag{8.84}$$

The identity of Eq. (8.84) allows the inner sum of Eq. (8.83) to be written as

$$x_2(n_0, n_1, k_{m-3}, \ldots, k_0)$$

$$= \sum_{k_{m-2}} x_1(n_0, k_{m-2}, \ldots, k_0) W^{(n_1 r_1 + n_0) k_{m-2}(r_3 r_4 \ldots r_m)} \tag{8.85}$$

We can now rewrite Eq. (8.83) in the form

$$X(n_{m-1}, n_{m-2}, \ldots, n_1, n_0) = \sum_{k_0} \sum_{k_1} \cdots \sum_{k_{m-3}} x_2(n_0, n_1, k_{m-3}, \ldots, k_0)$$

$$\times W^{n[k_{m-3}(r_4 r_5 \ldots r_m) + \cdots + k_0]} \tag{8.86}$$

If we continue reducing Eq. (8.86) in this manner, we obtain a set of recursive equations of the form

$$x_i(n_0, n_1, \ldots, n_{i-1}, k_{m-i-1}, \ldots, k_0)$$

$$= \sum_{k_{m-i}} x_{i-1}(n_0, n_1, \ldots, n_{i-2}, k_{m-i}, \ldots, k_0)$$

$$\times W^{[n_{i-1}(r_1 r_2 \ldots r_{i-1}) + \cdots + n_0]k_{m-i}(r_{i+1} \ldots r_m)} \qquad i = 1, 2, \ldots, m \tag{8.87}$$

The expression of Eq. (8.87) is valid provided we define $(r_{i+1} \ldots r_m) = 1$ for $i > m - 1$ and $k_{-1} = 0$.

The final results are given by

$$X(n_{m-1}, \ldots, n_0) = x_m(n_0, \ldots, n_{m-1}) \tag{8.88}$$

The expression of Eq. (8.87) is an extension due to Bergland [10] of the original Cooley-Tukey algorithm. We note that there are N elements in array x_1, each requiring r_1 operations (one complex multiplication and one complex addition), giving a total of Nr_1 operations to obtain x_1. Similarly, it takes Nr_2 operations to calculate x_2 from x_1. Thus, the computation of x_m requires $N(r_1 + r_2 + \cdots + r_m)$ operations. This bound does not take into account the symmetries of the complex exponential that can be exploited as developed in the following discussions.

Example 8.4   Base-4, $N = 16$, Twiddle Factor Algorithm

Recall from Eqs. (8.66) and (8.67) that the recursive equations for the base-4 FFT algorithm for $N = 16$ are given by

$$x_1(n_0, k_0) = \sum_{k_1=0}^{3} x_0(k_1, k_0) W^{4n_0 k_1}$$

$$x_2(n_0, n_1) = \sum_{k_0=0}^{3} x_1(n_0, k_0) W^{(4n_1 + n_0)k_0} \tag{8.89}$$

$$X(n_1, n_0) = x_2(n_0, n_1)$$

To illustrate the *twiddle factor* concept, let us rewrite Eq. (8.89) as

$$X(n_1, n_0) = \sum_{k_0=0}^{3} \left[\sum_{k_1=0}^{3} x_0(k_1, k_0) W^{4n_0 k_1} \right] W^{4n_1 k_0} W^{n_0 k_0} \tag{8.90}$$

Note that the term $W^{n_0 k_0}$ has been arbitrarily grouped with the outer sum and could have just as easily been grouped with the inner sum. By regrouping, Eq. (8.90) becomes

$$X(n_1, n_0) = \sum_{k_0=0}^{3} \left[\left\{ \sum_{k_1=0}^{3} x_0(k_1, k_0) W^{4n_0 k_1} \right\} W^{n_0 k_0} \right] W^{4n_1 k_0} \tag{8.91}$$

or in recursive form

$$x_1(n_0,k_0) = \left[\sum_{k_1=0}^{3} x_0(k_1,k_0)W^{4n_0k_1} \right] W^{n_0k_0} \tag{8.92}$$

$$x_2(n_0,n_1) = \left[\sum_{k_0=0}^{3} x_1(n_0,k_0)W^{4n_1k_0} \right] \tag{8.93}$$

$$X(n_1,n_0) = x_2(n_0,n_1) \tag{8.94}$$

The form of the algorithm given by Eq. (8.92) exploits the symmetries of the sine and cosine functions. To illustrate this point, consider the term $W^{4n_0k_1}$ in brackets in Eq. (8.92). Because $N = 16$, then

$$W^{4n_0k_1} = (W^4)^{n_0k_1} = (e^{-j2\pi(4)/16})^{n_0k_1} = (e^{-j\pi/2})^{n_0k_1} \tag{8.95}$$

Thus, $W^{4n_0k_1}$ only takes on the values $\pm i$ and ± 1, depending on the integer n_0k_1. As a result, the four-point transform in brackets in Eq. (8.92) can be evaluated without multiplications. These results are then *referenced* or *twiddled* [9] by the factor $W^{n_0k_0}$, which is outside the brackets in Eq. (8.92). Note that by similar arguments, Eq. (8.93) can be evaluated without multiplications. The total computations required to evaluate the base-4 algorithm have been reduced by this regrouping.

Cooley-Tukey Twiddle Factor Algorithm

We now develop a general formulation of the *twiddle factor* concept. The original Cooley-Tukey formulation is given by the set of recursive Eqs. (8.87). If we regroup these equations, the first array takes the form

$$\bar{x}_1(n_0,k_{m-2}, \ldots ,k_0)$$

$$= [\sum_{k_{m-1}} x_0(k_{m-1}, \ldots ,k_0)W^{n_0k_{m-1}(N/r_1)}]W^{(n_0k_{m-2})(r_3 \cdots r_m)} \tag{8.96}$$

and the succeeding equations are given by

$$\bar{x}_i(n_0,n_1, \ldots ,n_{i-1},k_{m-i-1}, \ldots .k_0)$$

$$= [\sum_{k_{m-1}} \bar{x}_{i-1}(n_0, \ldots ,n_{i-2},k_{m-i}, \ldots ,k_0)W^{n_i - 1k_{m-i}(N/r_i)}] \tag{8.97}$$

$$\times W^{[n_{i-1}(r_1r_2\cdots r_{i-1}) + \cdots + n_1r_1 + n_0]k_{m-i-1}(r_{i+2}\cdots r_m)}$$

We have used the notation $\bar{x}$ to indicate that these results have been obtained by *twiddling*. Equation (8.92) is valid for $i = 1, 2, \ldots , m$ if we interpret the case $i = 1$ in the sense of Eq. (8.96) and if we define $(r_{i+2} \ldots r_m) = 1$ for $i > m - 2$ and $k_{-1} = 0$.

Each iteration of Eq. (8.97) requires the evaluation of an r_i-point Fourier transform followed by a referencing or twiddling operation. The importance of this formation is that the bracketed r_i-point Fourier transforms can be computed with a minimum number of multiplications. For example, if r_i = 8 (that is, a base-8 transform), then the W^p factor in brackets only takes on the values ± 1, $\pm j$, $\pm e^{j\pi/4}$, and $\pm e^{-j\pi/4}$. Because the first two factors require no multiplications and the product of a complex number and either of the last two factors requires only two real multiplications, a total of only four real multiplications are required in evaluating each eight-point transform. As we see, the twiddle factor algorithms allow us to take advantage of the properties of the sine and cosine functions.

Computations Required by Base-2, Base-4, Base-8, and Base-16 Algorithms

Let us consider the case $N = 2^{12} = 4096$. The real number of multiplications and additions required to evaluate the recursive Eq. (8.97) is given in Table 8.1. This summary of operations was first reported by Bergland [10]. In counting the number of multiplications and additions, it is assumed that each of the twiddling operations requires one complex multiplication except when the multiplier is W^0.

TABLE 8.1　Operations Required in Computing
Base-2, Base-4, Base-8, and Base-16 FFT
Algorithms for $N = 4096$

Algorithm	Number of real multiplications	Number of real additions
Base 2	81,924	139,266
Base 4	57,348	126,978
Base 8	49,156	126,978
Base 16	48,132	125,442

PROBLEMS

8.1. Let $x_0(k) = k$, where $k = 0, 1, 2$, and 3. Compute Eq. (8.1) and note the total number of multiplications and additions. Repeat the calculation following the procedure outlined by Eqs. (8.6) through (8.14) and again note the total number of multiplications and additions. Compare your results.

8.2. It has been shown that the matrix-factoring procedure introduces scrambled results. For $N = 8, 16$, and 32, show the order of $X(n)$ that results from the scrambling.

8.3. It is desired to convert Eq. (8.9) into a signal flow graph for the case $N = 8$.
  (a) How many computation arrays are there?
  (b) Define dual nodes for this case. What is the dual-node spacing for each array? Give a general expression and then identify each node with its duals for each array.

   (c) Write the equation pair (8-21) for each node for array 1. Repeat for the other arrays.

   (d) Determine W^p for each node and substitute these values into the equation determined in part c.

   (e) Draw the signal flow graph for this case.

   (f) Show how to unscramble the results of the last computational array.

   (g) Illustrate the concept of node skipping on the signal flow graph.

8.4. Verify the computer program flowchart illustrated in Fig. 8.6 by mentally observing that each of the arrays determined in Problem 8.3 is correctly computed.

8.5. Relate each statement of the BASIC program illustrated in Fig. 8.7 with the computer program flowchart shown in Fig. 8.6.

8.6. Write an FFT computer program based on the flowchart illustrated in Fig. 8.6. The program should be capable of accepting complex time functions and performing the inverse transform using the alternate inversion formula. Call this program FFT.

8.7. Let $h(t) = e^{-t}$, where $t > 0$. Sample $h(t)$ with $T = 0.01$ and $N = 1024$. Compute the discrete Fourier transform of $h(k)$ with both FFT and DFT. Compare computing times.

8.8. Derive the FFT algorithm for $N = r_1 r_2$ for the case where the components of n are separated, i.e., the Sande-Tukey algorithm.

8.9. Develop the signal flow graph for the base-4 Sande-Tukey algorithm for $N = 16$.

8.10. Develop the Sande-Tukey base-"4 + 2" algorithm for the case $N = 8$.

8.11. Develop fully the Sande-Tukey algorithm for the case $N = r_1, r_2, \ldots, r_m$.

8.12. Develop the Cooley-Tukey base-8 algorithm for the case $N = 64$.

8.13. Let $N = 16$. Develop the Sande-Tukey twiddle factor algorithm.

8.14. Let $N = 8$. Is there an advantage in using twiddle factors in computing the FFT by the Cooley-Tukey base-2 algorithm? Verify your conclusions by demonstrating the required number of multiplications in each case.

8.15. Repeat Problem 8.14 for the base-"4 + 2" algorithm.

8.16. Develop a FFT computer program for a base-"4 + 2" Cooley-Tukey algorithm and bit-reversed data.

8.17. Develop a FFT computer program for a base-"4 + 2" Sande-Tukey algorithm and the data in natural order.

8.18. Develop a FFT computer program for a base-"8 + 4 + 2" Sande-Tukey algorithm with data in natural order. The program should maximize the number of base-8 computations, then maximize the number of base-4 computations.

REFERENCES

1. BERGLAND, G. D. "A Guided Tour of the Fast Fourier Transform." *IEEE Spectrum* (July 1969), Vol. 6, No. 7, pp. 41–52.

2. BRIGHAM, E. O., AND R. E. MORROW. "The Fast Fourier Transform." *IEEE Spectrum* (December 1967), Vol. 4, pp. 63–70.

3. COOLEY, J. W., AND J. W. TUKEY. "An Algorithm for Machine Calculation of Complex Fourier Series." *Math. Computation* (April 1965), Vol. 19, pp. 297–301.

4. GENTLEMAN, W. M. "Matrix Multiplication and Fast Fourier Transforms." *Bell Syst. Tech. J.* (July–August 1968), Vol. 47, pp. 1099–1103.

5. OPPENHEIM, A. V., AND R. W. SCHAFER. *Digital Signal Processing.* Englewood Cliffs, NJ: Prentice Hall, 1975.

6. PELED, A., AND B. LIU. *Digital Signal Processing.* New York: Wiley, 1976.

7. BURRIS, C. S., AND T. W. PARKS. *DFT-FFT & Convolution Algorithms & Implementation.* New York: Wiley, 1985.

8. G-AE Subcommittee on Measurement Concepts. "What is the Fast Fourier Transform?" *IEEE Trans. Audio and Electroacoustics.* (June 1967), Vol. AU-15, pp. 45–55. Also *Proc. IEEE* (Oct. 1967), Vol. 55, pp. 1664–1674.

9. GENTLEMAN, W. M., AND G. SANDE. "Fast Fourier Transform for Fun and Profit." *AFIPS Proc.,* 1966 Fall Joint Computer Conf., Vol. 29, pp. 563–678, Washington, DC: Spartan, 1966.

10. BERGLAND, G. D. "The Fast Fourier Transform Recursive Equations for Arbitrary Length Records." *Math. Computation* (April 1967), Vol. 21, pp. 236–238.

11. DUHAMEL, P. "Implementation of 'Split-Radix' FFT Algorithm for Complex, Real, and Real-Symmetric Data." *IEEE Trans. on Audio, Speech and Signal Processing* (April 1986), Vol. ASSP-34, No. 2, pp. 285–295.

12. KUMARESAN, R., AND P. K. GUPTA. "Prime Factor FFT Algorithm with Real Valued Arithmetic." *Proc. IEEE,* (July 1985), Vol. 73, No. 7, pp. 1241–1243.

13. PREUSS, R. D. "Very Fast Computation of the Radix 2 Discrete Fourier Transform." *IEEE Trans. on Audio, Speech and Signal Processing* (Aug. 1982), Vol. ASSP-30, No. 4, pp. 595–607.

14. SKINNER, D. P. "Prunning the Decimation-in-Time FFT Algorithm." *IEEE Trans. on Audio, Speech and Signal Processing* (April 1976), Vol. ASSP-24, No. 2, pp. 193–194.

9

FFT TRANSFORM
APPLICATIONS

A fundamental application of the FFT is to the multifaceted areas of transform analysis. In Chapter 6, we developed the relationship between discrete and continuous Fourier transforms. Because the discrete Fourier transform yields a close approximation to the continuous Fourier transform, we expect significant usage of the FFT in computing Fourier and inverse Fourier transforms. In this chapter, we explore the mechanics of applying the FFT to the computation of Fourier transforms, Fourier series, inverse Fourier transforms and Laplace transforms. As we show, differences between continuous transforms and FFT results arise because of the discrete transform requirements for sampling and truncation.

9.1 FOURIER TRANSFORM APPLICATIONS

To illustrate the application of the FFT to the computation of Fourier transforms, consider Fig. 9.1. We show in Fig. 9.1(a) the function e^{-t}. We wish to compute by means of the FFT an approximation to the Fourier transform of this function.

The first step in applying the discrete transform is to choose the number of samples N and the sample interval T. For $N = 32$ and $T = 0.25$, we show the samples of e^{-t} in Fig. 9.1(a). Note that we have defined the sample value at $t = 0$ to be consistent with Eq. (2.47), which states that the value of the function at a discontinuity must be defined to be the midvalue if the inverse Fourier transform is to hold.

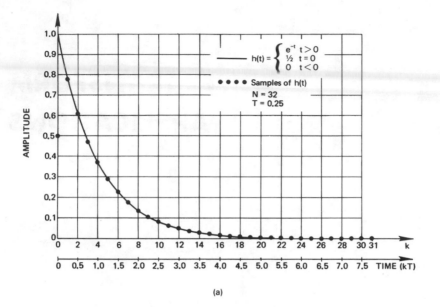

(a)

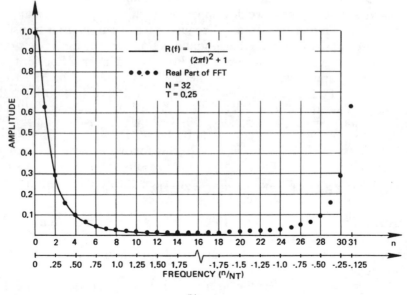

(b)

Figure 9.1 Example of Fourier transform computation via the FFT.

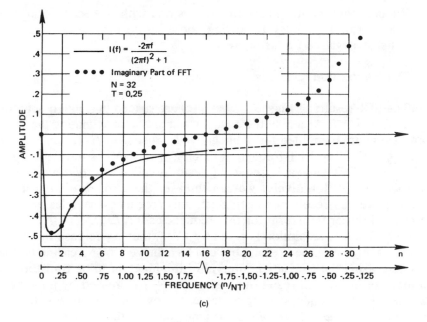

Figure 9.1   (cont.)

We next compute the discrete Fourier transform using the FFT:

$$H\left(\frac{n}{NT}\right) = T \sum_{k=0}^{N-1} [e^{-kT}]e^{-j2\pi nk/N} \qquad n = 0, 1, \ldots, N - 1 \qquad (9.1)$$

Note that the scale factor T is introduced to produce equivalence between the continuous and discrete transforms. These results are shown in Figs. 9.1(b) and (c). In Fig. 9.1(b), we show the real part of Fourier transform as determined in Ex. 2.1 and as computed by Eq. (9.1). Note that the discrete transform is symmetrical about $n = N/2$. This follows because the real part of the transform is even [Eq. (6.35)] and the results for $n > N/2$ are simply negative frequency results. This latter point is emphasized by plotting a true frequency scale beneath the scale for parameter n.

We could have graphed the data of Fig. 9.1(b) in the manner conventionally used to display the continuous Fourier transform, that is, from $-f_0$ to $+f_0$. However, conventional FFT computer programs provide results as a function of the parameter n. As long as we remember that those results for $n > N/2$ actually relate to negative frequency results, then we should encounter no interpretation problems.

In Fig. 9.1(c), we illustrate the imaginary part of the Fourier transform (Ex. 2.1) and the discrete transform. As shown, the FFT approximates rather poorly the continuous transform for the higher frequencies. To reduce this error, it is necessary to decrease the sample interval T and increase N.

We note that the imaginary function is odd with respect to $n = N/2$. This follows from Eq. (6.38). Repeating, those results for $n > N/2$ are to be interpreted as negative frequency results.

FFT Resolution

The FFT results of Figs. 9.1(b) and (c) are spaced in frequency by the interval $f_0 = 1/NT$. As a result, frequency samples approximating the Fourier transform are computed for nonnegative frequencies $0/NT$, $1/NT$, $2/NT$, . . . , $(N/2)/NT$. The FFT frequency spacing $f_0 = 1/NT$ is termed the *resolution* of the FFT and each frequency result is called a *resolution element* or *resolution cell*. Intuitively, we can think of the term *resolution* in the sense that we can resolve or distinguish only frequency samples approximating the Fourier transform for the frequencies $0/NT$, $1/NT$, $2/NT$, . . . , $(N/2)/NT$. Because resolution is given by $1/NT$, then a decrease in the frequency spacing (increased resolution) can be achieved by increasing N, that is, by increasing the truncation interval of the function to be transformed. (An increase in T could result in aliasing.) If N is increased by a factor of two, then the frequency spacing is decreased by a factor of two. Beware that the term *increase in resolution* is ambiguous in that one is not sure if a larger or smaller resolving power is implied.

Recall from Fig. 6.1 that the frequency spacing (resolution) in the discrete Fourier transform is determined by the width of the rectangle that multiplies and truncates the function to be transformed. This truncation in the time domain corresponds to convolution of the $[\sin(f)]/f$ function with the Fourier transform of the original time waveform. Convolution with the $[\sin(f)]/f$ function produces a smearing or blurring of the Fourier transform. The wider the time-domain truncation function, the narrower the $[\sin(f)]/f$ function and the less the frequency smear. The less the frequency smear, the better the frequency-resolving power that is possible. Hence, increased frequency-resolving power of the FFT is established by the degree of frequency smearing that is achieved by increasing the width of the rectangular truncation function.

A common mistake made by FFT users is to increase N by appending zeros to the sampled and truncated function and to interpret the results as having enhanced resolution. This is not the case, as can be seen from an examination of Figs. 6.2 and 9.2. We show in Fig. 9.2(a) the discrete Fourier transform resullts, Fig. 6.2(g), from the development of Fig. 6.2. We wish to observe the effect of appending zeros to the time function of Fig. 9.2(a). Assume that the number of zeros to be appended is N. This can be achieved by multiplication by the periodic time function illustrated in Fig. 9.2(b); the corresponding frequency function is also shown. Multiplication yields a periodic function of length $2N$, where the nonzero values are defined by the N samples of Fig. 9.2(a). Multiplication in time implies convolution of the

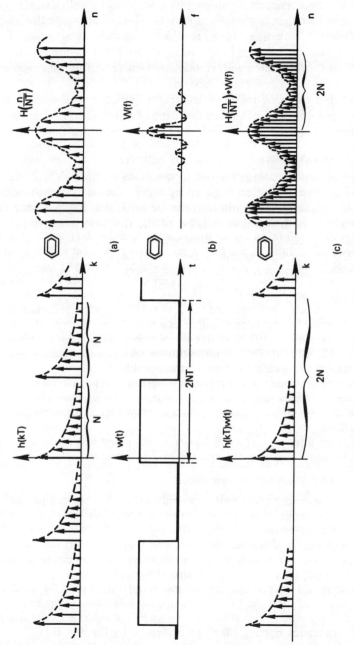

Figure 9.2 Example illustrating false FFT resolution enhancement by appending zeros.

frequency function of Figs. 9.2(a) and (b). Note that the frequency resolution has already been *set* in Fig. 9.2(a) and the convolution operation merely provided additional frequency samples by interpolating with a $[\sin(f)]/f$ function the original frequency transform results. Hence, although the frequency spacing of the FFT results are more closely spaced as a result of adding zeros, the resolution has not been changed. *FFT resolution cannot be enhanced by appending zeros unless the function is in fact zero-valued over the interval where the zeros are appended.*

Recall that this discussion simply reinforces the well-known fact that resolution is determined by the time duration of a signal. In FFT applications, the time duration of the signal is set by the truncation interval duration.

Example 9.1   FFT Aliasing

One problem encountered when computing the Fourier transform with the FFT is that of aliasing. As stated in Sec. 5.3, aliasing occurs if samples of the time function are not taken sufficiently close together. The result is that the frequency function *folds*, or *overlaps*, on itself. Figure 9.3 graphically illustrates this point.

In Figs. 9.3(a) to (c), we have sampled the function $h(t) = e^{-t}$, $t > 0$, with sample intervals $T = 1.0$, 0.5, and 0.25 s, respectively. N is set to 32 for each case. The magnitude of the Fourier transforms as computed by the FFT are also shown in Figs. 9.3(a) to (c). Note that the FFT results are severely aliased for sample interval $T = 1.0$ s. (The magnitude of the continuous Fourier transform is shown in Fig. 9.3(d).) As shown, aliasing is reduced for $T = 0.5$ s. A further reduction in sample interval to $T = 0.25$ s yields results that compare favorably with the theoretical frequency function. Figure 9.3 illustrates the principle that aliasing is reduced as the sample interval is reduced. There is no truncation effect as T is reduced because NT is always significantly greater than the nonzero width of $h(t)$.

Experimentally, one can choose an appropriate sample interval by performing a series of computations similar to those of Fig. 9.3. By successively reducing the sample interval, one notices negligible change in the FFT results when an approximately small sample interval T has been chosen. However, care must be exercised to ensure that the effect of decreasing the truncation interval NT does not enter into the results. If required, N can be increased as T is decreased.

Example 9.2   FFT Time-Domain Truncation

Another error commonly encountered in applying the FFT to computation of the Fourier transform arises from time-domain truncation. This error occurs when the total number of samples chosen to characterize the time function truncates the original time waveform. To illustrate this point, consider Figs. 9.4(a) to (c), where we have truncated $h(t)$ at $NT = 1$, 2, and 5 s, respectively. The magnitude of the Fourier transforms as computed by the FFT are also illustrated.

Truncation at 1.0 s produces considerable rippling in the FFT results. For a truncation interval of 2.0 s, the FFT results display a decrease in rippling. A further increase in the truncation interval to 5.0 s yields FFT results that have no apparent time-domain truncation rippling effect, as evidenced by Fig. 9.4(d).

Figure 9.4 illustrates an experimental approach for determining a suitable trun-

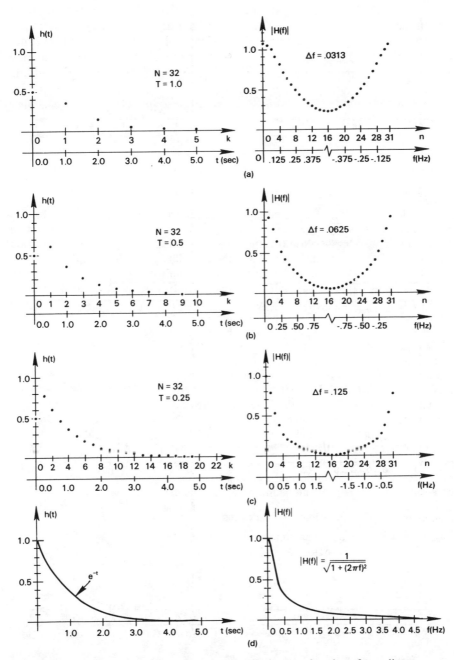

Figure 9.3   Illustration of frequency-domain aliasing as a function of sampling rate.

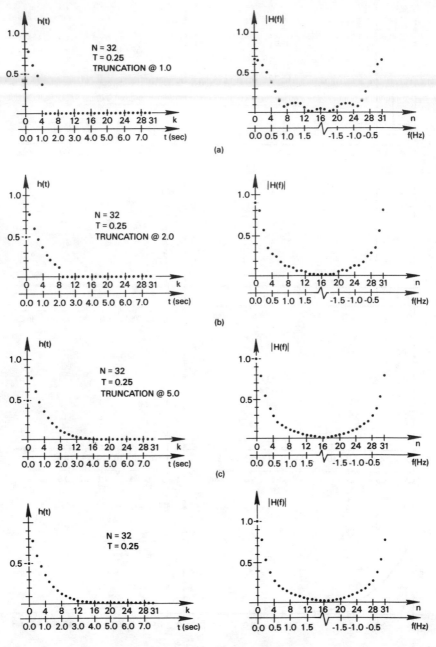

Figure 9.4 Illustration of time-domain truncation.

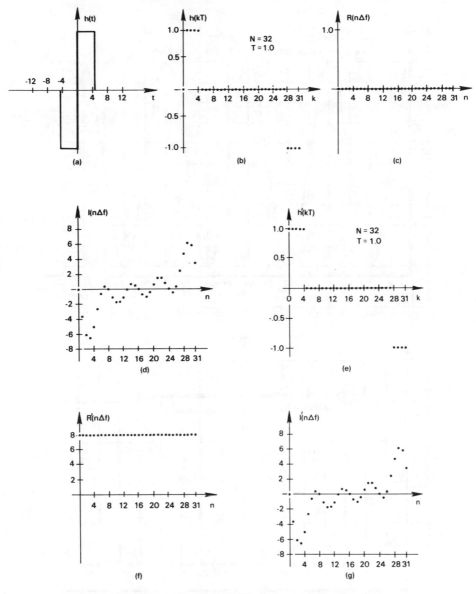

Figure 9.5   Illustration of the FFT of noncausal time functions.

cation interval. By successively increasing the truncation interval, we can see the gradual reduction in the rippling effect.

Example 9.3   FFT of Noncausal Time Functions

Because the discrete Fourier transform requires periodicity in the time domain, care must be exercised when computing the FFT of a time function defined for both

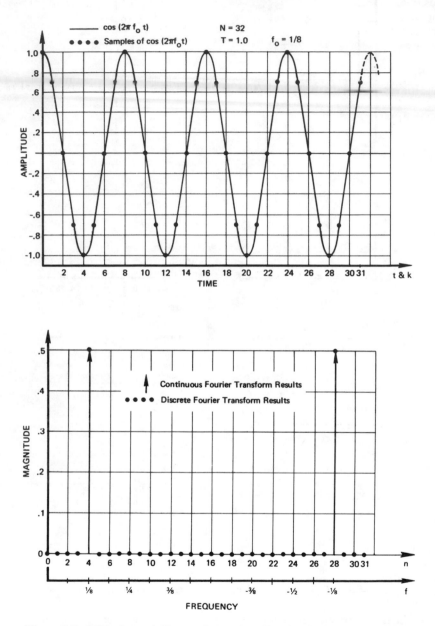

Figure 9.6 FFT of a periodic waveform: the truncation interval is equal to a multiple of the period.

positive and negative time (a noncausal function). To develop this point, consider the time function illustrated in Fig. 9.5(a). The appropriate technique for sampling such a function to maintain the time origin and to observe the periodicity constraint is shown in the sampled time function of Fig. 9.5(b). Although the one period shown does not closely resemble the original function, the periodic function does accurately reproduce the original time function. Because the time function is real and odd, from Eq. (6.38), we expect the frequency function to be purely imaginary and odd, as illustrated in Figs. 9.5(c) and (d).

Figures 9.5(e) to (g) illustrate a common mistake when applying the FFT to time functions of this type. Note that the FFT results for this example are both real and imaginary. The real frequency component results from the time sample at $t = 0$ being defined as 1 rather than 0, the midpoint of the discontinuity. As a result, the sampled time function of Fig. 9.5(e) is equal to the time function of Fig. 9.5(b) plus an impulse function of unity amplitude at the origin. The Fourier transform of an impulse is a constant real function of frequency, as illustrated in Fig. 9.5(f).

Example 9.4   FFT of Periodic Functions

To compute the FFT of a periodic function, we again must concern ourselves with choosing the sample interval T and the truncation interval. As before, T must be chosen to reduce aliasing to an acceptable level. The truncation interval for a periodic function presents a new problem in that the function does not decay as in the previous examples. However, recall that the N samples of the discrete Fourier transform results represent one period of a periodic time function. Hence, we choose the truncation interval to be one period (or multiple periods) of the time waveform. In this way, our sampled function accurately represents the original periodic waveform.

To illustrate this point, we have computed the FFT of the cosine function illustrated in Fig. 9.6(a). For sample interval $T = 1.0$ s and the number of samples $N = 32$, we also show samples of the cosine waveform in Fig. 9.6(a). Note that the 32 samples define exactly an integer multiple of the period of the waveform. In Fig. 9.6(b), we illustrate the magnitude of the FFT of these samples. As shown, the results are zero except at the desired frequency. Section 9.2 discusses FFT results for the case where the truncation interval is not chosen equal to an integer multiple of the period.

Summary

In applying the FFT to the computation of Fourier transforms, keep in mind the most important concept that the discrete Fourier transform implies is periodicity in both the time and frequency domains. If one always remembers that the N sample values of the time function represent one period of a periodic function, the application of the FFT should result in few surprises.

The previous discussion and examples have shown that application of the FFT to the computation of the Fourier transform requires that we exercise care in the choice of parameters T and N. Parameter T controls the level of aliasing, and parameters N and T control the width of the truncation function. If the bandwidth is roughly known, then T can be chosen readily.

Otherwise, an experimental procedure as outlined in Exs. 9.1 and 9.2 should be pursued. For an appropriate T and for N chosen sufficiently large so that truncation of the function being transformed does not occur, the FFT will yield an accurate approximation to the Fourier transform. For periodic functions with known periods, we choose NT equal to a period (or integer multiple of a period). For those cases where it is impossible to choose N sufficiently large or where the period of a periodic function is not known, the concept of a data-weighting function or data window must be employed.

9.2 FFT DATA-WEIGHTING FUNCTIONS

As discussed previously, time-domain truncation can lead to an unacceptable approximation to the Fourier transform. For those cases where data-processing constraints limit the value of N or for those cases where periodic functions of unknown period are considered, it is necessary to utilize data windows or data-weighting functions. In this section, we will investigate data-weighting functions, a technique for minimizing the undesired effects of time-domain truncation.

Rectangular Weighting Function

For review, reconsider the graphical development technique shown in Fig. 6.5. Recall that we first sample the sinusoid by multiplication with the infinite sequence of impulse functions, as illustrated in Fig. 6.5(b). This result, Fig. 6.5(c), must then be multiplied by the rectangular truncation function shown in Fig. 6.5(d) in order to limit the number of sample values to N. We refer to time-domain truncation as weighting the data by a rectangular weighting function. The result of time-domain truncation is evidenced clearly in Fig. 6.5(e). Note that the original frequency-domain impulse functions have been replaced by $[\sin(f)]/f$ functions because of the convolution that results from time-domain truncation. This convolution introduces additional frequency-domain components because of the side-lobe characteristics of the $[\sin(f)]/f$ function. These additional components are termed *leakage*. This terminology arises because the original frequency impulse function has *leaked* through the side lobes of the $[\sin(f)]/f$ function.

Note that even though our original time waveform is sinusoidal, the sampled time waveform is not. This is because the truncation interval is not equal to a period (or integer multiple of a period) and hence the convolution of the time functions in Figs. 6.5(e) and (f) does not yield the original periodic function. Rather, this convolution yields a periodic function with an envelope that is discontinuous. With this discontinuity, one expects the rippling effect in the frequency domain, which is illustrated in Fig. 6.5(g).

To further demonstrate the effect of a rectangular weighting function, we have computed the FFT of the cosine function illustrated in Fig. 9.7(a) for $T = 1.0$ s and $N = 32$. In Fig. 9.7(b), we show the magnitude of the discrete Fourier transform of the samples of Fig. 9.7(a). Note that the FFT produces nonzero frequency components at all discrete frequencies of the discrete transform. As stated previously, the additional frequency components are termed leakage and are a result of the side-lobe characteristics of the $[\sin(f)]/f$ function.

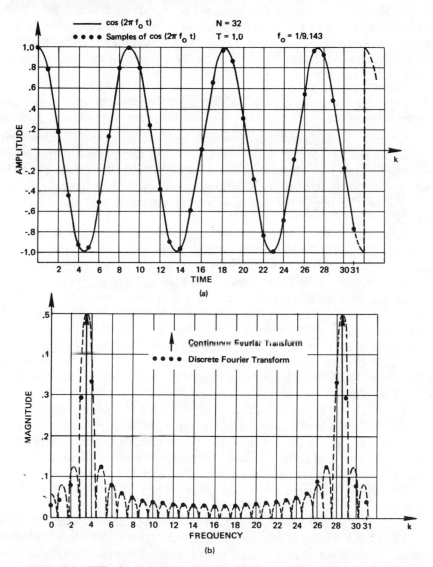

Figure 9.7 FFT of a periodic waveform: the truncation interval is not equal to a multiple of the period.

To reduce leakage, it is necessary to employ a time-domain truncation or weighting function that has frequency-domain side-lobe characteristics that are of smaller magnitude than those of the $[\sin(f)]/f$ function. The smaller the side lobes, the less leakage affects the results of the FFT. To clarify this point, consider Fig. 6.5(d) again. The Fourier transform of the rectangular weighting function is the $[\sin(f)]/f$ function shown. We know that we could choose, without change to the graphical development of Fig. 6.5, an alternate truncation function in Fig. 6.5(d) that would have lower side-lobe characteristics. This is the approach one takes to improve the FFT approximation to the Fourier transform. Data-weighting functions that truncate and weight the data are applied to the N-point sampled function before the FFT is computed.

Weighting Function Characteristics

Several popular truncation or weighting functions that have been employed with the FFT are shown in Fig. 9.8(a). The corresponding frequency-response functions are illustrated in Fig. 9.8(b). Table 9.1 lists the defining relationships for each of these weighting functions in both the time domain (centered at the origin for convenience of notation) and the frequency domain.

As shown in Fig. 9.8(b), all weighting functions have side lobes in the frequency domain of lower amplitude than those of the rectangular function and hence produce less leakage. However, all of the weighting functions also have a broader main lobe. Recall from Figs. 6.5(d) and (e) that the effect of time-domain truncation (weighting) is a frequency-domain convolution, with the respective frequency function shown in Fig. 9.8(b). Hence, the broader the main lobe, the more smeared the results of the FFT, that is, the broader the main lobe of the weighting function, the less the capability of the FFT to distinguish or resolve frequencies. In general, the more one reduces side lobes or leakage, the broader or more smeared the results of the FFT appear.

The trade-off between leakage (side-lobe level) and resolution (main-lobe bandwidth) is well-known in many scientific fields. Table 9.1 defines the highest side-lobe level and the 3-dB bandwidth for each weighting function. For general experimental work, we prefer to use the Hanning function because of its implementation simplicity. One should choose that weighting function whose characteristics are best suited to the problem being addressed.

It should be noted that irrespective of the weighting function chosen, the FFT yields results at frequency intervals $f_0 = 1/NT$. But the actual frequency resolution of FFT results is a function of the data-weighting function's bandwidth (see Fig. 9.8(b)). Hence, the commonly used FFT resolution definition $1/NT$ should be used with care as it describes only the

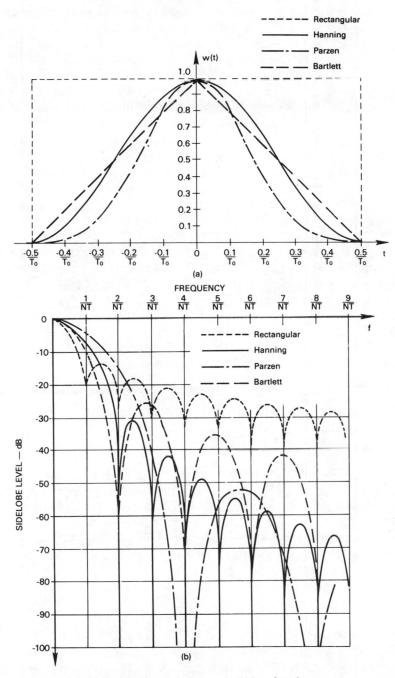

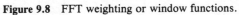

Figure 9.8 FFT weighting or window functions.

TABLE 9.1 Data Weighting Functions ($T_0 = NT$)

Weighting Function Nomenclature	Time Domain	Frequency Domain	Highest Side-Lobe Level (db)	3-dB Bandwidth	Asymptotic Rolloff (dB/Octave)
Rectangular	$w_R(t) = 1 \quad \|t\| \le \dfrac{T_0}{2}$ $\quad\quad\quad = 0 \quad \|t\| > \dfrac{T_0}{2}$	$W_R(f) = \dfrac{T_0 \sin(\pi f T_0)}{\pi f T_0}$	-13	$\dfrac{0.85}{T_0}$	6
Bartlett (triangle)	$w_B(t) = \left[1 - \dfrac{2\|t\|}{T_0}\right] \quad \|t\| < \dfrac{T_0}{2}$ $\quad\quad\quad = 0 \quad\quad\quad\quad\quad \|t\| > \dfrac{T_0}{2}$	$W_B(f) = \dfrac{T_0}{2}\left[\dfrac{\sin\left(\dfrac{\pi}{2} f T_0\right)}{\dfrac{\pi}{2} f T_0}\right]^2$	-26	$\dfrac{1.25}{T_0}$	12
Hanning (cosine)	$w_H(t) = \cos^2\left(\dfrac{\pi t}{T_0}\right)$ $\quad\quad = \dfrac{1}{2}\left[1 + \cos\left(\dfrac{2\pi t}{T_0}\right)\right] \quad \|t\| \le \dfrac{T_0}{2}$ $\quad\quad = 0 \quad\quad\quad\quad\quad\quad\quad \|t\| > \dfrac{T_0}{2}$	$W_H(f) = \dfrac{T_0}{2} \dfrac{\sin(\pi f T_0)}{\pi f T_0 [1 - (f T_0)^2]}$	-32	$\dfrac{1.4}{T_0}$	18
Parzen	$w_P(t) = 1 - 24\left(\dfrac{t}{T_0}\right)^2 + 48\left|\dfrac{t}{T_0}\right|^3 \quad \|t\| < \dfrac{T_0}{4}$ $\quad\quad = 2\left[1 - \dfrac{2\|t\|}{T_0}\right]^3 \quad \dfrac{T_0}{4} < \|t\| < \dfrac{T_0}{2}$ $\quad\quad = 0 \quad\quad\quad\quad\quad\quad \|t\| \ge \dfrac{T_0}{2}$	$W_P(f) = \dfrac{3T_0}{8}\left[\dfrac{\sin(\pi f T_0/4)}{\pi f T_0/4}\right]^4$	-52	$\dfrac{1.92}{T_0}$	24

frequency spacing of the FFT results and is independent of the window used. In Chapter 13, where the FFT in the context of filters is discussed, we return to this point.

Because of the low side-lobe characteristics of the illustrated data-weighting functions, we expect that their utilization significantly reduces the leakage that results from time-domain truncation. In Fig. 9.9(a), we show the cosine waveform illustrated in Fig. 9.7(a) multiplied by the Hanning weighting function illustrated in Fig. 9.8(a).

Figure 9.9(b) illustrates the FFT of the samples of Fig. 9.9(a). As expected, leakage is significantly reduced. Note that the majority of the frequency components are considerably *broadened* or *smeared* with respect to the desired impulse function. Recall that this is expected because the effect of time-domain truncation or weighting is to convolve the frequency-impulse function with the Fourier transform of the weighting function.

Example 9.5   FFT Signal Detection

A practical application of the utilization of the Hanning weighting function (or any other good truncation function) is in signal detection. To illustrate, consider the frequency function in Fig. 9.10(a), which has been computed by the FFT using the standard rectangular weighting function. A cursory comparison of this illustration with that of Fig. 9.7(b) leads to the conclusion that the time function consists of a single sinusoid.

Now consider Fig. 9.10(b), where we have recomputed the FFT of the original time function but have used the Hanning weighting function. Note that a second sinusoidal component is clearly visible. We now conclude that the time function is composed of two sinusoids. Leakage resulting from the rectangular truncation function almost completely obscures the second, smaller frequency component in Fig. 9.10(a). The detection of signals buried in noise is addressed in Sec. 14.3.

Example 9.6   Dolph-Chebyshev Weighting Functions

As discussed previously, low side lobes are achieved at the expense of main-lobe bandwidth or broadening. Even though the Hanning function is a weighting function that has low side lobes and is convenient to use, a weighting function that has better side-lobe characteristics is the Dolph-Chebyshev function. This window, Refs. [7] and [8], minimizes the bandwidth or smearing while forcing all side lobes to be below a specified level. For some FFT applications, the additional complexity of this weighting function may be warranted.

The Dolph-Chebyshev weighting function can be calculated [8] from the relationship

$$w_N(i) = \frac{N-1}{N-i} \sum_{k=0}^{M} \binom{i-2}{k} \binom{N-i}{k+1} \beta^{k+1} \qquad i \neq 1 \text{ or } N \qquad (9.2)$$

where

$$M = i - 2 \qquad i \leq (N+1)/2$$

$$= N - i - 1 \qquad i \geq (N+1)/2$$

and

$$w_N(1) = w_N(N) = 1$$

and β is the desired side-lobe level in dB. Figure 9.11 shows a BASIC program for computing $w_N(i)$. The required inputs are the number of weights N, (i.e., the number of data samples) and the desired side-lobe level (SSL) in decibels. SSL is input as a positive number. The logarithm function used in the program is the natural loga-

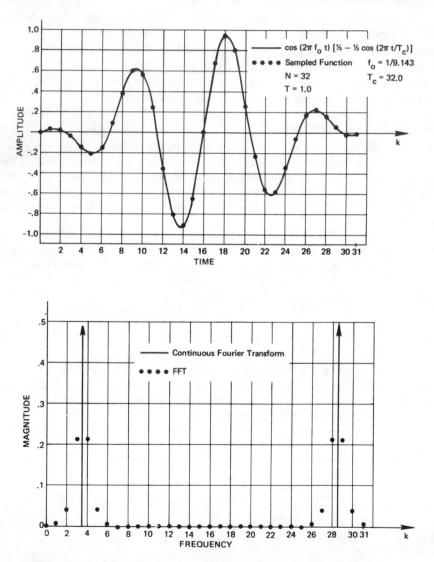

Figure 9.9 Example of applying the Hanning function to reduce leakage in computation of the FFT.

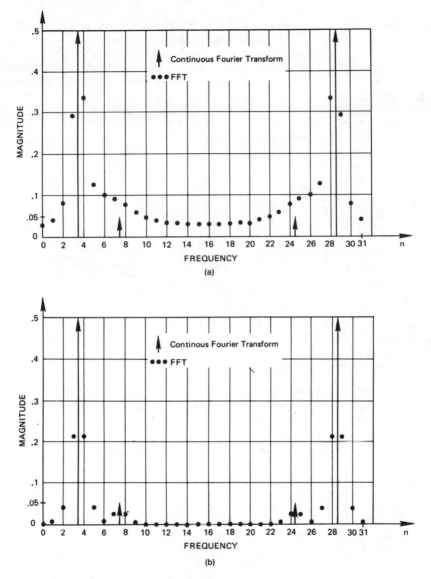

Figure 9.10   (a) Example signal obscured by side-lobe leakage, and (b) signal detection by application of the Hanning weighting function.

rithm. Exercise caution when computing Dolph-Chebyshev weights because of the precision required if N is large [13, 14].

When the number of data samples is 10 or larger, the 3-dB bandwidth of the main lobe is essentially independent of N and depends only on the side-lobe level. Figure 9.12(a) shows a plot of the increase in bandwidth as the side-lobe level parameter varies. Hence, side lobes can be reduced to any desired value but with a

```
8500   REM:   DOLPH-CHEBYSHEV WEIGHTING FUNCTION SUBROUTINE
8510   REM:   THE CALLING PROGRAM SHOULD DIMENSION
8520   REM:   THE WEIGHTING FUNCTION ARRAY W(I%), THE NUMBER
8530   REM:   OF WEIGHTING FUNCTION VALUES N%, AND THE DESIRED
8540   REM:   SIDE-LOBE LEVEL IN DB SHOULD BE INITIALIZED.
8550   AN=N%
8560   N1%=(N%+1)\2
8570   S=10!^(SSL/20!)
8580   A=2!*LOG(S+SQR(S*S-1!))/AN-1!)
8590   B=(EXP(A)-1!)*(EXP(A)-1!)
8600   C=(EXP(A)+1!)*(EXP(A)+1!)
8620   D=B/C
8630   FOR I%=2 TO N1%
8640   AI=I%
8650   I1=I%-1
8660   E=0!
8670   FOR K%=1 TO I1
8680   K1%=K%-1
8690   AK=K%
8700   G=1!
8710   H=1!
8720   IF (K%-1)=0 THEN 8770 ELSE 8730
8730   FOR J%=1 TO K1%
8740   AJ=J%
8750   G=G*(AI-1!-AJ)/AJ
8760   NEXT J%
8770   FOR L%=1 TO K%
8780   AL=L%
8790   H=H*(AN-AI+1!-AL)/AL
8795   NEXT L%
8800   E=E+G*H*(D^AK)
8810   NEXT K%
8820   W(I%)=(AN-1!)*E/(AN-AI)
8830   W(N%-I%+1)=W(I%)
8840   NEXT I%
8850   W(N%)=1!
8860   W(1)=1!
8870   RETURN
8880   END
```

Figure 9.11   Computer subroutine in BASIC for computing the Dolph-Chebyshev
weighting function.

corresponding increase in bandwidth. The normalized data plotted in Fig. 9.12(a) is
adequate to evelute the degree of increased bandwidth as a function of side-lobe
level. Figure 9.12(b) illustrates Dolph-Chebyshev weighting functions for several
choices of side-lobe levels. The rectangular and Hanning functions are also shown
for comparison.

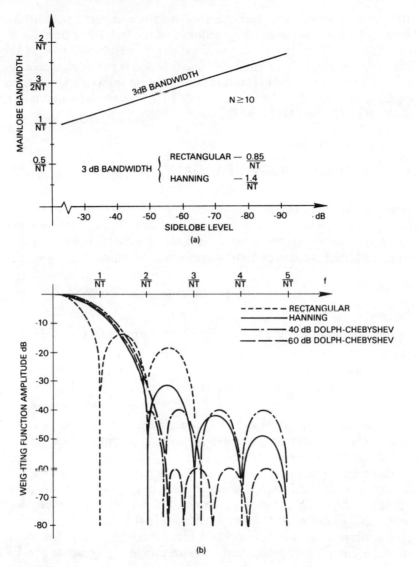

Figure 9.12 (a) Bandwidth vs. side-lobe level for the Dolph-Chebyshev weighting function, and (b) Dolph-Chebyshev frequency-domain functions for 40- and 60-dB side-lobe levels.

Summary

The reader should not infer from the previous discussion that the FFT is of little utility for computing the Fourier transform of periodic functions. If you know the period, then you should take advantage of this knowledge by selecting the truncation interval equal to an integer multiple of the period.

If the period is unknown, then the FFT results computed with a Hanning weighting function are as good an estimate of the frequency function as any other estimate. The only reason one questions the results is if *truth* is known. If the experimenter believes that the time function is periodic, then a succession of FFTs with sequentially longer truncation intervals should identify the period. A catalog and comparison of FFT weighting functions is given in Refs. [9], [10], and [11].

9.3 FFT ALGORITHMS FOR REAL DATA

In applying the FFT, we often consider only real functions of time, whereas the frequency functions are, in general, complex. Thus, a single computer program written to determine both the discrete transform and its inverse is written such that a complex time waveform is assumed:

$$H(n) = \frac{1}{N} \sum_{k=0}^{N-1} [h_r(k) + jh_i(k)]e^{-j2\pi nk/N} \tag{9.3}$$

This follows because the alternate inversion formula of Eq. (6.33) is given by

$$h(k) = \frac{1}{N} \left[\sum_{n=0}^{N-1} [H_r(n) + jH_i(n)]^* e^{-j2\pi nk/N} \right]^* \tag{9.4}$$

and because both Eqs. (9.3) and (9.4) contain the common factor $e^{-j2\pi nk/N}$, then a single computer program can be used to compute both the discrete transform and its inverse.

If the time function being considered is real, we must set to zero the imaginary part of the complex time function in Eq. (9.3). This approach is inefficient in that the computer program still performs the multiplications involving $jh_i(k)$ in Eq. (9.4) even though $jh_i(k)$ is zero.

In this section, we develop two techniques for using this imaginary part of the complex time function to more efficiently compute the FFT of real functions.

FFT of Two Real Functions Simultaneously

It is desired to compute the discrete Fourier transform of the real time functions $h(k)$ and $g(k)$ from the complex function

$$y(k) = h(k) + jg(k) \tag{9.5}$$

That is, $y(k)$ is constructed to be the sum of two real functions, where one of these real functions is taken to be imaginary. From the linearity property of Eq. (6.25), the discrete Fourier transform of $y(k)$ is given by

$$Y(n) = H(n) + jG(n)$$

$$= [H_r(n + jH_i(n)] + j[G_r(n) + jG_i(n)]$$

$$= [H_r(n) - G_i(n)] + j[H_i(n) + G_r(n)] \tag{9.6}$$

$$= R(n) + jI(n)$$

By means of the frequency-domain equivalent of Eq. (6.39), we decompose both $R(n)$, the real part of $Y(n)$, and $I(n)$, the imaginary part of $Y(n)$, into even and odd components:

$$Y(n) = \left(\frac{R(n)}{2} + \frac{R(N-n)}{2}\right) + \left(\frac{R(n)}{2} - \frac{R(N-n)}{2}\right)$$
$$+ j\left(\frac{I(n)}{2} + \frac{I(N-n)}{2}\right) + j\left(\frac{I(n)}{2} - \frac{I(N-n)}{2}\right) \tag{9.7}$$

From Eqs. (6.45) and (6.46),

$$H(n) = R_e(n) + jI_0(n)$$

$$= \left(\frac{R(n)}{2} + \frac{R(N-n)}{2}\right) + j\left(\frac{I(n)}{2} - \frac{I(N-n)}{2}\right) \tag{9.8}$$

Similarly, from Eqs. (6.47) and (6.48),

$$jG(n) = R_0(n) + jI_e(n)$$

or

$$G(n) = I_e(n) - jR_0(\dot{n})$$

$$= \left(\frac{I(n)}{2} + \frac{I(N-n)}{2}\right) - j\left(\frac{R(n)}{2} - \frac{R(N-n)}{2}\right) \tag{9.9}$$

Thus, if the real and imaginary parts of the discrete transform of a complex time function are decomposed according to Eqs. (9.8) and (9.9), then the simultaneous discrete transform of two real time functions can be accomplished. This procedure results in a *two-series* capability with only the requirement for *sorting* the results. For ease of reference, the necessary steps to simultaneously compute the FFT of two real functions are listed in Fig. 9.13. Note that Step 4 is written in terms of $R(N)$ and $I(N)$. By periodicity, we know that $R(N) = R(0)$ and $I(N) = I(0)$.

A BASIC computer program according to the procedure defined in

Fig. 9.13 is listed in Fig. 9.14. The two-input real data arrays are X1REAL(I%) and X2REAL(I%), where each is N points. Parameters N% and NU% should be initialized in the main program; N% $= N$. Transform results for the real data array X1REAL(I%) are returned from the subroutine with the real part of the transform stored in X1REAL(I%) and the imaginary part of the transform stored in X1IMAG(I%). Similarly, the real part of the transform of the X2REAL(I%) data array is returned in array X2REAL(I%) and the imaginary part in X2IMAG(I%). The subroutine completely sorts the results so that the output is in exactly the same form as two independent FFTs. The program uses the FFT subroutine listed in Fig. 8.7 and hence XREAL(I%) and XIMAG(I%) must be dimensioned. For clarity of presentation, we have utilized additional arrays that are not required if memory is

---

1. Functions $h(k)$ and $g(k)$ are real $k = 0, 1, \ldots, N - 1$
2. Form the complex function:

$$y(k) = h(k) + jg(k) \qquad k = 0, 1, \ldots, N - 1$$

3. Compute

$$Y(n) = \sum_{k=0}^{N-1} y(k)e^{-j2\pi nk/N}$$

$$= R(n) + jI(n) \qquad n = 0, 1, \ldots, N - 1$$

where $R(n)$ and $I(n)$ are the real and imaginary parts of $Y(n)$, respectively.

4. Compute

$$H(n) = \left[\frac{R(n)}{2} + \frac{R(N - n)}{2}\right] + j\left[\frac{I(n)}{2} - \frac{I(N - n)}{2}\right]$$

$$G(n) = \left[\frac{I(n)}{2} + \frac{I(N - n)}{2}\right] - j\left[\frac{R(n)}{2} - \frac{R(N - n)}{2}\right]$$

$$n = 0, 1, \ldots, N - 1$$

where $R(N) = R(0)$, $I(N) = I(0)$, and $H(n)$ and $G(n)$ are the discrete transforms of $h(k)$ and $g(k)$, respectively.

5. Scale the results by the sample interval T.

Figure 9.13 Computation procedure for simultaneous FFT of two real functions.

```
11000 REM:    SUBROUTINE FOR SIMULTANEOUS FFT OF TWO REAL FUNCTIONS
11002 REM:    STORED IN ARRAYS X1REAL(I%) AND X2REAL(I%). RESULTS ARE
11004 REM:    RETURNED IN ARRAYS X1REAL(I%), X1IMAG(I%), X2REAL(I%), AND
11006 REM:    X2IMAG(I%). THESE ARRAYS AND XREAL(I%), XIMAG(I%) MUST BE
11008 REM:    DIMENSIONED IN THE MAIN PROGRAM. N% AND NU% MUST BE
11010 REM:    INITIALIZED. THIS PROGRAM CALLS THE FFT SUBROUTINE
11012 REM:    BEGINNING AT LINE 10000 (FIG. 8-7).
11014 REM:
11020 FOR I%=1 TO N%
11030     XREAL(I%)=X1REAL(I%)
11040     XIMAG(I%)=X2REAL(I%)
11050 NEXT I%
11060     GOSUB 10000
11070 N2%=N%/2
11080     X1REAL(1)=XREAL(1)
11090     X1IMAG(1)=0
11100     X2REAL(1)=XIMAG(1)
11110     X2IMAG(1)=0
11120 FOR I%=2 TO N%
11130     X1REAL(I%)=(XREAL(I%)+XREAL(N%+2-I%))/2
11140     X1IMAG(I%)=(XIMAG(I%)-XIMAG(N%+2-I%))/2
11150     X2REAL(I%)=(XIMAG(I%)+XIMAG(N%+2-I%))/2
11160     X2IMAG(I%)=-(XREAL(I%)-XREAL(N%+2-I%))/2
11170 NEXT I%
11180 RETURN
11190 END
```

Figure 9.14 Computer subroutine in BASIC for simultaneous FFT of two real functions.

a constraint. Output results must be scaled by the sample interval T to obtain equivalence to the continuous Fourier transform.

Transform of 2 *N* Samples with an *N*-Sample Transform

The imaginary part of the complex time function can also be used to compute more efficiently the discrete transform of a single real time function. Consider a function $x(k)$ that is described by 2N samples. It is desired to compute the discrete transform of this function using Eq. (9.3). That is, we wish to break the 2N-point function $x(k)$ into two N-sample functions. Function $x(k)$ cannot simply be divided in half; instead, we divide $x(k)$ as follows:

$$\left.\begin{array}{l} h(k) = x(2k) \\[2mm] g(k) = x(2k + 1) \end{array}\right\} k = 0, 1, \ldots, N - 1 \qquad (9.10)$$

That is, function $h(k)$ is equal to the even-numbered samples of $x(k)$, and

$g(k)$ is equal to the odd-numbered samples. Eq. (9.3) can then be written as

$$X(n) = \sum_{k=0}^{2N-1} x(k)e^{-j2\pi nk/2N}$$

$$= \sum_{k=0}^{N-1} x(2k)e^{-j2\pi n(2k)/2N} + \sum_{k=0}^{N-1} x(2k+1)e^{-j2\pi n(2k+1)/2N}$$

$$= \sum_{k=0}^{N-1} x(2k)e^{-j2\pi nk/N} + e^{-j\pi n/N} \sum_{k=0}^{N-1} x(2k+1)e^{-j2\pi nk/N} \qquad (9.11)$$

$$= \sum_{k=0}^{N-1} h(k)e^{-j2\pi nk/N} + e^{-j\pi n/N} \sum_{k=0}^{N-1} g(k)e^{-j2\pi nk/N}$$

$$= H(n) + e^{-j\pi n/N} G(n)$$

To efficiently compute $H(n)$ and $G(n)$, use the previously discussed technique. Let

$$y(k) = h(k) + jg(k) \qquad (9.12)$$

then

$$Y(n) = R(n) + jI(n)$$

From Eqs. (9.8) and (9.9),

$$H(n) = R_e(n) + jI_0(n) \qquad (9.13)$$
$$G(n) = I_e(n) - jR_0(n)$$

Substitution of Eq. (9.13) into Eq. (9.11) yields

$$X(n) = R_e(n) + jI_0(n) + e^{-j\pi n/N}[I_e(n) - jR_0(n)]$$

$$= \left[R_e(n) + \cos\left(\frac{\pi n}{N}\right)I_e(n) - \sin\left(\frac{\pi n}{N}\right)R_0(n) \right]$$

$$\qquad + j\left[I_0(n) - \sin\left(\frac{\pi n}{N}\right)I_e(n) - \cos\left(\frac{\pi n}{N}\right)R_0(n) \right] \qquad (9.14)$$

$$= X_r(n) + jX_i(n)$$

Hence, the real part of the $2N$-sample function $x(k)$ is

$$X_r(n) = \left[\frac{R(n)}{2} + \frac{R(N-n)}{2} \right] + \cos\left(\frac{\pi n}{N}\right)\left[\frac{I(n)}{2} + \frac{I(N-n)}{2} \right]$$

$$\qquad - \sin\left(\frac{\pi n}{N}\right)\left[\frac{R(n)}{2} - \frac{R(N-n)}{2} \right] \qquad (9.15)$$

1. Function $x(k)$ is real $k = 0, 1, \ldots, 2N - 1$
2. Divide $x(k)$ into two functions:

$$\left.\begin{array}{l} h(k) = x(2k) \\ g(k) = x(2k + 1) \end{array}\right\} \quad k = 0, 1, \ldots, N - 1$$

3. Form the complex function:

$$y(k) = h(k) + jg(k) \qquad k = 0, 1, \ldots, N - 1$$

4. Compute

$$Y(n) = \sum_{k=0}^{N-1} y(k)e^{-j2\pi nk/N}$$

$$= R(n) + jI(n) \qquad n = 0, 1, \ldots, N - 1$$

where $R(n)$ and $I(n)$ are the real and imaginary parts of $Y(n)$, respectively.

5. Compute

$$X_r(n) = \left[\frac{R(n)}{2} + \frac{R(N - n)}{2}\right] + \cos\left(\frac{\pi n}{N}\right)\left[\frac{I(n)}{2} + \frac{I(N - n)}{2}\right]$$

$$- \sin\left(\frac{\pi n}{N}\right)\left[\frac{R(n)}{2} - \frac{R(N - n)}{2}\right]$$

$$n = 0, 1, \ldots, N - 1$$

$$X_i(n) = \left[\frac{I(n)}{2} - \frac{I(N - n)}{2}\right] - \sin\left(\frac{\pi n}{N}\right)\left[\frac{I(n)}{2} + \frac{I(N - n)}{2}\right]$$

$$- \cos\left(\frac{\pi n}{N}\right)\left[\frac{R(n)}{2} - \frac{R(N - n)}{2}\right]$$

$$n = 0, 1, \ldots, N - 1$$

where $R(N) = R(0)$, $I(N) = I(0)$, $X_r(n)$ and $X_i(n)$ are the real and imaginary parts of the $2N =$ point discrete transform of $x(k)$, respectively.
6. Scale the results by the sample interval T.

Figure 9.15 Computation procedure for the FFT of a $2N$-point function by means of a N-point FFT.

```
12000 REM:   SUBROUTINE FOR EFFICIENT COMPUTATION OF THE FFT OF REAL
12002 REM:   FUNCTIONS. THE CALLING PROGRAM SHOULD DIMENSION XREAL(I%)
12004 REM:   AND XIMAG(I%) AND THE COMPUTATION ARRAYS X1REAL(I%) AND
12006 REM:   X1IMAG(I%). N%=2N AND NU% MUST BE INITIALIZED.
12008 REM:   THIS PROGRAM CALLS THE FFT SUBROUTINE BEGINNING AT
12010 REM:   LINE 10000 (FIG. 8-7).
12012 REM:
12020     N%=N%/2
12030     NU%=NU%-1
12040 REM:   PLACE ODD NUMBERED SAMPLES IN XREAL( ), EVEN IN XIMAG( ).
12050 FOR I%=1 TO N%
12060       X1REAL(I%)=XREAL(2*I%-1)
12070       X1IMAG(I%)=XREAL(2*I%)
12080 NEXT I%
12090 FOR I%=1 TO N%
12100       XREAL(I%)=X1REAL(I%)
12110       XIMAG(I%)=X1IMAG(I%)
12120 NEXT I%
12130 REM:   COMPUTE THE FFT.
12140     GOSUB 10000
12150 ARG=3.145926#/N%
12160 INIT=ARG
12170 FOR I%=2 TO N%
12180   S=SIN(ARG)
12190   C=COS(ARG)
12200       A1=(XREAL(I%)+XREAL(N%+2-I%))/2
12210       A2=(XREAL(I%)-XREAL(N%+2-I%))/2
12220       B1=(XIMAG(I%)+XIMAG(N%+2-I%))/2
12230       B2=(XIMAG(I%)-XIMAG(N%+2-I%))/2
12240       X1REAL(I%)=A1+C*B1-S*A2
12250       X1IMAG(I%)=B2-S*B1-C*A2
12260       ARG=ARG+INIT
12270 NEXT I%
12280       XREAL(N%+1)=XREAL(1)-XIMAG(1)
12290       XIMAG(N%+1)=0!
12300       XREAL(1)=XREAL(1)+XIMAG(1)
12310       XIMAG(1)=0
12320       XIMAG(N%+1)=0
12330 FOR I%=2 TO N%
12340       XREAL(I%)=X1REAL(I%)
12350       XIMAG(I%)=X1IMAG(I%)
12360       K%=2*N%
12370       XREAL(K%+2-I%)=XREAL(I%)
12380       XIMAG(K%+2-I%)=-XIMAG(I%)
12390 NEXT I%
12400     N%=2*+%
12410     NU%=NU%+1
12420   RETURN
12430   END
```

Figure 9.16   Computer subroutine in BASIC for the FFT of 2*N* data samples with a *N*-point FFT.

and, similarly, the imaginary part is

$$X_i(n) = \left[\frac{I(n)}{2} - \frac{I(N-n)}{2}\right] - \sin\left(\frac{\pi n}{N}\right)\left[\frac{I(n)}{2} + \frac{I(N-n)}{2}\right]$$
$$- \cos\left(\frac{\pi n}{N}\right)\left[\frac{R(n)}{2} - \frac{R(N-n)}{2}\right] \quad (9.16)$$

Thus, the imaginary part of the complex time function can be used advantageously to compute the transform of a function defined by $2N$ samples by using a discrete transform that sums only over N values. We normally speak of this computation as performing a $2N$-point transform by means of a N-point transform. For reference, an outline of the computation approach is given in Fig. 9.15. Note that Step 5 is written in terms of $R(N)$ and $I(N)$. By periodicity, $R(N) = R(0)$ and $I(N) = I(0)$.

A BASIC computer program that implements the outline of Fig. 9.15 is listed in Fig. 9.16. The $2N$-point real input function is stored in XREAL(I%). The main program should initialize N% and NU%, where N% = $2N$. For clarity of presentation, dummy array DREAL(I%) and DIMAG(I%) are used. All arrays should be of dimension $2N$. Because this program cuts FFT computation time approximately in half, readers with a personal computer should find it very useful. Output results must be scaled by the sample interval T to obtain equivalence to the continuous Fourier transform.

9.4 INVERSE FOURIER TRANSFORM APPLICATIONS

Assume that we are given the continuous real and imaginary frequency functions considered in Figs. 9.1(b) and (c) and that we wish to determine the corresponding time function by means of the inverse FFT:

$$h(kT) = \Delta f \sum_{n=0}^{N-1} [R(n\Delta f)] + jI(n\Delta f)e^{j2\pi nk/N}$$

$$k = 0, 1, \ldots, N-1 \quad (9.17)$$

where Δf is the frequency sample interval frequency. Assume $N = 32$ and $\Delta f = 1/8$.

Because we know that $R(f)$, the real part of the complex frequency function, must be an even function, then we *fold* $R(f)$ about the frequency $f = 2.0$, which corresponds to the sample point $n = N/2$. As shown in Fig. 9.17(a), we simply sample the frequency function up to the point $n = N/2$ and then *fold* these values about $n = N/2$ to obtain the remaining samples.

In Figure 9.17(b), we illustrate the method for determining the N samples of the imaginary part of the frequency function. Because the imaginary

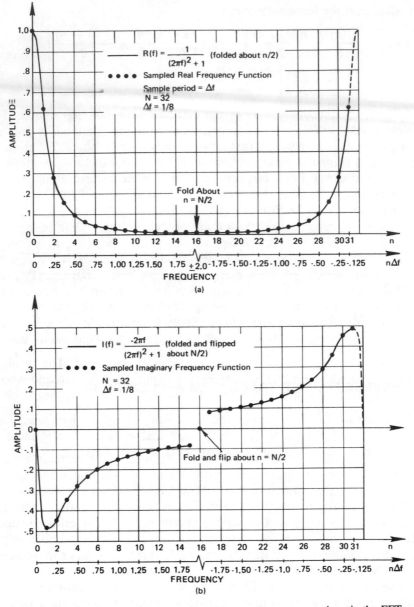

Figure 9.17 Example of the inverse Fourier transform computation via the FFT.

frequency function is odd, we must not only *fold* about the sample value $N/2$, but also *flip* the results. To preserve symmetry, we set the sample at $n = N/2$ to zero.

Computation of Eq. (9.17) with the sampled function illustrated in Figs.

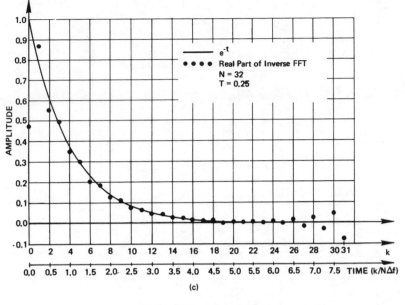

Figure 9.17 (*continued*)

9.17(a) and (b) yields the inverse discrete Fourier transform. The result is a complex function whose imaginary part is approximately zero and whose real part is as shown in Fig. 9.17(c). We note that at $k = 0$, the result is approximately equal to the correct midvalue and reasonable agreement is obtained for all except the results for k large. Improvement can be obtained by reducing Δf and increasing N.

Note that there can exist a requirement for a frequency-domain weighting function analogous to the previously discussed time-domain weighting function. The slightly oscillating results of Fig. 9.17(c) occur because the imaginary frequency function has been truncated. This truncation effect can be reduced by increasing N, the number of data points, or by using a weighting function. To use the Hanning weighting function in this example, apply it so that it is unity at $f = 0$ Hz, zero at $f = 2$ Hz, and returns to unity for the negative frequencies.

The key to using the inverse FFT for obtaining an approximation to continuous results is to specify the sampled frequency functions correctly. Figures 9.17(a) and (b) illustrate this correct method. One should observe the scale factor Δf, which is required to give a correct approximation to continuous inverse Fourier transform results.

It is not necessary to write a special FFT program to compute the inverse transform relation of Eq. (9.17). Rather, we use a direct transform FFT algorithm and employ the alternate inversion formula of Eq. (6.33). To

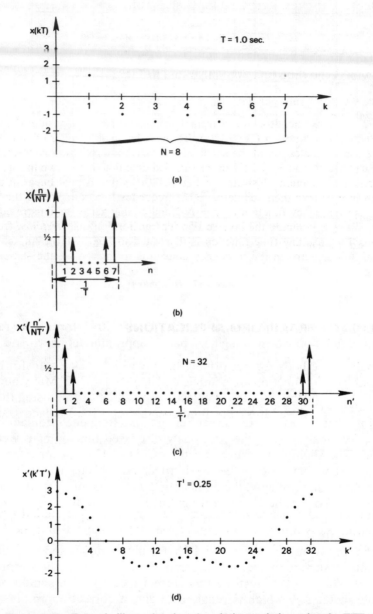

Figure 9.18　Example illustrating time-domain interpolation using the FFT.

use this relationship, we first conjugate the complex frequency function, that is, the imaginary sampled function illustrated in Fig. 9.17(b) is multiplied by -1. Because the resulting time function is real,

$$h(kT) = \Delta f \sum_{n=0}^{N-1} [R(n\Delta f) + j(-1)I(n\Delta f)]e^{-j2\pi nk/N} \qquad (9.18)$$

which yields the time function illustrated in Fig. 9.17(c).

Example 9.7   FFT Interpolation

The FFT is a convenient tool for interpolating a time function. Consider the sampled time function illustrated in Fig. 9.18(a). Because the small number of samples does not give a good indication of the shape of the curve, interpolation is desired.

First, we compute the FFT of the sampled time function shown in Fig. 9.18(a). The frequency function illustrated in Fig. 9.18(b) is purely real because the time function is even. We then add zeros to the frequency function. We do this by *separating* the frequency function at $n = N/2$ and adding zeros, as illustrated in Fig. 9.18(c). We next compute the inverse Fourier transform of this *stretched* frequency function. The resulting time function is illustrated in Fig. 9.18(d). As shown, the result of adding zeros in the frequency domain is to interpolate the sampled time function.

9.5 LAPLACE TRANSFORM APPLICATIONS

The analytic methods for performing the inverse Laplace transform of an irrational transfer function are complicated and incomplete. Many numerical methods are available, but the simplest to implement is that using the FFT. In this section, we address the fundamentals of FFT Laplace transform inversion.

The Laplace transform of a real function of time is given by

$$G(s) = \int_0^\infty g(t)e^{-st}\, dt \qquad (9.19)$$

The Fourier and Laplace transforms are very closely related. In general, the Fourier transform is a function of the real variable f and the Laplace transform is a function of the complex variable s. If we let $s = c + j2\pi f$, then Eq. (9.19) becomes

$$G(c + j2\pi f) = \int_0^\infty [g(t)e^{-ct}]e^{-j2\pi ft}\, dt \qquad (9.20)$$

If $g(t) = 0$ for $t < 0$, then the lower limit of Eq. (9.19) can be replaced by $-\infty$ and Eq. (9.20) becomes the Fourier transform relationship. Hence, the

Laplace transform can be cast in terms of the Fourier transform:

$$g(t)e^{-ct} \iff G(c + j2\pi f) \tag{9.21}$$

From Eq. (9.21), we can formulate a procedure for inversion of the Laplace transform. First, rewrite the transform function $G(s)$ by replacing the variable s by $c + j2\pi f$. This inversion yields the time function $g(t)e^{-ct}$. Multiplication of this function by e^{ct} yields the desired function $g(t)$. Recall from Laplace transform theory that the parameter c must be chosen larger than the real part of the poles for the transform function $G(s)$.

Although c can be chosen to be any value larger than the real part of the poles of $G(x)$, it should be noted that the effect of a large value of c is to attenuate $g(t)$ and thereby reduce the effects of time-domain aliasing. However, if c is chosen too large, then the product $[g(t)e^{-ct}]e^{ct}$ for large t results in errors due to the rounding errors that occur in the computation of $g(t)e^{-ct}$. Cooley [3] has determined a procedure for optimally choosing c to balance aliasing and rounding errors. Unless one is overly concerned with accuracy (1 part in 10^4), then an optimum choice for c is not warranted.

Example 9.8   Inverse Laplace Transform: $C = 0$

To illustrate the computation of the inverse Laplace transform, consider the transfer function $G(s) = 1/(s + 1)$. Replace parameter s by $c + j2\pi f$ to obtain $G(c + j2\pi f)$ $= 1/(j2\pi f + 1 + c)$. Because the pole is located at $s = -1$, then c can be chosen as any value greater than -1, say 0. Thus,

$$G(c + j2\pi f) = G(j2\pi f) = 1/(j2\pi f + 1)$$
$$= 1/[(2\pi f)^2 + 1] - j2\pi f/[(2\pi f)^2 + 1] \tag{9.22}$$

which is exactly the example considered in Sec. 9.4. The procedure for computing the inverse Fourier transform of Eq. (9.20) using the FFT is then identical to those described in the previous section and Figs. 9.17(a) to (c) apply. The time function of Fig. 9.17(c) is $g(t)e^{-ct}$, but because $c = 0$, then $e^{-ct} = 1$ and it is not necessary to multiply by e^{ct}. The desired time function is then given by Fig. 9.17(c).

Example 9.9   Inverse Laplace Transform: $C = 1$

Consider $G(s) = 1/s$. The pole is located at $s = 0$ and, hence, c must be chosen greater than zero. Let $c = 1$. Replacing s by $1 + j2\pi f$ yields $G(1 + j2\pi f) = 1/(j2\pi f + 1)$, which is the frequency function considered in Eq. (9.22). Therefore, the inverse Fourier transform yields the time function illustrated in Fig. 9.17(c). To obtain the desired function $g(t)$, we multiply Fig. 9.17(c) by $e^{ct} = e^t$; the result is shown in Fig. 9.19. The result approximates the theoretically correct step function. Recall from Sec. 9.4 that the error in this computation is due to frequency-domain truncation. Note that Fig. 9.19 is periodic, as required by the FFT. One must take this into account in interpreting inverse Laplace transform results.

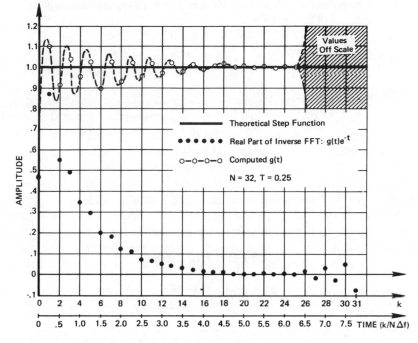

Figure 9.19 Example inverse Laplace transform computation using the FFT.

PROBLEMS

9.1 By means of the FFT, compute the amplitude spectrum $|H(f)|$ and phase $\theta(f)$ for the functions illustrated in Fig. 2.12.

9.2 Let $h(t) = e^{-t}$. Sample $h(t)$ with $T = 0.25$. Compute the FFT of $h(kT)$ for $N = 8, 16, 32,$ and 64. Compare these results and explain the differences. Repeat for $T = 0.1$ and $T = 1.0$ and discuss the results.

9.3 Let $h(t) = \cos(2\pi t)$. Sample $h(t)$ with $T = \pi/8$. Compute the FFT for $N = 16$. Repeat for $N = \pi/9$. Compare these results with those of Figs. 9.6 and 9.7.

9.4 Consider $h(t)$ illustrated in Fig. 6.7(a). Let $T_0 = 1.0$. Sample $h(t)$ with $T = 0.1$ and $N = 16$. Compute the FFT. Repeat for $T = 0.2$ and $N = 4$, and for $T = 0.01$ and $N = 128$. Compare and explain these results.

9.5 Let $h(t) = te^{-t}$, $t > 0$. Compute the FFT. Give reasons for your choice of T and N.

9.6 Let $h(k)$ be defined according to Problem 9.5. Let

$$g(k) = \cos(2\pi k/1024) \qquad k = 0, \ldots, 1023$$

Simultaneously, compute the discrete Fourier transform of $h(k)$ and $g(k)$ using the procedure described in Fig. 9.13.

9.7 Demonstrate the procedure illustrated in Fig. 9.15 on the function defined in Problem 9.5. Let $2N = 1024$.

9.8 Find the inverse FFT of the following functions:

(a) $\dfrac{[\sin(2\pi)f)]^2}{2\pi f}$

(b) $\dfrac{1}{(1 + j2\pi f)^2}$

9.9 By means of the FFT, compute the inverse Laplace transform of the following functions. The theoretical inverse is given as a means for checking your answer.

(a) $G(s) = \dfrac{e^{-s}}{s^2}$ $\qquad$ $g(t) = 0;\ 0 < t < 1$
$\qquad\qquad\qquad\qquad\qquad = t;\ t > 1$

(b) $G(s) = \dfrac{e^{-\pi s}}{s^2 + 1}$ $\qquad$ $g(t) = 0;\ 0 < t < \pi$
$\qquad\qquad\qquad\qquad\qquad = -\sin(t);\ t > \pi$

(c) $G(s) = 1/[(s + 4)^2 + 1]$ $\qquad$ $g(t) = e^{-t}\sin(t);\ t > 0$

REFERENCES

1. COOLEY, J. W., P. A. W. LEWIS, AND P. D. WELCH. "The Finite Fourier Transform." *IEEE Trans. Audio and Electroacoust.* (June 1969), Vol. AU-17, No. 2, pp. 77–85.

2. COOLEY, J. W., P. A. W. LEWIS, AND P. D. WELCH. "The Fast Fourier Transform and its Applications." *IEEE Trans. Education* (March 1969), Vol. 12, No. 1, pp. 27–34.

3. COOLEY, J. W., P. A. W. LEWIS, AND P. D. WELCH. "The Fast Fourier Transform Algorithm: Programming Considerations in the Calculation of Sine, Cosine and Laplace Transforms." *J. Sound Vib.* (July 1970), Vol. 12, No. 3, pp. 315–337.

4. DUBNER, H., AND J. ABATE. "Numerical Inversion of Laplace Transforms by Relating Them to the Finite Fourier Cosine Transform." *J. Assoc. Comput. Mach.* (January 1968), Vol. 15, No. 1, pp. 115–123.

5. PRASAD, K. P. "Fast Interpolation Algorithm Using FFT." *Elec. Lett.* (February 1986), Vol. 22, No. 4, pp. 185–187.

6. SINGHAL, K. "Interpolation Using the Fast Fourier Transform." *Proc. IEEE* (December 1972), Vol. 60, No. 12, p. 1558.

7. DOLPH, C. L. "A Current Distribution For Broadside Arrays Which Optimize the Relationship Between Beam Width and Sidelobe Level," *Proc. IRE* (June 1946), Vol. 34, pp. 335–348.

8. WARD, H. R. "Properties of Dolph-Chebyshev Weighting Functions." *IEEE Trans. Aerospace and Elec. Syst.* (September 1973), Vol. AE5-9, No. 5, pp. 785–786.

9. HARRIS, F. J. "On the Use of Windows for Harmonics Analysis with the Discrete Fourier Transform." *Proc. IEEE* (January 1978), Vol. 66, No. 1, pp. 51–83.

10. CHILDERS, D., AND A. DURLING. *Digital Filtering and Signal Processing*. St Paul, MN: West Publishing, 1975.

11. GECKINLI, N. C., AND D. YARRIS. "Some Novel Windows and a Concise Tutorial Comparison of Window Families." *IEEE Trans. Acoust. Speech Sig. Proc.* (December 1978), ASSP-26, pp. 501–507.

12. RAMIREZ, R. W. *The FFT Fundamentals and Concepts.* Englewood Cliffs, NJ: Prentice-Hall, 1985.

13. DIDERICH, R. "Calculating Chebyshev Shading Coefficients via the Discrete Fourier Transform." *IEEE Proc. Lett.* (October 1974), Vol. 62, No. 10, pp. 1395–1396.

14. NUTTAL, A. H. "Generation of Dolph-Chebyshev Weights via a Fast Fourier Transform." *IEEE Proc. Lett.* (October 1974), Vol. 62, No. 10, p. 1936.

10

FFT CONVOLUTION
AND CORRELATION

FFT applications such as matched filtering, digital signal processing, simulation, systems analysis, and time-interval measurements are based on an implementation of the discrete convolution or correlation integral. In general, a straightforward computation of the discrete integral relationships is not practical because of the excessive number of required multiplications. However, as discussed in Chapter 6, both integrals can be computed by means of the discrete Fourier transform. With the tremendous increase in computational speed that can be achieved using the FFT, it is more efficient to compute the convolution and correlation integrals by means of the discrete Fourier transform.

In this chapter, we develop the techniques for applying the FFT to high-speed convolution and correlation.

10.1 FFT CONVOLUTION OF FINITE-DURATION WAVEFORMS

The discrete convolution relationship is given by Eq. (7.1) as

$$y(k) = \sum_{i=0}^{N-1} x(i)h(k - i) \qquad (10.1)$$

where both $x(k)$ and $h(k)$ are periodic functions with period N. As discussed in Chapter 7, discrete convolution, if correctly performed, produces a replica of the continuous convolution, provided that both the functions $x(t)$ and $h(t)$

are of finite duration. We now extend that discussion to include efficient computation by means of the FFT.

Consider the finite-duration, or *aperiodic*, waveforms $x(t)$ and $h(t)$ illustrated in Fig. 10.1(a). Continuous convolution of these functions is also shown. By means of discrete convolution, it is desired to produce a replica of the continuous convolution. Recall from Chapter 7 that discrete convolution requires that we sample both $x(t)$ and $h(t)$ and form periodic functions with period N, as illustrated in Fig. 10.1(b). The resulting discrete convolution [Fig. 10.1(c)] is periodic; however, each period is a replica of the desired finite duration, or aperiodic waveform. Scaling constant T (sample

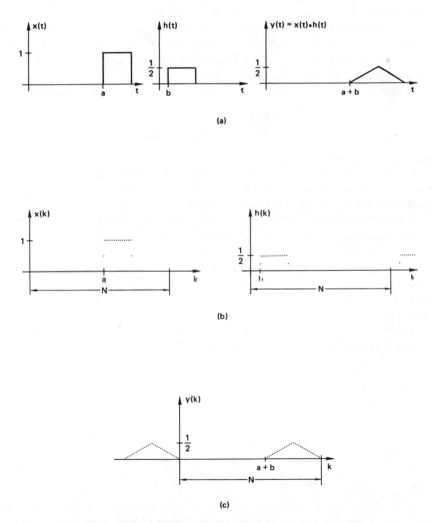

Figure 10.1   Example illustrating inefficient discrete convolution.

period) has been introduced to obtain results comparable with continuous convolution. Note that because both $x(t)$ and $h(t)$ are shifted from the origin, a large N is required to produce a period sufficiently large to eliminate the *overlap* or *end effect* described in Chapter 7. Computationally, the discrete convolution illustrated in Fig. 10.1(c) is very inefficient because of the large number of zeros produced in the interval $[0, a + b]$. To perform the discrete convolution more efficiently, we simply restructure the data.

Restructuring the Data

As illustrated in Fig. 10.2, we shift each sampled function [Fig. 10.1(b)] to the origin; from Eq. (7.6), we choose the period $N > P + Q - 1$ to eliminate *overlap* effects. Because we ultimately desire to use the FFT to perform the convolution, we also require that $N = 2^\gamma$, where γ is integer valued; we assume that a radix-2 algorithm is used. Our results are easily extended for the case of other algorithms.

Functions $x(k)$ and $h(k)$ are required to have a period N satisfying

$$N > P + Q - 1$$

$$N = 2^\gamma \qquad \gamma \text{ integer valued}$$

(10.2)

Discrete convolution for this choice of N is illustrated in Fig. 10.2(b); the results differ from that of Fig. 10.1(c) only in a shift of origin. But this shift is known a priori. From Fig. 10.1(a), the shift of the convolution $y(t)$ is simply the sum of the shifts of the functions being convolved. Consequently, no information is lost if we shift each function to the origin prior to convolution.

To compute the identical waveform of Fig. 10.2(b) by means of the

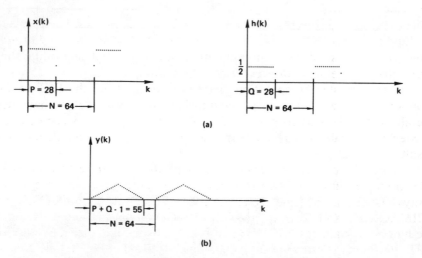

Figure 10.2   Discrete convolution of restructured data.

FFT, we first shift both $x(t)$ and $h(t)$ to the origin. Let the shifts of $x(t)$ and $h(t)$ be a and b, respectively. Both functions are then sampled. Next, N is chosen to satisfy Eq. (10.2). The resulting sampled periodic functions are defined by the relationships

$$
\begin{aligned}
x(k) &- x(kT + a) & k &= 0, 1, \ldots, P - 1 \\
x(k) &= 0 & k &= P, P + 1, \ldots, N - 1 \\
h(k) &= h(kT + b) & k &= 0, 1, \ldots, Q - 1 \\
h(k) &= 0 & k &= Q, Q + 1, \ldots, N - 1
\end{aligned}
\tag{10.3}
$$

The same notation is used to emphasize that our discussions should assume only sampled periodic functions shifted to the origin. We now compute the discrete convolution by means of the discrete convolution theorem of Eq. (6.50). The discrete Fourier transforms of $x(k)$ and $h(k)$ are computed:

$$
X(n) = \sum_{k=0}^{N-1} x(k)e^{-j2\pi nk/N}
\tag{10.4}
$$

$$
H(n) = \sum_{k=0}^{N-1} h(k)e^{-j2\pi nk/N}
\tag{10.5}
$$

Next, the product

$$
Y(n) = X(n)H(n)
\tag{10.6}
$$

is formed, and finally we compute the inverse discrete transform of $Y(n)$ and obtain the discrete convolution $y(k)$:

$$
y(k) = \frac{1}{N} \sum_{n=0}^{N-1} Y(n)e^{j2\pi nk/N}
\tag{10.7}
$$

Note that the single discrete convolution of Eq. (10.1) has now been replaced by Eqs. (10.4) to (10.7). This gives rise to the term *the long way around*. However, because of the computing efficiency of the FFT algorithm, these four equations define *a shortcut by the long way around*.

A step-by-step computation procedure for applying the FFT to convolution of discrete functions is given in Fig. 10.3. Note that we have used the alternate inversion formula of Eq. (6.33) in Step 7 and scaled by $1/N$. In Step 8, we scale by the sample interval T for comparison with continuous results.

A BASIC computer program following the procedure of Fig. 10.3 is shown in Fig. 10.4. The two real functions to be convolved are stored in arrays X1REAL(I%) and X2REAL(I%). These arrays, X1IMAG(I%), X2IMAG(I%), XREAL(I%), and XIMAG(I%), should be dimensioned by the number of samples N%. N% and NU% must be initialized. We use the FFT subroutine starting at line 10000, which is listed in Fig. 8.7. The reader is responsible for implementing Steps 2 to 4 to eliminate overlap effects.

1. Let $x(t)$ and $h(t)$ be finite-length functions shifted from the origin by a and b, respectively.
2. Shift $x(t)$ and $h(t)$ to the origin and sample

$$x(k) = x(kT + a) \qquad k = 0, 1, \ldots, P - 1$$

$$h(k) = h(kT + b) \qquad k = 0, 1, \ldots, Q - 1$$

3. Choose N to satisfy the relationships

$$N \geq P + Q - 1$$

$$N = 2^\gamma \qquad \gamma \text{ integer valued}$$

where P is the number of samples defining $x(t)$, and Q is the number of samples defining $h(t)$.

4. Augment with zeros the sampled functions of Step 2:

$$x(k) = 0 \qquad k = P, P + 1, \ldots, N - 1$$

$$h(k) = 0 \qquad k = Q, Q + 1, \ldots, N - 1$$

5. Compute the FFT of $x(k)$ and $h(k)$:

$$X(n) = \sum_{k=0}^{N-1} x(k)e^{-j2\pi nk/N}$$

$$H(n) = \sum_{k=0}^{N-1} h(k)e^{-j2\pi nk/N}$$

6. Compute the product

$$Y(n) = X(n)H(n)$$

7. Compute the inverse FFT using the forward FFT (note scaling by $1/N$):

$$y(k) = \sum_{n=0}^{N-1} \left(\frac{1}{N} Y^*(n) \right) e^{-j2\pi nk/N}$$

8. Scale the results by the sample interval T.

Figure 10.3 Computation procedure for FFT convolution of finite-length functions.

Convolution results are returned stored in XREAL(I%) and must be scaled by the sample interval T to obtain results equivalent to continuous convolution. XIMAG(I%) results should be approximately zero. Note that the factor $1/N$, shown in Step 7, is incorporated in the program.

```
13000 REM:    SUBROUTINE FOR CONVOLVING TWO REAL FUNCTIONS STORED
13002 REM:    IN ARRAYS X1REAL(I%) AND X2REAL(I%). N% AND NU% MUST BE
13004 REM:    INITIALIZED. DIMENSION X1REAL(I%),X1IMAG(I%),X2REAL(I%),
13006 REM:    X2IMAG(I%),XREAL(I%) AND XIMAG(I%). USER IS RESPONSIBLE FOR
13008 REM:    PREVENTING END EFFECTS. CONVOLUTION RESULTS ARE
13010 REM:    RETURNED IN ARRAY XREAL(I%). THIS PROGRAM CALL THE FFT
13012 REM:    SUBROUTINE STARTING AT LINE 10000 (FIG. 8-7).
13020 FOR I%=1 TO N%
13030       XREAL(I%)=X1REAL(I%)
13040       XIMAG(I%)=0
13050 NEXT I%
13060       GOSUB 10000
13070 FOR I%=1 TO N%
13080       X1REAL(I%)=XREAL(I%)
13090       X1IMAG(I%)=XIMAG(I%)
13100       XREAL(I%)=X2REAL(I%)
13110       XIMAG(I%)=0
13120 NEXT I%
13130       GOSUB 10000
13140 FOR I%=1 TO N%
13150       X2REAL(I%)=XREAL(I%)
13160       X2IMAG(I%)=XIMAG(I%)
13170       XREAL(I%)=(X1REAL(I%)*X2REAL(I%)-X1IMAG(I%)*X2IMAG(I%))/N%
13180       XIMAG(I)=-(X1REAL(I%)*X2IMAG(I%)+X1IMAG(I%)*X2REAL(I%))/N%
13190 NEXT I%
13200       GOSUB 10000
13210 RETURN
13220 END
```

Figure 10.4   BASIC subroutine for FFT convolution.

Example 10.1   FFT Convolution

The application of the FFT to convolution computation is illustrated in Fig. 10.5. The sampled function $x(kT)$, with $N = 32$, is shown in Fig. 10.5(a). Results of applying Eq. (10.4) using the FFT is also shown in Fig. 10.5(a). Note that FFT results are complex and we show a magnitude function. The sampled function $h(kT)$ and its FFT as computed from Eq. (10.5) are shown in Fig. 10.5(b). Because $P = 16$ and $Q = 16$, then $N = 32 > P + Q - 1$ and there is no overlap.

We next form the product frequency function of Eq. (10.6). This result is shown in Fig. 10.5(c) in magnitude form. This complex frequency function is input to the inverse FFT, Eq. (10.7), or is conjugated and input to the forward FFT (Step 7, Fig. (10.3)). All results have been scaled to approximate continuous results.

Computational Efficiency of FFT Convolution

Evaluation of the N samples of the convolution result $y(k)$ by means of Eq. (10.1) requires a computation time proportional to N^2, the number

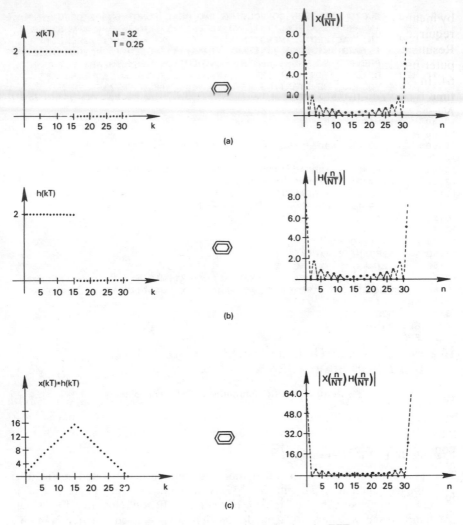

Figure 10.5 Example convolution using the FFT.

of multiplications. From Sec. 8.2, the computation time of the FFT is proportional to $N \log_2 N$; computation time of Eqs. (10.4) to (10.6) is then proportional to $3N \log_2 N$ and the computation time of Eq. (10.7) is proportional to N. It is generally faster to use the FFT and Eqs. (10.4) through (10.7) to compute the discrete convolution rather than computing Eq. (10.1) directly.

Exactly how much faster the FFT approach is than the conventional approach depends not only on the number of points but also on the details of the FFT and convolution programs being employed. To indicate the point at which FFT convolution is faster and the time savings that can be obtained by means of FFT convolution, we have observed as a function of N the time required to compute Eq. (10.1) by both the direct and FFT approaches. Results of this simulation are given in Table 10.1. As shown, with our computer programs, it is faster to employ the FFT for convolution if N exceeds 64. In Sec. 10.3, we describe a technique for reducing the FFT computing time by an additional factor of two; as a result, the breakeven point is for $N = 32$.

TABLE 10.1   Computing Times (Seconds)

N	Direct Method	FFT Method	Speed Factor
16	0.0008	0.003	0.27
32	0.003	0.007	0.43
64	0.012	0.015	0.8
128	0.047	0.033	1.4
256	0.19	0.073	2.6
512	0.76	0.16	4.7
1024	2.7	0.36	7.5
2048	11.0	0.78	14.1
4096	43.7	1.68	26.0

10.2 FFT CONVOLUTION OF INFINITE- AND FINITE-DURATION WAVEFORMS

We have discussed to this point only the class of functions for which both $x(t)$ and $h(t)$ are of finite duration. Further, we have assumed that $N = 2^\gamma$ was sufficiently small so that the number of samples did not exceed our computer memory. When either of these two assumptions is false, it is necessary to use the concept of *sectioning*.

Consider the waveforms $x(t)$, $h(t)$, and their convolution $y(t)$, as illustrated in Fig. 10.6. We assume that $x(t)$ is of infinite duration or that the number of samples representing $x(t)$ exceeds the memory of the computer. As a result, it is necessary to decompose $x(t)$ into sections and compute the discrete convolution as many smaller convolutions. Let NT be the time duration of each section of $x(t)$ to be considered; these sections are illustrated in Fig. 10.6(a). As shown in Fig. 10.7(a), we form the periodic sampled function $x(k)$, where a period is defined by the first section of $x(t)$; $h(t)$ is sampled and zeros are added to obtain the same period. Convolution $y(k)$ of these functions is also illustrated in Fig. 10.7(a). Note that we do not show the first $Q - 1$ points of the discrete convolution; these samples are incorrect because of the end effect. Recall from Sec. 7.3 for $h(k)$ defined by Q samples that the first $Q - 1$ samples of $y(k)$ have no relationship to the desired continuous convolution and should be discarded.

In Fig. 10.7(b), we illustrate the discrete convolution of the second

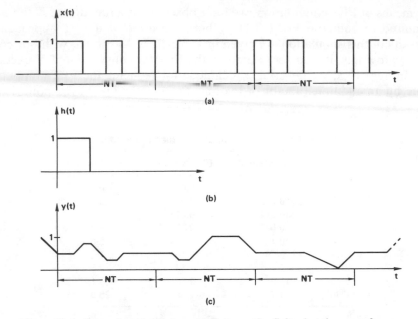

Figure 10.6 Example convolution of infinite- and a finite-duration waveforms.

section of duration NT illustrated in Fig. 10.6(a). As described in Sec. 10.1, we have shifted this section to the origin for purposes of efficient convolution. The section is then sampled and forced to be periodic; functions $h(k)$ and the resulting convolution $y(k)$ are also shown. Again, the first $Q - 1$ samples of the convolution function are deleted because of the end effect.

The final section of $x(t)$ is shifted to the origin and sampled, as illustrated in Fig. 10.7(c); discrete convolution results with the first $Q - 1$ samples deleted are also shown.

Each of the discrete convolution sections of Figs. 10.7(a) to (c) is reconstructed in Figs. 10.8(a) to (c), respectively. We have replaced the shift from the origin, which was removed for efficient convolution. Note that with the exception of the *holes* created by the addition of these sectioned results, Fig. 10.8(d) approximates closely the desired continuous convolution of Fig. 10.8(e). By simply overlapping the sections of $x(t)$ by a duration $(Q - 1)T$, we can eliminate these holes entirely.

Overlap-Save Sectioning

In Fig. 10.9(a), we show the identical waveform $x(t)$ of Fig. 10.6(a). However, note that the sections of $x(t)$ are now overlapped by $(Q - 1)T$, the duration of the function $h(t)$ minus T.

We shift each section of $x(t)$ to the origin, sample the section, and form

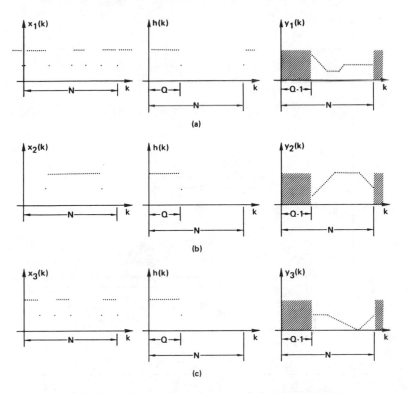

Figure 10.7   Discrete convolution of each section of Fig. 10.6(a).

a periodic function. Figures 10.9(b) to (e) illustrate the discrete convolution resulting from each section. Note that as a result of the overlap, additional sections are necessary. The first $Q - 1$ samples of each section are again eliminated because of the end effect.

As illustrated in Fig. 10.10, we add each section of the discrete convolution. The appropriate shift is added to each section. We do not have holes as before because the end effect occurs for a duration of the convolution that was computed by the previous section. Combination of each of the sections yields over the entire range the desired continuous convolution [Fig. 10.6(c)]. The only end effect that cannot be compensated is the first one, as illustrated. All illustrations have been scaled by the factor T for convenience of comparison with continuous results. It remains to specify mathematically the relationships that have been developed graphically.

Refer to Fig. 10.9(a). Note that we choose the first section to be of duration NT. To use the FFT, we require that

$$N = 2^\gamma \qquad \gamma \text{ integer valued} \tag{10.8}$$

and obviously, we require $N > Q$ (the optimum choice of N is discussed

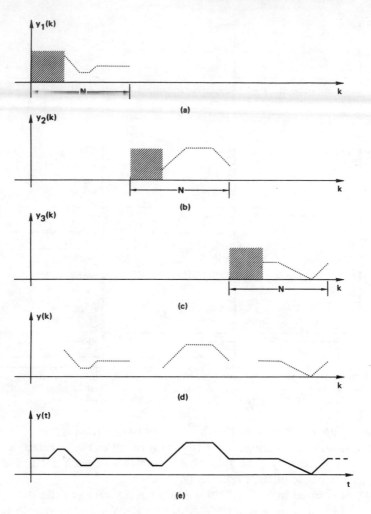

Figure 10.8 Reconstructed results of the discrete convolution of Fig. 10.7.

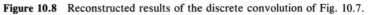

later). We form the sampled periodic function

$$x_1(k) = x(kT) \qquad k = 0, 1, \ldots, N - 1$$

and by means of the FFT compute

$$X_1(n) = \sum_{k=0}^{N-1} x_1(k)e^{-j2\pi nk/N} \tag{10.9}$$

Next, we take the Q sample values defining $h(t)$ and assign zero to the

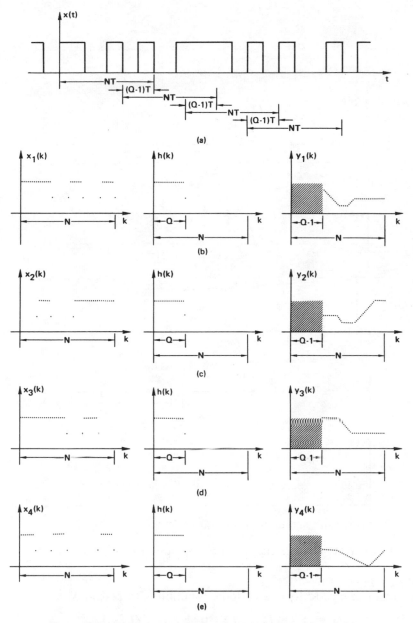

Figure 10.9   Discrete convolution of overlapped sections of data.

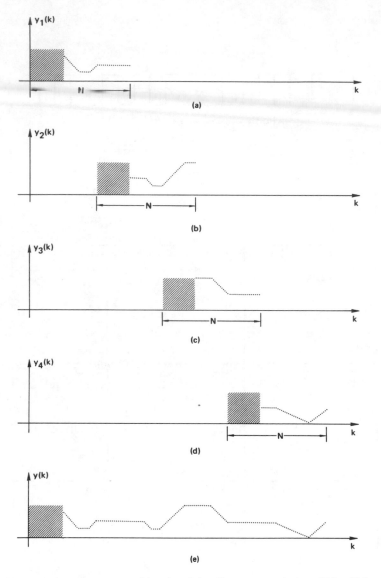

Figure 10.10   Reconstructed results of the discrete convolution of Fig. 10.9.

remaining samples to form a periodic function with period N:

$$h(k) = \begin{cases} h(kT) & k = 0, 1, \ldots, Q - 1 \\ 0 & k = Q, Q + 1, \ldots, N - 1 \end{cases} \tag{10.10}$$

If $h(t)$ is not shifted to the origin, as illustrated in Fig. 10.6(b), then $h(t)$ is first shifted to the origin and Eq. (10.10) is applied. Using the FFT, we compute

$$H(n) = \sum_{k=0}^{N-1} h(k)e^{-j2\pi nk/N} \qquad (10.11)$$

and then the product

$$Y_1(n) = X_1(n)H(n) \qquad (10.12)$$

Finally, we compute the inverse discrete transform of $Y_1(n)$:

$$y_1(k) = \frac{1}{N} \sum_{n=0}^{N-1} Y_1(n)e^{j2\pi nk/N} \qquad (10.13)$$

and because of the end effect, delete the first $Q - 1$ samples of $y(k)$: $y(0)$, $y(1), \ldots, y(Q - 2)$. The remaining samples are identical to those illustrated in Fig. 10.10(a) and should be saved for future combination.

The second section of $x(t)$, illustrated in Fig. 10.9(a), is shifted to the origin and sampled:

$$x_2(k) = x[(k + [N - Q + 1])T] \qquad k = 0, 1, \ldots, N - 1 \qquad (10.14)$$

Equations (10.11) through (10.13) are then repeated. From Eq. (10.11), the frequency function $H(n)$ has previously been determined and need not be recomputed. Multiplication, as indicated in Eq. (10.12), and subsequent inverse transformation, as indicated in Eq. (10.13), yield the waveform $y_2(k)$, illustrated in Fig. 10.10(b). Again, the first $Q - 1$ samples of $y_2(k)$ are deleted because of the end effect. All remaining sectioned convolution results are determined similarly.

The method of combining the sectioned results is as illustrated in Fig. 10.10(e):

$$
\begin{aligned}
y(k) \text{ undefined} \qquad & k = 0, 1, \ldots, Q - 2 \\
y(k) = y_1(k) \qquad & k = Q - 1, \\
& Q, \ldots, N - 1 \\
y(k + N) = y_2(k + Q - 1) \qquad & k = 0, 1, \ldots, N - Q \\
y(k + 2N) = y_3(k + Q - 1) \qquad & k = 0, 1, \ldots, N - Q \\
y(k + 3N) = y_4(k + Q - 1) \qquad & k = 0, 1, \ldots, N - Q
\end{aligned}
\qquad (10.15)
$$

The terms *select-saving* and *overlap-save* are given in the literature [2, 3] for this technique of sectioning.

Overlap-Add Sectioning

An alternate technique for sectioning has been termed the *overlap-add* [2, 3] method. Consider the illustrations of Fig. 10.11. We assume that the finite-length function $x(t)$ is of a duration such that the samples representing $x(t)$ exceed the memory of our computer. As a result, we show the sections of length $(N - Q)T$, as illustrated in Fig. 10.11(a). The desired convolution is illustrated in Fig. 10.11(c). To implement this technique, we first sample the first section of Fig. 10.11(a); these samples are illustrated in Fig. 10.12(a). The samples are augmented with zeros to form one period of a periodic function. In particular, we choose $N = 2^\gamma$, $N - Q$ samples of the function $x(t)$:

$$x_1(k) = x(kT) \qquad k = 0, 1, \ldots, N - Q \tag{10.16}$$

and $Q - 1$ zero values:

$$x_1(k) = 0 \qquad k = N - Q + 1, \ldots, N - 1 \tag{10.17}$$

Note that the addition of $Q - 1$ zeros ensures that there will be no end effect. Function $h(t)$ is sampled to form a function $h(k)$ with period N, as illustrated; the resulting convolution is also shown.

The second section of $x(t)$, illustrated in Fig. 10.11(a), is shifted to zero and then sampled:

$$x_2(k) = x[(k + N - Q + 1)T] \qquad k = 0, \ldots, N - Q$$

$$= 0 \qquad\qquad k = N - Q + 1, \ldots, N - 1$$

$$\tag{10.18}$$

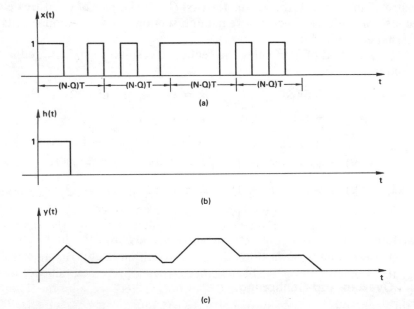

(a)

(b)

(c)

Figure 10.11 Example illustrating proper sectioning for overlap-add discrete convolution.

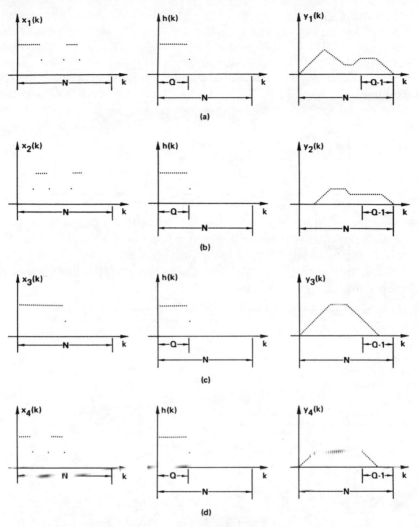

Figure 10.12   Discrete convolution of each section of Fig. 10.11.

As before, we add $Q - 1$ zeros to the sampled function $x(t)$. Convolution with $h(k)$ yields the function $y_2(k)$, as illustrated in Fig. 10.12(b). Convolution of each of the additional sequences is obtained similarly; the results are illustrated in Figs. 10.12(c) and (d).

We now combine these sectioned convolution results, as illustrated in Fig. 10.13. Each section has been shifted to the appropriate value. Note that the resulting addition yields a replica of the desired convolution. The trick of this technique is to add sufficient zeros to eliminate any end effects. These convolution results are then *overlapped* and *added* at identically those samples where zeros were added. This gives rise to the term *overlap-add*.

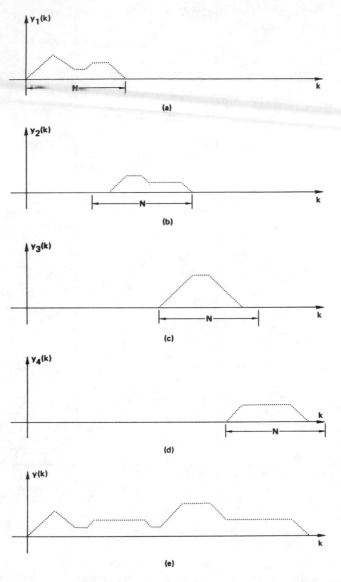

Figure 10.13 Reconstructed results of the discrete convolution of Fig. 10.12.

Computational Efficiency of FFT Sectioned Convolution

In both of the sectioning techniques described, the choice of N seems to be rather arbitrary as long as $N = 2^\gamma$. This choice determines the number of sections that must be computed, and thus the computing time. If an M-

1. Refer to Figs. 10.9 and 10.10 for a graphical interpretation of the algorithm.
2. Let Q be the number of samples representing $h(t)$.
3. Choose N according to Table 10.2.
4. Form the sampled periodic function $h(k)$:

$$h(k) = h(kT) \qquad k = 0, 1, \ldots, Q - 1$$
$$= 0 \qquad\qquad k = Q, Q + 1, \ldots, N - 1$$

5. Compute the FFT of $h(k)$:

$$H(n) = \sum_{k=0}^{N-1} h(k)e^{-j2\pi nk/N}$$

6. Form the sampled periodic function:

$$x_i(k) = x(kT) \qquad k = 0, 1, \ldots, N - 1$$

7. Compute the FFT of $x_i(k)$:

$$X_i(n) = \sum_{k=0}^{N-1} x_i(k)e^{-j2\pi nk/N}$$

8. Compute the product

$$Y_i(n) = X_i(n)H(n)$$

9. Compute the inverse FFT of $Y_i(n)$ (note scaling by $1/N$):

$$y_i(k) = \sum_{n=0}^{N-1} \left(\frac{1}{N} Y_i^*(n) \right) e^{-j2\pi nk/N}$$

10. Delete samples $y_i(0)$, $y_i(1)$, $\ldots$, $y_i(Q - 2)$, and save the remaining samples.
11. Repeat Steps 6 to 10 until all sections are computed.
12. Combine the sectioned results by the relationships

$$y(k) \text{ undefined} \qquad\qquad k = 0, 1, \ldots, Q - 2$$
$$y(k) = y_1(k) \qquad\qquad k = Q - 1, Q, \ldots, N - 1$$
$$y(k + N) = y_2(k + Q - 1) \qquad k = 0, 1, \ldots, N - Q$$
$$y(k + 2N) = y_3(k + Q - 1) \qquad k = 0, 1, \ldots, N - Q$$

$$\vdots$$

13. Scale the results by the sample interval T.

Figure 10.14 Computation procedure for FFT convolution: select-savings method.

1. Refer to Figs. 10.12 and 10.13 for a graphical interpretation of the algorithm.
2. Let Q be the number of samples representing $h(t)$.
3. Choose N according to Table 10.2.
4. Form the sampled periodic function $h(k)$:

$$h(k) = h(kT) \qquad k = 0, 1, \ldots, Q - 1$$
$$= 0 \qquad k = Q, Q + 1, \ldots, N - 1$$

5. Compute the FFT of $h(k)$:

$$H(n) = \sum_{k=0}^{N-1} h(k)e^{-j2\pi nk/N}$$

6. Form the sampled periodic function:

$$x_i(k) = x(kT) \qquad k = 0, 1, \ldots, N - Q$$
$$= 0 \qquad k = N - Q + 1, \ldots, N - 1$$

7. Compute the FFT of $x_i(k)$:

$$X_i(n) = \sum_{k=0}^{N-1} x_i(k)e^{-j2\pi nk/N}$$

8. Compute the product

$$Y_i(n) = X_i(n)H(n)$$

9. Compute the inverse FFT of $Y_i(n)$ (note scaling by $1/N$):

$$y_i(k) = \sum_{n=0}^{N-1} \left(\frac{1}{N} Y_i^*(n) \right) e^{-j2\pi nk/N}$$

10. Repeat Steps 6 to 9 until all sections are computed.
11. Combine the sectioned results by the relationships

$$y(k) = y_1(k)$$
$$k = 0, 1, \ldots, N - Q$$
$$y(k + N - Q + 1) = y_1(k + N - Q + 1) + y_2(k)$$
$$k = 0, 1, \ldots, N - Q$$
$$y[k + 2(N - Q + 1)] = y_2(k + N - Q + 1) + y_3(k)$$
$$k = 0, 1, \ldots, N - Q$$

.
.
.

12. Scale the results by sample interval T.

Figure 10.15 Computation procedure for FFT convolution: overlap-add method.

point convolution is desired, then approximately $M/(N - Q + 1)$ sections must be computed. If it is assumed that M is sufficiently greater than $N - Q + 1$, then the time required to compute $H(n)$ via the FFT can be ignored. Each section requires a forward and an inverse transform; hence, the FFT must be repeated approximately $2M/(N - Q + 1)$ times. We have experimentally determined the optimum value of N; the results of this investigation are given in Table 10.2. One can depart from the values of N shown without greatly increasing the computing time.

TABLE 10.2 Optimum
Value of N for FFT
Convolution

Q	$N = 2^\gamma$
≤ 10	32
11– 19	64
20– 29	128
30– 49	256
50– 99	512
100–199	1024
200–299	2048
300–599	4096
600–999	8192

We describe the step-by-step computational procedure for the *select saving* and the *overlap-add* methods of sectioning in Figs. 10.14 and 10.15, respectively. Both algorithms are approximately equivalent with respect to computational efficiency.

If the functions $x(t)$ and $h(t)$ are real, then we can use additional techniques to more efficiently compute the FFT. In the next section, we describe exactly how this is accomplished.

10.3 EFFICIENT FFT CONVOLUTION

We have to this point in the discussion considered that the functions being convolved are real functions of time. As a result, we have not utilized the full capabilities of the FFT. In particular, the FFT algorithm is designed for complex input functions; thus, if we only consider real functions, then the imaginary part of the algorithm is wasted. In this section, we describe how to divide a single real waveform into two parts, calling one part real, one part imaginary, and how to compute the convolution in one-half the normal FFT computing time. Alternately, our technique can be used to convolve two signals with an identical function simultaneously.

Consider the real periodic sampled functions $g(k)$ and $s(k)$. It is desired to convolve simultaneously these two functions with the real function $h(k)$ by means of the FFT. We accomplish this task by applying the technique of efficient discrete transforms, which was discussed in Sec. 9.3. First, we compute the discrete Fourier transform of $h(k)$, setting the imaginary part of $h(k)$ to zero:

$$H(n) = \sum_{k=0}^{N-1} h(k)e^{-j2\pi nk/N}$$
$$= H_r(n) + jH_i(n) \tag{10.19}$$

Next, we form the complex function

$$p(k) = g(k) + js(k) \qquad k = 0, 1, \ldots, N-1 \tag{10.20}$$

and compute

$$P(n) = \sum_{k=0}^{N-1} p(k)e^{-j2\pi nk/N}$$
$$= R(n) + jI(n) \tag{10.21}$$

Using the discrete convolution theorem, Eq. (6.50), we compute

$$y(k) = y_r(k) + jy_i(k) = p(k) * h(k) = \frac{1}{N}\sum_{k=0}^{N-1} P(n)H(n)e^{j2\pi nk/N}$$
$$\tag{10.22}$$

From Eqs. (9.6) and (9.7), the frequency function $P(n)$ can be expressed as

$$P(n) = R(n) + jI(n)$$
$$= [R_e(n) + R_0(n)] + j[I_e(n) + I_0(n)] \tag{10.23}$$
$$= G(n) + jS(n)$$

where

$$G(n) = R_e(n) + jI_0(n) \tag{10.24}$$
$$S(n) = I_e(n) - jR_0(n)$$

Product $P(n)H(n)$ is then given by

$$P(n)H(n) = G(n)H(n) + jS(n)H(n) \tag{10.25}$$

and thus the inversion formula yields

$$y(k) = y_r(k) + jy_i(k) = \frac{1}{N}\sum_{n=0}^{N-1} P(n)H(n)e^{j2\pi nk/N} \tag{10.26}$$

where

$$y_r(k) = \frac{1}{N} \sum_{k=0}^{N-1} G(n)H(n)e^{j2\pi nk/N}$$

$$jy_i(k) = \frac{1}{N} \sum_{k=0}^{N-1} jS(n)H(n)e^{j2\pi nk/N}$$

(10.27)

which is the desired result. That is, $y_r(k)$ is the convolution of $g(k)$ and $h(k)$, and $y_i(k)$ is the convolution of $s(k)$ and $h(k)$. If $g(k)$ and $s(k)$ represent successive sections, as described in the previous section, then we have reduced the computing time by a factor of two by using this technique. One still must combine the results as appropriate for the method of sectioning being employed.

Now consider the case where it is desired to perform the discrete convolution of $x(k)$ and $h(k)$ in one-half the time by using the imaginary part of the complex time function, as discussed in Sec. 9.3. Assume $x(k)$ is described by $2N$ points; define

$$g(k) = x(2k) \qquad k = 0, 1, \ldots, N - 1$$

$$s(k) = x(2k + 1) \qquad k = 0, 1, \ldots, N - 1$$

(10.28)

and let

$$p(k) = g(k) + js(k) \qquad k = 0, 1, \ldots, N - 1 \qquad (10.29)$$

But Eq. (10.29) is identical to Eq. (10.20); therefore,

$$z(k) = z_r(k) + jz_i(k) = \frac{1}{N} \sum_{n=0}^{N-1} P(n)H(n)e^{j2\pi nk/N}$$

where the desired convolution $y(k)$ is given by

$$y(2k) = z_r(k) \qquad k = 0, 1, \ldots, N - 1$$

$$y(2k + 1) = z_i(k) \qquad k = 0, 1, \ldots, N - 1$$

(10.30)

As in the previous method, we must still combine the results as appropriate for the method of sectioning being considered.

10.4 FFT CORRELATION OF FINITE-DURATION WAVEFORMS

Application of the FFT to discrete correlation is very similar to FFT convolution. As a result, our discussion on correlation will only point out the differences in the two techniques.

Consider the discrete correlation relationship

$$z(k) = \sum_{i=0}^{N-1} h(i)x(k + i) \tag{10.31}$$

where both $x(k)$ and $h(k)$ are periodic functions with period N. Figure 10.16(a) illustrates the same periodic functions $x(k)$ and $h(k)$ considered in Fig. 10.1(b). Correlation of these two functions according to Eq. (10.31) is shown in Fig. 10.16(b). Scaling factor T has been introduced for ease of comparison with continuous results. Note from Fig. 10.16(b) that the shift from the origin of the resultant correlation function is given by the difference between the leading edge of $x(k)$ and the trailing edge of $h(k)$. Recall that a positive shift for $h(k)$ is to the left.

In convolution, either function can be folded and shifted. The results are unchanged. This is not the case for correlation. Figure 10.16(c) illustrates the correlation function resulting from the shift $x(k)$ rather than $h(k)$. Note that the results give the same waveform but the waveform is shifted to the right by $a - d$ in Fig. 10.16(b) and shifted to the left by $a - d$ in Fig. 10.16(c). Care should be exercised in interpreting the correlation results of Fig. 10.16(c) to ensure that the correct shift from the origin has been determined. As in our convolution example, the correlation computation illustrated in Fig. 10.16(b) is inefficient because of the number of zeros included in the N points defining one period of the periodic correlation function. Restructuring of the data is again the solution we choose for efficient computation.

If we shift both functions to the origin, as shown in Fig. 10.17(a), then

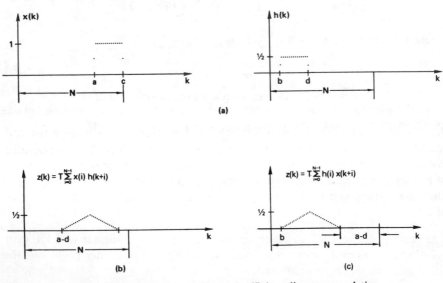

Figure 10.16 Example illustrating inefficient discrete correlation.

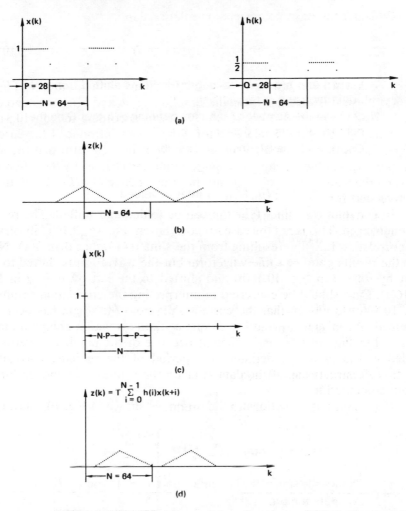

Figure 10.17   Discrete correlation of restructured data.

the resulting correlation is as illustrated in Fig. 10.17(b). Although the correlation waveform is correct, it must be *unraveled* before it is meaningful. We can remedy this situation by restructuring the waveform $x(k)$, as shown in Fig. 10.17(c). For this condition, the resulting correlation waveform is as illustrated in Fig. 10.17(d). This is the desired waveform with the exception of a known time shift.

To apply the FFT to the computation of Eq. (10.31), we choose the period N to satisfy the relationships

$$N \geq P + Q - 1$$

$$N = 2^\gamma \qquad \gamma \text{ integer valued}$$

(10.32)

1. Let $x(t)$ and $h(t)$ be finite-length functions shifted from the origin by a and b, respectively.

2. Let P be the number of samples defining $x(t)$ and Q be the number of samples defining $h(t)$.

3. Choose N to satisfy the relationships

$$N \geq P + Q - 1$$

$$N = 2^\gamma \quad \gamma \text{ integer valued}$$

4. Define $x(k)$ and $h(k)$ as follows:

$$x(k) = 0 \qquad\qquad k = 0, 1, \ldots, N - P$$

$$x(k) = x(kT + a) \qquad k = N - P + 1,$$
$$N - P + 2, \ldots, N - 1$$

$$h(k) = h(kT + b) \qquad k = 0, 1, \ldots, Q - 1$$

$$h(k) = 0 \qquad\qquad k = Q, Q + 1,$$
$$\ldots, N - 1$$

5. Compute the FFT $x(k)$ and $h(k)$:

$$X(n) = \sum_{k=0}^{N-1} x(k)e^{-j2\pi nk/N}$$

$$H(n) = \sum_{k=0}^{N} h(k)e^{-j2\pi nk/N}$$

6. Change the sign of the imaginary part of $H(n)$ to obtain $H^*(n)$.

7. Compute the product

$$Z(n) = X(n)H^*(n)$$

8. Compute the inverse FFT using the forward FFT: (note scaling by $1/N$):

$$z(k) = \sum_{n=0}^{N-1} \left(\frac{1}{N} Z^*(n)\right) e^{-j2\pi nk/N}$$

9. Scale the results by sample interval T.

Figure 10.18 Computation procedure for FFT correlation of finite-length functions.

We shift and sample $x(t)$ as follows:

$$x(k) = 0 \qquad\qquad k = 0, 1, \ldots , N - P$$

$$x(k) = x[kT + a] \qquad k = N - P + 1, N - P + 2, \ldots , N - 1$$

$$(10.33)$$

That is, we shift the P samples of $x(k)$ to the extreme right of the N samples defining a period. Function $h(t)$ is shifted and sampled according to the relations

$$h(k) = h(kT + b) \qquad k = 0, 1, \ldots , Q - 1 \tag{10.34}$$

$$h(k) = 0 \qquad\qquad k = Q, Q + 1, \ldots , N - 1$$

Based on the discrete correlation theorem, Eq. (7.13), we compute the following:

$$X(n) = \sum_{k=0}^{N-1} x(k)e^{-j2\pi nk/N} \tag{10.35}$$

$$H(n) = \sum_{k=0}^{N-1} h(k)e^{-j2\pi nk/N} \tag{10.36}$$

$$Z(n) = X(n)H^*(n) \tag{10.37}$$

$$z(k) = \frac{1}{N} \sum_{n=0}^{N-1} Z(n)e^{j2\pi nk/N} \tag{10.38}$$

The resulting $z(k)$ is identical to the illustration of Fig. 10.17(d).

Computing times of Eqs. (10.35) through (10.38) are essentially the same as the convolution Eqs. (10.4) through (10.7) and the results of the previous section are applicable. The computations leading to Eq. (10.38) are outlined in Fig. 10.18 for easy reference.

The key to carrying one's knowledge of FFT convolution techniques to FFT correlation is to remember that in correlation *there is no folding operation and that a shift to the left is positive*. This latter factor is probably responsible for the majority of errors in interpreting FFT correlation results.

PROBLEMS

10.1 Given the functions $h(t)$ and $x(t)$ illustrated in Fig. 10.19, determine the optimum choice of N to eliminate overlap effects during convolution and correlation. Assume a sample period of $T = 0.1$ and a base-2 FFT algorithm. Graphically show how to restructure the data for efficient convolution computation.

10.2 Consider the functions $x(t)$ and $h(t)$ of Fig. 10.19. Graphically show how to apply the overlap-save and overlap-add sectioning techniques for computing the convolution of $x(t)$ and $h(t)$.

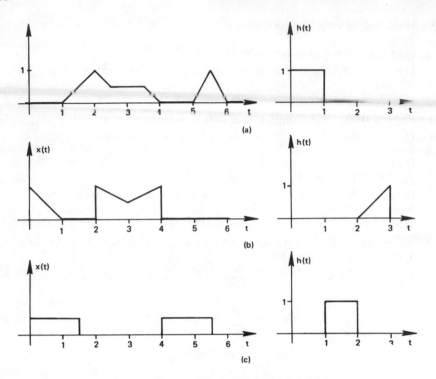

Figure 10.19 Functions for Probs. 10.1 to 10.4.

10.3 Repeat Problem 10.2 for the correlation of $x(t)$ and $h(t)$.

10.4 Repeat Problem 10.3 for the functions $x(t)$ and $h(t)$ illustrated in Fig. 10.6.

10.5 Use the FFT to duplicate the results shown in Figs. 10.7, 10.9, 10.10, 10.12, and 10.13. Apply the efficient FFT convolution techniques described in Sec. 10.4.

10.6 Develop graphically the overlap-save and overlap-add sectioning techniques for discrete correlation.

10.7 Repeat Problem 10.5 for the case of correlation of the two waveforms.

REFERENCES

1. COOLEY, J. W., P. A. W. LEWIS, and P. D. WELCH. "Application of the Fast Fourier Transform to Computation of Fourier Integrals, Fourier Series, and Convolution Integrals." *IEEE Trans. Audio and Electroacoust.* (June 1967), Vol. AU-15, No. 2, pp. 79–84.

2. HELMS, H. D. "Fast Fourier Transform Method of Computing Difference Equations and Simulating Filters." *IEEE Trans. Audio and Electroacoust.* (June 1967), Vol. AU-15, No. 2, pp. 85–90.

3. STOCKHAM, T. G. "High-Speed Convolution and Correlation." *AFIPS Proc.*

(1966 Spring Joint Computer Conf.), Vol. 28, pp. 229–233. Washington, DC: Spartan.

4. GENTLEMAN, W. M., and G. SANDE. "Fast Fourier Transforms for Fun and Profit." *AFIPS Proc.* (1966 Spring Joint Computer Conf.), Vol. 29, pp. 563–578, Washington, DC: Spartan.

5. COOLEY, J. W., P. A. W. LEWIS, and P. D. WELCH. "The Finite Fourier Transform." *IEEE Trans. Audio and Electroacoust.* (June 1969), Vol. AU-17, No. 2, pp. 77–85.

6. AGARWAL, R. C., and J. W. COOLEY. "New Algorithms for Digital Convolution." *IEEE Trans. Acoust. Speech Sig. Proc.* (October 1977), Vol. ASSP-25, No. 5, pp. 392–410.

7. BORGIOLI, R. C. "Fast Fourier Transform Correlation versus Direct Discrete Time Correlation." *Proc. IEEE* (September 1968), Vol. 56, No. 9, pp. 1602–1604.

8. NUSSBAUMER, H. J. *Fast Fourier Transforms and Convolution Algorithms*. New York: Springer-Verlag, 1982.

11

TWO-DIMENSIONAL
FFT ANALYSIS

In previous chapters, we applied the FFT to the analysis and processing of one-dimensional waveforms. Many of the techniques, procedures, and concepts discussed can be readily extended to two-dimensional FFT signal processing. A two-dimensional signal is a function $h(x,y)$ of two variables x and y. Two-dimensional FFTs are of considerable computational importance in the digital processing of two-dimensional waveforms such as images, geophysical arrays, gravity and magnetic data, and antenna analysis. Our approach is to develop the fundamental principles on which these applications of the FFT are based.

We will discuss in this chapter the concepts and techniques for applying the FFT to two-dimensional forward and inverse Fourier transforms. Applications of the FFT to two-dimensional convolution and correlation integrals are also addressed. As we will see, these applications are an extension of the previously developed one-dimensional case. However, because the two-dimensional Fourier transform is generally a less familiar analysis tool than the one-dimensional transform, we have chosen to develop our results from two-dimensional definitions rather than generalizing one-dimensional results.

11.1 TWO-DIMENSIONAL FOURIER TRANSFORMS

A two-dimensional function $h(x,y)$ has a two-dimensional Fourier transform $H(u,v)$ given by

$$H(u,v) \doteq \int_{-\infty}^{\infty} \int_{-\infty}^{\infty} h(x,y)e^{-j2\pi(ux+vy)} \, dx \, dy \tag{11.1}$$

Analogous to the one-dimensional case, Eq. (11.1) describes the analysis of the two-dimensional function $h(x,y)$ into components of the form $\cos[2\pi(ux + vy)]$ and $\sin[2\pi(ux + vy)]$.

An example of a two-dimensional waveform is illustrated in Fig. 11.1(a). The function shown represents a cosinusoidally corrugated two-dimensional surface. If a section is made through the corrugation in the y-h plane, the sectioned function oscillates with a frequency of v_0 cycles per unit of y (i.e., analogous to cycles per second). To distinguish between frequencies associated with functions of time and functions of length, the terms *temporal* and *spatial* are used, respectively. The two-dimensional Fourier transform of Fig. 11.1(a) as determined from Eq. (11.1) is the pair of impulse functions shown in Fig. 11.1(b).

The concept of a two-dimensional waveform is further illustrated in Fig. 11.2(a). For this example, a section is made through the corrugations in the x-h plane. The waveform oscillates with a spatial frequency of $v_0 \sin(\theta)$ cycles per unit of x. Similarily, a section made through the corrugations in the y-h plane oscillates with a frequency of $v_0 \cos(\theta)$ cycles per unit of y. Figure 11.2(a) is simply that of Fig. 11.1(a) rotated through an angle θ.

The two-dimensional Fourier transform of Fig. 11.2(a) is illustrated in Fig. 11.2(b). As shown, the spacial frequency at which the corrugation oscillates in a section perpendicular to the lines of zero phase is given by $[v_0 \cos^2(\theta) + v_0 \sin^2(\theta)]^{1/2} = v_0$. Note that the frequency impulse functions are located on an axis rotated through an angle θ with respect to the results of Fig. 11.1(b). A comparison of Figs. 11.1 and 11.2 shows that if a function $h(x,y)$ is rotated through an angle θ, then its two-dimensional Fourier transform is also rotated through an angle θ.

Example 11.1   Two-Dimensional Pulse Waveform

Find the two-dimensional Fourier transform of the function illustrated in Fig. 11.3(a). From Fig. 11.3,

$$h(x,y) = 1 \qquad -1 < x < 1; \, -1 < y < 1 \tag{11.2}$$
$$= 0 \qquad \text{otherwise}$$

Substitution of Eq. (11.2) into Eq. (11.1) yields

$$H(u,v) = \int_{-1}^{1} e^{-j2\pi vy} \, dy \int_{-1}^{1} e^{-j2\pi ux} \, dx$$

$$= \int_{-1}^{1} [\cos(2\pi vy) - j \sin(2\pi vy)] \, dy \tag{11.3}$$

$$\times \int_{-1}^{1} [\cos(2\pi ux) - j \sin(2\pi ux)] \, dx$$

Because the $\sin(\xi)$ term integrates to zero over the interval $(-1,1)$, then integration

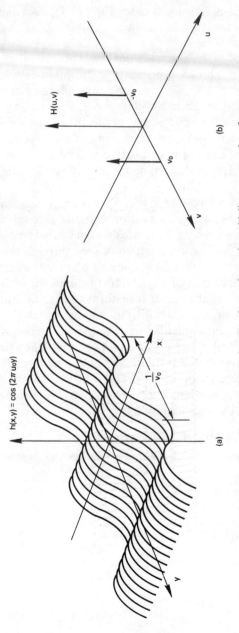

Figure 11.1 Two-dimensional Fourier transform of a cosinusoidally corrugated surface.

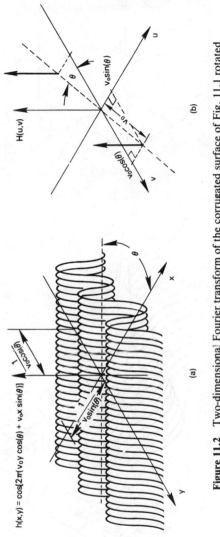

$h(x,y) = \cos\{2\pi[v_0 y \cos(\theta) + v_0 x \sin(\theta)]\}$

(a)

(b)

Figure 11.2 Two-dimensional Fourier transform of the corrugated surface of Fig. 11.1 rotated θ degrees.

235

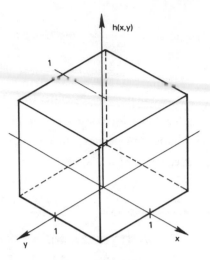

(a)

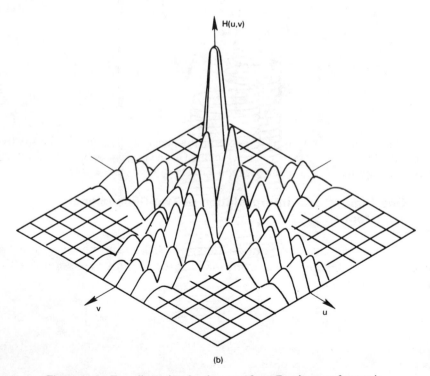

(b)

Figure 11.3 Two-dimensional pulse-waveform Fourier transform pair.

of Eq. (11.3) yields

$$H(u,v) = \frac{\sin(2\pi u)}{\pi u} \int_{-1}^{1} \cos(2\pi vy) \, dy$$

$$= \frac{\sin(2\pi u) \sin(2\pi v)}{\pi^2 uv} \tag{11.4}$$

Figure 11.3(b) illustrates this two-dimensional Fourier transform result.

Example 11.2   Two-Dimensional Fourier Transform: Separable Functions

Find the two-dimensional Fourier transform of the function

$$h(x,y) = \cos(2\pi u_0 x) \cos(2\pi v_0 y) \tag{11.5}$$

From Eq. (11.1),

$$H(u,v) = \int_{-\infty}^{\infty} \int_{-\infty}^{\infty} \cos(2\pi u_0 x) \cos(2\pi v_0 y) e^{-j2\pi(ux + vy)} \, dx \, dy$$

$$= \int_{-\infty}^{\infty} \cos(2\pi v_0 y) e^{-j2\pi vy} \, dy \int_{-\infty}^{\infty} \cos(2\pi u_0 x) e^{-j2\pi ux} \, dx$$

$$= \tfrac{1}{2} \int_{-\infty}^{\infty} (e^{j2\pi v_0 y} + e^{-j2\pi v_0 y}) e^{-j2\pi vy} \, dy$$

$$\times \left[\tfrac{1}{2} \int_{-\infty}^{\infty} (e^{j2\pi u_0 x} + e^{-j2\pi u_0 x}) e^{-j2\pi ux} \, dx \right] \tag{11.6}$$

$$= \tfrac{1}{2} \int_{-\infty}^{\infty} (e^{-j2\pi y(v - v_0)} + e^{-j2\pi y(v + v_0)}) \, dy$$

$$+ \tfrac{1}{2} \int_{-\infty}^{\infty} \{e^{-j2\pi x[u - u_0]} + e^{-j2\pi x[u + u_0]}\} \, dx$$

$$= \tfrac{1}{2} \delta(u,v - v_0) + \tfrac{1}{2} \delta(u,v + v_0)$$

$$+ \tfrac{1}{2} \delta(u - u_0,v) + \tfrac{1}{2} \delta(u + u_0,v)$$

The relationship of Eq. (11.5) is termed a *separable function* in that its two-dimensional Fourier transform can be computed as the product of two single-variable integrals. Note that the function given in Eq. (11.2) is also separable.

One-Dimensional Interpretation of Two-Dimensional Transforms

The two-dimensional Fourier transform $H(u,v)$ can be viewed as two successive one-dimensional transforms. To develop this viewpoint, we first rewrite Eq. (11.1) as

$$H(u,v) = \int_{-\infty}^{\infty} e^{-j2\pi vy} \left[\int_{-\infty}^{\infty} h(x,y) e^{-j2\pi ux} \, dx \right] dy \tag{11.7}$$

Note that the term in brackets is simply the one-dimensional Fourier transform of $h(x,y)$ with respect to x, that is,

$$Z(u,y) = \int_{-\infty}^{\infty} h(x,y) e^{-j2\pi ux} \, dx \tag{11.8}$$

Equation (11.7) can then be rewritten as

$$H(u,v) = \int_{-\infty}^{\infty} Z(u,y)e^{-j2\pi vy} \, dy \tag{11.9}$$

where we have substituted $Z(u,y)$ for the term in brackets in Eq. (11.7). An examination of Eq. (11.9) reveals that $H(u,v)$ is the one dimensional transform of $Z(u,y)$ with respect to y. Hence, the two-dimensional Fourier transform $H(u,v)$ can be interpreted as the two successive one-dimensional transforms given by Eqs. (11.8) and (11.9).

Analytical evaluation of a two-dimensional integral can be implemented by simply determining the two successive one-dimensional integrals of Eqs. (11.8) and (11.9). These single-dimension integrals are evaluated exactly by the procedures followed in the one-dimensional Fourier transform case. Hence, the two-dimensional Fourier transform can be evaluated using the methods and fundamentals developed in previous discussions. As we will see, this interpretation is of considerable importance in applying the FFT to the computation of two-dimensional Fourier transforms.

Example 11.3   Two-Dimensional Fourier Transforms: Successive One-Dimensional Transforms

In Fig. 11.2, we illustrated the two-dimensional Fourier transform of the function

$$h(x,y) = \cos\{2\pi[v_0 y \cos(\theta) + v_0 x \sin(\theta)]\} \tag{11.10}$$

To compute the transform analytically, we substitute Eq. (11.10) into Eq. (11.1):

$$H(u,v) = \int_{-\infty}^{\infty} \int_{-\infty}^{\infty} \cos\{2\pi[v_0 y \cos(\theta)$$
$$+ v_0 x \sin(\theta)]\}e^{-j2\pi(ux+vy)} \, dx \, dy \tag{11.11}$$

We next apply the principles developed in Eqs. (11.8) and (11.9) by rewriting Eq. (11.12) as two successive single-dimension Fourier transforms:

$$H(u,v) = \int_{-\infty}^{\infty} e^{-j2\pi vy} \left(\int_{-\infty}^{\infty} \cos\{2\pi[v_0 y \cos(\theta) \right.$$
$$\left. + v_0 x \sin(\theta)]\}e^{-j2\pi ux} \, dx \right) dy$$

$$= \frac{1}{2}\int_{-\infty}^{\infty} e^{-j2\pi vy} \left[\int_{-\infty}^{\infty} (e^{j2\pi[v_0 y \cos(\theta) + v_0 x \sin(\theta)]})e^{-j2\pi ux} \, dx \right.$$
$$+ \int_{-\infty}^{\infty} (e^{-j2\pi[v_0 y \cos(\theta) + v_0 x \sin(\theta)]})e^{-j2\pi ux} \, dx \Bigg] dy$$

$$= \frac{1}{2}\int_{-\infty}^{\infty} e^{-j2\pi vy} \left[\int_{-\infty}^{\infty} (e^{j2\pi[v_0 y \cos(\theta) - x(u-v_0)\sin(\theta)]}) \, dx \right.$$
$$+ \int_{-\infty}^{\infty} (e^{-j2\pi[v_0 y \cos(\theta) + x(u+v_0)\sin(\theta)]}) \, dx \Bigg] dy$$

$$= \frac{1}{2} \int_{-\infty}^{\infty} e^{-j2\pi vy} \left(e^{j2\pi v_0 y \cos(\theta)} \int_{-\infty}^{\infty} e^{-j2\pi x (u - v_0)\sin(\theta)} \, dx \right.$$

$$\left. + e^{-j2\pi v_0 y \cos(\theta)} \int_{-\infty}^{\infty} e^{-j2\pi x (u + v_0) \sin(\theta)} \, dx \right) dy \qquad (11.12)$$

$$= \frac{1}{2} \int_{-\infty}^{\infty} e^{-j2\pi vy} \{ e^{j2\pi v_0 y \cos(\theta)} \, \delta[u - v_0 \sin(\theta), y]$$

$$+ e^{-j2\pi v_0 y \cos(\theta)} \, \delta[u + v_0 \sin(\theta), y] \} \, dy$$

$$= \frac{1}{2} \int_{-\infty}^{\infty} \delta[u - v_0 \sin(\theta), y] e^{-j2\pi y[v - v_0 \cos(\theta)]} \, dy$$

$$+ \frac{1}{2} \int_{-\infty}^{\infty} \delta[u + v_0 \sin(\theta), y] e^{-j2\pi y[v - v_0 \cos(\theta)]} \, dy$$

$$= \frac{1}{2} \delta[u - v_0 \sin(\theta), v - v_0 \cos(\theta)]$$

$$+ \frac{1}{2} \delta[u + v_0 \sin(\theta), v + v_0 \cos(\theta)]$$

The two-dimensional frequency function of Eq. (11.12) is illustrated in Fig. 11.2(b).

Inverse Fourier Transform

The two-dimensional inverse Fourier transform is given by

$$h(x,y) = \int_{-\infty}^{\infty} \int_{-\infty}^{\infty} H(u,v) e^{j2\pi(ux + vy)} \, du \, dv \qquad (11.13)$$

Analogous to the one-dimensional inverse Fourier transform, Eq. (11.13) implies that corrugations of appropriate frequencies, orientations, phases, and amplitudes can be summed to produce the original two-dimensional waveform. However, it is recognized that the two-dimensional inverse Fourier transform is much more difficult to visualize than the one-dimensional transform.

Example 11.4   Two-Dimensional Inverse Fourier Transform

Find the inverse two-dimensional Fourier transform of the frequency function

$$H(u,v) = \Omega \qquad -a \le u \le a, \; -b \le v \le b \qquad (11.14)$$
$$= 0 \qquad \text{otherwise}$$

From Eq. (11.13),

$$h(x,y) = \int_{-\infty}^{\infty} \int_{-\infty}^{\infty} \Omega e^{j2\pi(ux + vy)} \, du \, dv$$

$$= \Omega \int_{-b}^{b} e^{j2\pi vy} \, dv \left[\int_{-a}^{a} e^{j2\pi ux} \, du \right] \qquad (11.15)$$

$$= \Omega \int_{-b}^{b} \cos(2\pi vy) \, dv \left[\int_{-a}^{a} \cos(2\pi ux) \, du \right]$$

$$= \Omega \left[\frac{\sin(2\pi by)}{\pi y} \right] \left[\frac{\sin(2\pi ax)}{\pi x} \right]$$

Example 11.4 demonstrates the property that if the frequency function $H(u,v)$ can be decomposed into a product of a function of the variable u and a function of the variable v, then the function $h(x,y)$ can be decomposed into a product of a function of the variable x and a function of the variable y.

Summary

We normally find two-dimensional Fourier transform relationships much more difficult to picture than one-dimensional relationships. This generally follows from the emphasis of one's formal training and experience in the analysis and synthesis of single-dimension functions. As with any new area of study, a thorough and fundamental understanding comes only with considerable exposure and practical experience. The previously developed basic principles should form the foundation for such an investigative endeavor. A comprehensive treatment of two-dimensional Fourier transform properties is given in Ref. [1].

11.2 TWO-DIMENSIONAL FFTs

Recall from the development of Eqs. (11.8) and (11.9) and Ex. 11.3 that the two-dimensional Fourier transform can be written as two successive single-dimension Fourier transforms. This interpretation of the two-dimensional transform is also readily seen in the two-dimensional discrete Fourier transform. We assume that the two-dimensional function $h(x,y)$ has been sampled in the x dimension with sample interval T_x and sampled in the y dimension with sample interval T_y. The resulting sampled function is $h(pT_x,qT_y)$, where $p = 0, 1, \ldots, N - 1$ and $q = 0, 1, \ldots, M - 1$.

Analytical Development

Analogous to the one-dimensional case, the two-dimensional discrete Fourier transform is defined as

$$H(n/NT_x,m/MT_y) = \sum_{q=0}^{M-1} \left[\sum_{p=0}^{N-1} h(pT_x,qT_y)e^{-j2\pi np/N} \right] e^{-j2\pi mq/M}$$

$$p = 0, 1, \ldots, N - 1 \qquad n = 0, 1, \ldots, N - 1$$

$$q = 0, 1, \ldots, M - 1 \qquad m = 0, 1, \ldots, M - 1$$

$$(11.16)$$

Note that the term in brackets is simply a one-dimensional discrete Fourier transform along the data array defined by parameter p. To evaluate the term in brackets, we compute M one-dimensional transforms: one for each q, where $q = 0, 1, \ldots, M - 1$, along the data array defined by p. If we call each of these FFT results $Z(n/NT_x, qT_y)$, then Eq. (11.16) can be rewritten as

$$H(n/NT_r, m/MT_y) = \sum_{q=0}^{M-1} Z(n/NT_x, qT_y)e^{-j2\pi mq/M} \qquad (11.17)$$

Equation (11.17) is evaluated by N one-dimensional discrete Fourier transforms, each along the data array defined by parameter q. As shown analytically, the two-dimensional discrete Fourier transform can be implemented straightforwardly by computing one-dimensional discrete Fourier transforms: first on the function $h(pT_x, qT_y)$, where $p = 0, 1, \ldots, N - 1$ for each q; and then a second one-dimensional transform on the function $Z(n/NT_x, qT_y)$, where $q = 0, 1, \ldots, M - 1$ for each $n = 0, 1, \ldots, N - 1$. Equations (11.16) and (11.17) must be multiplied by the scale factor $T_x T_y$ to obtain equivalence between the continuous and discrete transforms.

Graphical Development

To further illustrate the one-dimensional computation of a two-dimensional Fourier transform, consider Fig. 11.4(a). As illustrated, we interpret the sampled data of the two-dimensional waveform as a data matrix with $M = 8$ rows, where $q = 0, 1, \ldots, M - 1$, and $N = 8$ columns, where $p = 0, 1, \ldots, N - 1$. Because the terms in brackets in Eq. (11.16) sum on the parameter p, then this summation corresponds to computing the one-dimensional discrete Fourier transform for each row of data, that is, a transform is computed for each $q = 0, 1, \ldots, M - 1$.

The discrete Fourier transform or FFT of row 0 is a frequency function defined by all zero values because the values of the sampled function represented by row 0 is a zero-valued function. In Fig. 11.4(b), row 0, we plot this FFT result. The sampled function defined by row 1 is also zero and, correspondingly, the FFT computed for this row is a zero-valued function, as shown in Fig. 11.4(b), row 1.

Now consider the sample values of row 2. These samples define a pulse or rectangular waveform that has a $| [\sin(f)]/f |$ transform. The magnitude of the FFT of row 2 is the frequency function illustrated in Fig. 11.4(b), row 2. Note that we display the FFT results in the standard one-dimensional format, that is, the first $N/2$ values represent positive frequency results and the remaining values represent negative frequency results. The sample values of rows 3 to 5 in Fig. 11.4(a) also define a pulse waveform and hence, the magnitude of the FFT of each of these rows is the $| [\sin(f)]/f |$ function shown in Fig. 11.4(b), rows 3 to 5. Rows 6 and 7 of Fig. 11.4(a) are zero-valued; rows 6 and 7 of Fig. 11.4(b) are then zero-valued.

To this point, we have computed the FFT of the sampled matrix of Fig. 11.4(a) for each row. The complex data matrix represented by the mag-

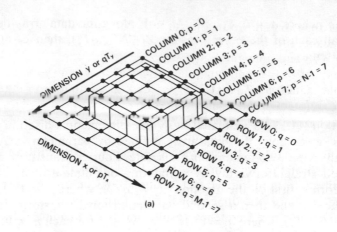

(a)

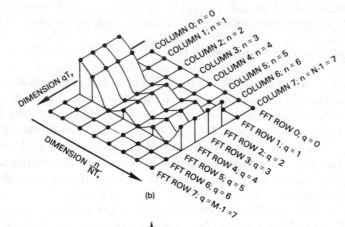

(b)

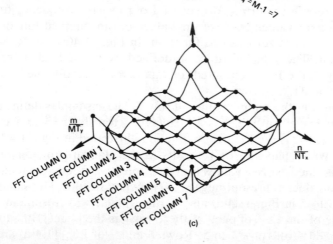

(c)

Figure 11.4 Graphical development of the two-dimensional FFT as a sequence of one-dimensional transforms.

nitude function shown in Fig. 11.4(b) corresponds to computing the terms in brackets in Eq. (11.16) for each value of q. That is, we have effectively set $q = 0$ and computed an FFT over the $p = 0, 1, \ldots, N - 1$ sample values; set $q = 1$ and computed the FFT over the $p = 0, 1, \ldots N - 1$ sample values; etc. Next, we proceed to compute the outer sum of Eq. (11.16). Note that this summation is on q, the row data values of Fig. 11.4(b) for each $n = 0, 1, \ldots, N - 1$. Hence, we compute the FFT of the complex sample values of each column of the matrix.

In Fig. 11.4(b), the sample values of column 0 define a pulse waveform. The FFT of this waveform is then the $|\,[\sin(f)]/f\,|$ function illustrated in magnitude form in Fig. 11.4(c), column 0. As before, we display the results in the standard one-dimensional FFT format, where the first $M/2$ values represent positive frequency results and the remaining values are negative frequency results.

Observe that the sampled functions defined by each nonzero column of Fig. 11.4(b) are of the same form, a pulse waveform with differing amplitude. Our input to the FFT in each case is the complex result determined in Fig. 11.4(b). Hence, the FFT of each nonzero column is a $|\,[\sin(f)]/f\,|$ function and the results for each column differ only in amplitude. Figure 11.4(c) illustrates the magnitude of the FFT for each column of Fig. 11.4(b).

As in the one-dimensional FFT, we must consider both the sampled data matrix that forms the input to the FFT and the two-dimensional FFT results to be one period of a two-dimensional periodic sequence with period (N,M). For this reason, we must interpret the illustrations of Figs. 11.4(a) to (c) as one period of a waveform that is periodic in both the row and column indices. This periodicity constraint is examined further in a later section.

Also analogous to the one-dimensional case, spacial frequency resolution in two-dimensional FFT results is given by

$$\Delta u = 1/(NT_x)$$
$$\Delta v = 1/(MT_y)$$

(11.18)

Computations Required for Two-Dimensional FFTs

Figure 11.4 readily illustrates the concept of computing the two-dimensional discrete Fourier transform by determining successive single-dimensional transforms. We first compute the FFT of each row of data, that is, M transforms of N samples each. We organize these results, as shown in Fig. 11.4(b), and then compute the FFT of each column of data, that is, N transforms of M samples each. Therefore, a data matrix of size $N \times M$ requires $N + M$ FFTs to be computed. From Chapter 8, the total number of computations is $NM \log_2 NM$.

Reorganizing Two-Dimensional FFTs
for Conventional Viewing

Recall in the one-dimensional FFT case that it was necessary to *rearrange* the FFT results if we wished to display them in a format for conventional viewing (Sec. 9.1). A similar situation is encountered for two-dimensional FFTs. Figure 11.4(c) must be *rearranged* or *reorganized* if the results are to be viewed conventionally. We repeat Fig. 11.4(c) in Fig. 11.5(a) and illustrate the required reorganization of Fig. 11.5(a) in Fig. 11.5(b). The same reorganization procedure is shown in Figs. 11.5(c) and (d), but from a data-matrix perspective to further clarify the required data restructuring.

Note that if we examine the data matrix in terms of quadrants, then the restructuring procedure is simply one of a right circular shift through two quadrants. An examination of Figs. 11.5(c) and (d) illustrates this point. The FFT output data in quadrant I ends up in quadrant III after restructuring. Quadrant III is a right circular shift through two quadrants from quadrant I. We repeat the Nyquist spacial frequency sample values, $H(n/NT_x,4)$ and $H(4,m/MT_y)$, in each quadrant. For the real array, quadrant III is a positive reflection of quadrant I, and quadrant IV is a positive reflection of II. For the imaginary array, quadrant III is a negative reflection of I and quadrant IV is a negative reflection of II.

Example 11.5   Two-Dimensional FFT Computation

To further demonstrate two-dimensional FFT computation, consider the cosinusoidally corrugated two-dimensional surface illustrated in Fig. 11.6(a). We first sample the surface with sample intervals T_x and T_y, resulting in 4 rows and 16 columns of data. Note that the row samples define exactly a multiple period of the cosine waveform surface.

We next compute the one-dimensional FFT of each row of sampled data. From Fig. 11.6(a), row 0 is a cosine waveform and hence the FFT of row 0 is the impulse functions shown in row 0 (columns 2 and 14) of Fig. 11.6(a). Recall that the impulse function in column 14 is a negative frequency result because columns 9 to 15 represent negative frequencies. Because each row of sampled data defines the same cosine waveform, the FFT results for each row are identical, as shown in Fig. 11.6(b). This data matrix is the input to the second series of one-dimensional FFTs. We next compute the FFT of each column of data in Fig. 11.6(b). Each column of data is a zero-valued function except for columns 2 and 14. Columns 2 and 14 are both constant-value functions whose FFTs are impulse functions at zero spacial frequency $(m/(MT_y),$ where $m = 0)$. These results are shown in Fig. 11.6(c).

Figure 11.6(c) illustrates the results of the two-dimensional FFT obtained by implementing successive single-dimension FFTs. However, it is necessary to *rearrange* these results according to the restructuring procedure illustrated in Fig. 11.5. Restructured two-dimensional FFT results are shown in Fig. 11.6(d). Note the restructuring procedure is one of a right circular shift through two quadrants. Nyquist spacial frequency data values are repeated in each quadrant.

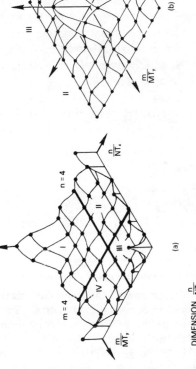

Figure 11.5 Graphical presentation of two-dimensional FFT reorganization required for conventional viewing.

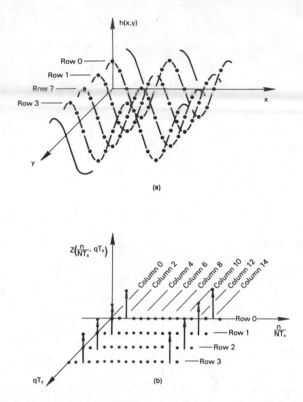

Figure 11.6 Two-dimensional FFT computation of a cosinusoidally corrugated surface: (a) sampled waveform, (b) FFT of row data, (c) FFT of columns of part (b), and (d) reorganization of part (c).

Example 11.6   Alternative Two-Dimensional FFT Computational Procedure

In Ex. 11.5, we first compute the FFT of each row of the input data matrix and then the FFT of each column of the intermediate computational matrix. Equivalent results could be obtained if we first compute the FFT of each column and subsequently compute the FFT of each row of the intermediate results. This simply states the fact that Eq. (11.16) can be rewritten in a form such that one first sums on q, which corresponds to computing the one-dimensional discrete Fourier transform for each column of data.

Figure 11.7 illustrates the alternate two-dimensional FFT computational procedure. In comparison to the previous example, the two-dimensional waveform shown is a corrugated sinusoidal surface. Note that the row samples define exactly a multiple period of the sinusoidal surface. Because each column of sampled data is a constant-value function, then the FFT of each column of data is an impulse at zero spacial frequency. The amplitude of each impulse is equal to the amplitude of the constant sample value for each column. FFT results of the column data are illustrated in Fig. 11.7(b).

We next compute the FFT of each row of the data matrix of Fig. 11.7(b). Row

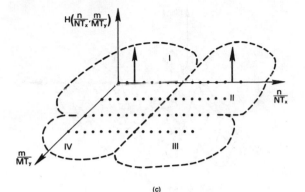

(c)

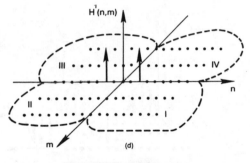

Figure 11.6 (*continued*)

0 data defines a sinusoidal waveform whose FFT is the set of impulse functions shown in Fig. 11.7(c). The impulse function of column 14 is a negative frequency value, as illustrated in the restructured two-dimensional results of Fig. 11.7(d).

Two-Dimensional Periodicity Constraints

In Chapter 6, we saw that the discrete Fourier transform is defined only for periodic sampled functions. A similar result can be shown for the two-dimensional discrete Fourier transform. A two-dimensional sampled function is periodic in the row index p with period N and in the column index q with period M if

$$h(pT_x,qT_y) = h[(p + cN)T_x,(q + dM)T_y] \qquad (11.19)$$

where c and d are arbitrary positive or negative integers.

Figure 11.8 illustrates the implications of Eq. (11.19). The 4×4 matrix within the dotted square is assumed to be the sampled surface. Note that we sampled the two-dimensional function only for positive values of x and y. However, due to the periodicity constraint of Eq. (11.19), we must in-

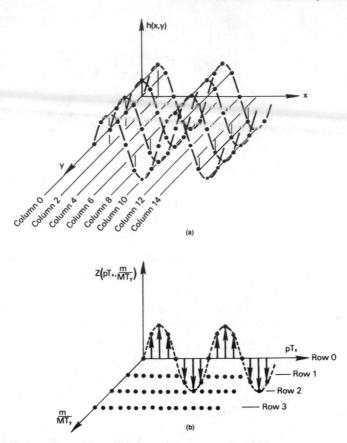

Figure 11.7 Example of an alternate two-dimensional FFT computation procedure: (a) sampled waveform, (b) FFT of column data, (c) FFT of rows of part (b), and (d) reorganization of part (c).

terpret this matrix of data as one period of a periodic two-dimensional function. Figure 11.8 shows four periods of this sampled waveform. It is important to observe the relationship of (row,column) indices for each period with the (row,column) indices within the dotted square.

Example 11.7 Periodicity in Two Dimensions

To further demonstrate the two-dimensional periodicity constraint, consider the surface shown in Fig. 11.9(a). Let us assume that it is desired to sample this surface and compute the two-dimensional FFT. Considerable care must be paid to the periodicity constraint. A review of Fig. 11.8 points out that if we wish to sample the surface shown in Fig. 11.9(a), then it is necessary to actually sample the function illustrated in Fig. 11.9(b). Although this two-dimensional surface looks significantly

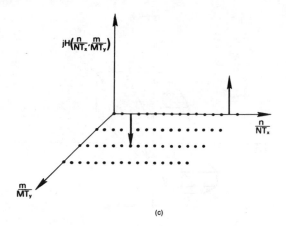

(c)

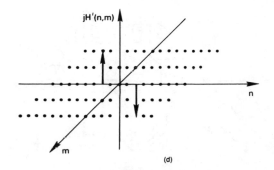

(d)

Figure 11.7   (*continued*)

(0,0)	(0,1)	(0,2)	(0,3)	(0,0)	(0,1)	(0,2)	(0,3)
(1,0)	(1,1)	(1,2)	(1,3)	(1,0)	(1,1)	(1,2)	(1,3)
(2,0)	(2,1)	(2,2)	(2,3)	(2,0)	(2,1)	(2,2)	(2,3)
(3,0)	(3,1)	(3,2)	(3,3)	(3,0)	(3,1)	(3,2)	(3,3)

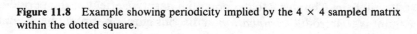

Figure 11.8   Example showing periodicity implied by the 4 × 4 sampled matrix within the dotted square.

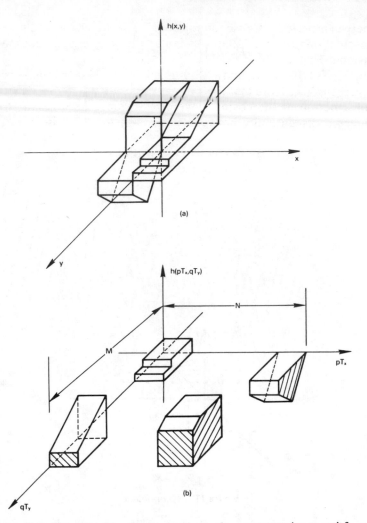

Figure 11.9 Example of two-dimensional waveform restructuring to satisfy periodicity constraints.

different from Fig. 11.9(a), the sampling periodicity constraint yields the original surface.

Care must be exercised when setting up the two-dimensional matrix of sampled values to ensure that the function being FFTed is that which is actually desired. Recall that this warning is the same as that given in Chapter 9.

Two-Dimensional Data Windows

Recall from Chapter 9 that a potential negative effect of the periodicity constraint is to introduce *truncation*, a discontinuity or abrupt change in the data between periods. Consequentially, FFT frequency-domain results exhibit oscillations or side lobes. A similar result is encountered with two-dimensional FFTs.

In Fig. 11.10(a), we show a sampled corrugated two-dimensional co-sinusoidal surface. Note that the sample values do not define exactly a multiple of a period of the cosine waveform surface. As a result, truncation introduces a discontinuity between periods in the pT_x dimension and the

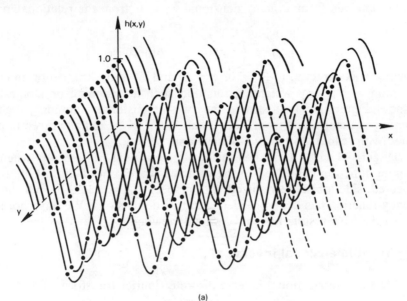

(a)

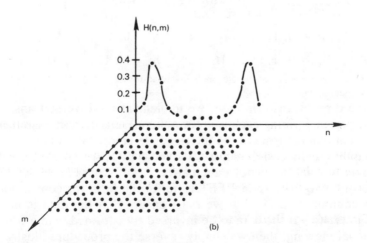

(b)

Figure 11.10 Graphical illustration of two-dimensional FFT results before a weighting function is applied.

two-dimensional FFT results shown in Fig. 11.10(b) exhibit large side lobes. As in the one-dimensional case, we use a weighting function to reduce the side lobes.

The weighting function techniques discussed in Chapter 9 are directly extendable to two dimensions. As expected, a rectangular two-dimensional weighting function exhibits large side lobes (see Fig. 11.3). Haung [9] has shown that a good two-dimensional symmetrical weighting function $w'(\cdot)$ can be obtained from a one-dimensional window from the relationship:

$$w'(x,y) = w[(x^2 + y^2)^{1/2}] \qquad |x^2 + y^2| < T'/2 \qquad (11.20)$$
$$= 0 \qquad\qquad\qquad \text{otherwise}$$

where $w(\cdot)$ is centered at $[x = 0, y = 0]$ and T' is the truncation interval. Function $w(\cdot)$ is any weighting function such as Hanning or Dolph-Chebyshev. Figure 11.11 illustrates the two-dimensional Hanning weighting function as determined from Eq. (11.20) for $M = N$ and as correctly positioned with respect to the two-dimensional period.

Figure 11.11(b) illustrates the two-dimensional FFT results when the Hanning function is applied to the sampled waveform of Fig. 11.10(a). As expected, side lobes are reduced with respect to Fig. 11.10(b), but the frequency function has been broadened in two dimensions. We have not rearranged the FFT results for conventional viewing.

Two-Dimensional Inverse FFTs

The two-dimensional inverse discrete Fourier transform is defined as

$$h(pT_x,qT_y) = \frac{1}{M} \sum_{m=0}^{M-1} \frac{1}{N} \left[\sum_{n=0}^{N-1} H(n/NT_x,m/\mathrm{MT}_y)e^{j2\pi np/N} \right] e^{j2\pi mq/M}$$

$$p = 0, 1, \ldots, N - 1 \qquad n = 0, 1, \ldots, N - 1$$
$$q = 0, 1, \ldots, M - 1 \qquad m = 0, 1, \ldots, M - 1$$

$$(11.21)$$

As in the direct transform case, we implement the inverse transform by first inverse transforming each row (or column) and then inverse transforming each column (or row) of the intermediate computation matrix.

When applying the two-dimensional inverse FFT, we must be careful in setting up the spacial frequency-data matrix. Recall from Fig. 11.4(c) that the output of the two-dimensional FFT is not in the form for conventional viewing. To compute Eq. (11.21), we must ensure that the data is in the format of Fig. 11.4(c). If the data to be inversed transformed is in a format for conventional viewing, then we simply reverse the procedure illustrated in Fig. 11.5 before inputting the data to Eq. (11.21). This process is analogous to that described in Chapter 9 for computing the inverse FFT of a one-

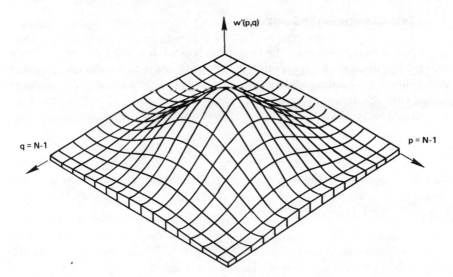

(a)

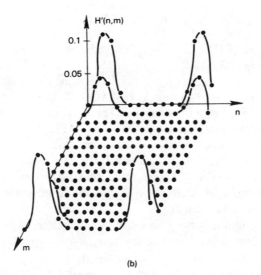

(b)

Figure 11.11 (a) Figure 11.10(a) multiplied by the two-dimensional Hanning weighting function, and (b) two-dimensional FFT of part (a).

dimensional frequency function. Equation (11.21) must be multiplied by the factor $\Delta u \Delta v$ to obtain results equivalent to the continuous inverse two-dimensional Fourier transform.

Two-Dimensional Sampling

In the development of the two-dimensional discrete Fourier transform, we did not discuss the requirements imposed by the two-dimensional sampling theorem. To develop the theorem, we simply extend the one-dimensional concept. If we are given a two-dimensional surface $h(x,y)$ whose Fourier transform is $H(u,v)$, where

$$H(u,v) = 0 \qquad u \geq u_c, \, v \geq v_c \qquad (11.22)$$

then we simply state the Nyquist criteria independently in the two dimensions. That is, we must sample $h(x,y)$ to obtain $h(pT_x,qT_y)$, such that T_x and T_y satisfy the relationships

$$T_x \leq 1/2u_c$$
$$T_y \leq 1/2v_c \qquad (11.23)$$

Summary

A BASIC computer program for the two-dimensional FFT of an array $W(n,m)$ is given in Fig. 11.12. The real component of the two-dimensional signal is placed in W1REAL(II%,JJ%) and the imaginary component is placed in W1IMAG(II%,JJ%). Parameters N%, NU%, M%, and MU% must be initialized. Real and imaginary spacial frequency-domain results are returned in W1REAL(II%,JJ%) and W1IMAG(II%,JJ%), respectively. Note that the program first computes the FFT of each column of the data array using the one-dimensional FFT program listed in Fig. 8.7. The program then computes the FFT of each row of the intermediate array, again using the one-dimensional FFT. XREAL(I%) and XIMAG(I%) should be dimensioned by the larger of N% or M%. Users must rearrange the output results for conventional viewing and scale by T_xT_y to obtain equivalence to the continuous two-dimensional transform.

Two-dimensional inverse FFTs can be computed with the program by first conjugating the spacial frequency function by exactly the procedure used in one-dimensional inverse transforms.

In this section, we have developed the basic fundamentals for applying the FFT to the computation of the two-dimensional Fourier and inverse Fourier transforms. The discussion by no means has been exhaustive but the fundamental principles necessary for further investigations have been established. If one carefully extends the concepts of one-dimensional FFT analysis to each of the FFTs computed in the two-dimensional transform, then few difficulties should be encountered. As developed, the mathematics of two-dimensional transform analysis is sufficiently close to that of the one-dimensional case to justify such a conclusion.

```
9000 REM:   TWO-DIMENSIONAL FFT SUBROUTINE- THE MAIN
9002 REM:   PROGRAM SHOULD DIMENSION THE DATA ARRAYS
9004 REM:   W1REAL(II%,JJ%) AND W1IMAG(II%,JJ%).
9006 REM:   N%,NU%,M%, AND MU% MUST BE INITIALIZED.
9008 REM:   XREAL(I%) AND XIMAG(J%) SHOULD BE DIMENSIONED
9010 REM:   THE LARGER OF N% OR M%. THIS PROGRAM
9012 REM:   CALLS THE FFT ROUTINE (FIG. 8-7) BEGINNING
9014 REM:   AT LINE 10000.
9026          NN%=N% : NNU%=NU% : MM%=M% : MMU%=MU%
9028 REM: COMPUTE THE FFT OF EACH COLUMN.
9030 FOR JJ%=1 TO MM%
9040         FOR II%=1 TO NN%
9050             XREAL(II%)=W1REAL(II%,JJ%)
9060             XIMAG(II%)=W1IMAG(II%,JJ%)
9070         NEXT II%
9080     GOSUB 10000
9090         FOR KK%=1 TO NN%
9100             W1REAL(KK%,JJ%)=XREAL(KK%)
9110             W1IMAG(KK%,JJ%)=XIMAG(KK%)
9120         NEXT KK%
9130 NEXT JJ%
9140 REM: COMPUTE THE FFT OF EACH ROW.
9150 FOR JJ%=1 TO NN%
9160         FOR II%=1 TO MM%
9170             XREAL(II%)=W1REAL(JJ%,II%)
9180             XIMAG(II%)=W1IMAG(JJ%,II%)
9190         NEXT II%
9200     N%=MM% : NU%=MMU%
9210     GOSUB 10000
9220         FOR KK%=1 TO MM%
9230             W1REAL(JJ%,KK%)=XREAL(KK%)
9240             W1IMAG(JJ%,KK%)=XIMAG(KK%)
9250         NEXT KK%
9260 NEXT JJ%
9270 N%=NN% : NU%=NNU%
9280 RETURN
9290 END
```

Figure 11.12   Subroutine in BASIC for computing the two-dimensional FFT.

11.3 TWO-DIMENSIONAL CONVOLUTION AND CORRELATION

The convolution integral for two-dimensional functions is defined as

$$g(x,y) = \int_{-\infty}^{\infty} \int_{-\infty}^{\infty} r(\tau_x,\tau_y)h(x-\tau_x,y-\tau_y)d\tau_x d\tau_y = r(x,y) ** h(x,y)$$

$$(11.24)$$

Interpretation of Eq. (11.24) is analogous to the one-dimensional case, as we demonstrate in the following graphical analysis.

Graphical Evaluation

Let $r(\tau_x, \tau_y)$ and $h(\tau_x, \tau_y)$ be given by the graphs shown in Figs. 11.13(a) and (b), respectively. For ease of presentation, we do not show the amplitudes of the two-dimensional functions. To evaluate Eq. (11.24) for the point $g(x', y')$, function $h(x' - \tau_x, y' - \tau_y)$ is required. From Fig. 11.13(c), note that $h(-\tau_x, -\tau_y)$ is obtained by rotating $h(\tau_x, \tau_y)$ 180° about the origin. Function $h(x' - \tau_x, y' - \tau_y)$ is obtained by displacing $h(-\tau_x, -\tau_y)$ by the amount x' along the τ_x axis and by the amount y' along the τ_y axis, as illustrated in Fig. 11.13(d). The volume (double integral) of the product $r(\tau_x, \tau_y) \times h(x' - \tau_x, y' - \tau_y)$ yields the convolution result $g(x', y')$.

Example 11.7   Two-Dimensional Convolution: Line Functions

An example of two-dimensional convolution is illustrated in Fig. 11.14. The two functions to be convolved are shown in Figs. 11.14(a) and (b). Because $h(x,y)$ is

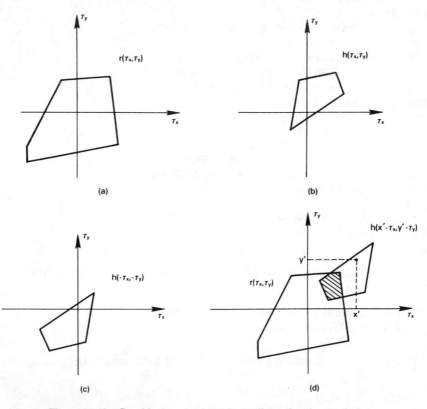

Figure 11.13   Graphical evaluation of two-dimensional convolution.

symmetrical about the τ_y axis, then the 180° rotation required by Eq. (11.23) yields the same function, $h(\tau_x,\tau_y) = h(-\tau_x,-\tau_y)$, as shown in Fig. 11.14(a). Figure 11.14(c) shows the displacement function $h(x'-\tau_x,y'-\tau_y)$; multiplication with $r(\tau_x,\tau_y)$ and integration yields the spacial point of the resulting two-dimensional convolution illustrated in Fig. 11.14(d).

Example 11.8   Two-Dimensional Convolution: Impulse Functions

To demonstrate two-dimensional convolution involving impulse functions, consider Fig. 11.15. In Fig. 11.15(a), we show a two-dimensional sequence of impulse functions. The impulse functions are separated by T_x in the x dimension and by T_y in the y dimension. The function to be convolved with these impulses is shown in Fig. 11.15(b).

For ease of presentation, we have deleted the amplitude information of Figs. 11.15(a) and (b) and show only the length, width, and displacement information in Figs. 11.15(c) and (d). To convolve Figs. 11.15(c) and (d), recall from Chapter 4 that convolution with an impulse function requires centering the function to be convolved on the impulse. The resulting convolution is shown in Figs. 11.15(e) and (f). Results of this example can be extended to graphically demonstrate the two-dimensional sampling theorem (Prob. 11.15).

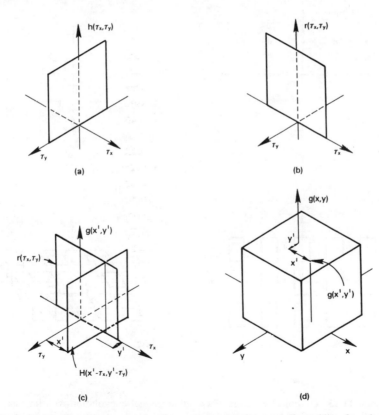

Figure 11.14   Graphical example of two-dimensional convolution of line functions.

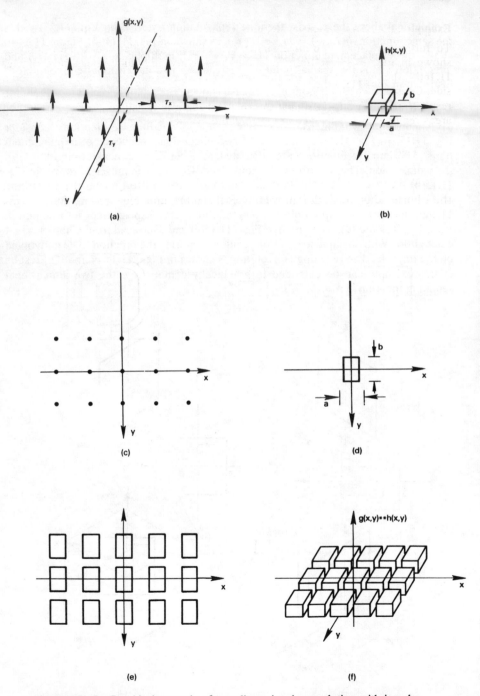

Figure 11.15 Graphical example of two-dimensional convolution with impulse functions.

Example 11.9   Two-Dimensional Convolution: Amplitude Determination

To further illustrate two-dimensional convolution, consider the square surfaces shown in Fig. 11.16(a). Both functions have unity amplitude. As shown in Fig. 11.16(b), the dimensions of the nonzero convolution result is a square $(a + b)$ on a side. Constant-amplitude contours are also shown. The amplitude of the convolution result along the x axis is shown in Fig. 11.16(c); appropriate shifts and integration areas required to evaluate example amplitude values on the x axis are also shown.

Two-Dimensional Convolution Theorem

We can compute the two-dimensional convolution by multiplication in the Fourier transform domain. Hence, if $r(x,y)$ and $h(x,y)$ are two waveforms

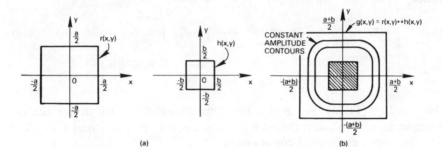

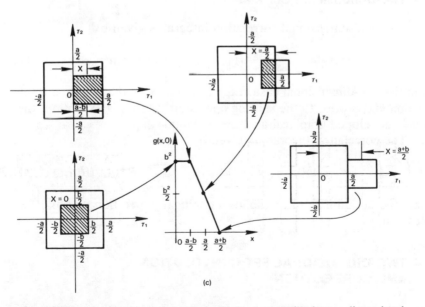

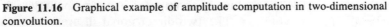

Figure 11.16 Graphical example of amplitude computation in two-dimensional convolution.

with transforms $R(u,v)$ and $H(u,v)$, respectively, then

$$r(x,y) ** h(x,y) \quad \Longleftrightarrow \quad R(u,v)H(u,v) \qquad (11.25)$$

To show the relationship of Eq. (11.25), we apply the one-dimensional transform interpretation of the two-dimensional Fourier transform, which was developed in Eqs. (11.7) to (11.9).

If $Z_r(u,y)$ and $Z_h(u,y)$ are the one-dimensional transforms of $r(x,y)$ and $h(x,y)$ with respect to x, respectively, we can write from the one-dimensional convolution theorem:

$$\int_{-\infty}^{\infty} r(\tau_x,\tau_y)h(x-\tau_x,y-\tau_y)d\tau_x \quad \Longleftrightarrow \quad Z_r(u,\tau_y)Z_h(u,y-\tau_y) \qquad (11.26)$$

Hence,

$$\int_{-\infty}^{\infty}\int_{-\infty}^{\infty} r(\tau_x,\tau_y)h(x-\tau_x,y-\tau_y)d\tau_x d\tau_y \quad \Longleftrightarrow \quad \int_{-\infty}^{\infty} Z_r(u,\tau_y)Z_h(u,y-\tau_y)d\tau_y$$

$$\Longleftrightarrow \quad R(u,v)H(u,v)$$

$$(11.27)$$

which is Eq. (11.25). As in the one-dimensional case, we utilize the two-dimensional convolution theorem as a means for applying the FFT to compute the two-dimensional convolution.

Two-Dimensional Correlation

The two-dimensional correlation integral is given by

$$p(x,y) = \int_{-\infty}^{\infty}\int_{-\infty}^{\infty} r(\tau_x,\tau_y)h(x+\tau_x,y+\tau_y)d\tau_x \, d\tau_y \qquad (11.28)$$

As in the one-dimensional case, Eq. (11.28) implies that we do not fold the function $h(x,y)$ prior to the shifting operation. With this exception, the previously developed graphical analysis techniques apply.

The correlation theorem is given by

$$\int_{-\infty}^{\infty}\int_{-\infty}^{\infty} r(\tau_x,\tau_y)h(x+\tau_x,y+\tau_y)d\tau_x \, d\tau_y \quad \Longleftrightarrow \quad R^*(u,v)H(u,v) \qquad (11.29)$$

where the notation $R^*(u,v)$ implies a conjugate operation on both the variables u and v.

11.4 TWO-DIMENSIONAL FFT CONVOLUTION AND CORRELATION

A two-dimensional FFT convolution is computed in exact analogy with one-dimensional FFT convolution. The two-dimensional discrete convolution relationship is given by

$$g(pT_x,qT_y) = \sum_{j=0}^{M-1} \sum_{i=0}^{N-1} r(iT_x,jT_y)h[(p - i)T_x,(q - j)T_y]$$

$$p = 0, 1, \ldots, N - 1 \qquad i = 0, 1, \ldots, N - 1$$

$$q = 0, 1, \ldots, M - 1 \qquad j = 0, 1, \ldots, M - 1$$

$$(11.30)$$

where $g(pT_x,qT_y)$, $r(pT_x,qT_y)$, and $h(pT_x,qT_y)$ are periodic functions with periods NT_x and MT_y in the x and y coordinates, respectively.

We use the frequency convolution theorem to compute the discrete convolution by means of the FFT. First, we compute the two-dimensional FFT of the functions $r(iT_x,jT_y)$ and $h(iT_x,jT_y)$:

$$R(n/NT_x,m/MT_y) = \sum_{q=0}^{M-1} \left[\sum_{p=0}^{N-1} r(pT_x,qT_y)e^{-j2\pi np/N} \right] e^{-j2\pi mq/M}$$

$$(11.31)$$

$$H(n/NT_x,m/MT_y) = \sum_{q=0}^{M-1} \left[\sum_{p=0}^{N-1} h(pT_x,qT_y)e^{-j2\pi np/N} \right] e^{-j2\pi mq/M}$$

$$n = 0, 1, \ldots, N - 1$$

$$p = 0, 1, \ldots, N - 1$$

$$m = 0, 1, \ldots, M - 1$$

$$(11.32)$$

$$q = 0, 1, \ldots, M - 1$$

Next, we compute the product $R(n/NT_x,m/MT_y)H(n/NT_x,m/MT_y)$ (taking into account that both functions are in general complex), and then we compute the inverse FFT of this product:

$$g(pT_x,qT_y) = (1/NM) \sum_{m=0}^{M-1} \left[\sum_{n=0}^{N-1} R(n/NT_x,m/MT_y) \right.$$

$$(11.33)$$

$$\left. \times\ H(n/NT_x,m/MT_y)e^{j2\pi np/N} \right] e^{j2\pi mq/M}$$

The convolution result appears in the real array of the FFT output.

Because Eq. (11.30) represents a periodic convolution of the periodic sampled function $r(pT_x,qT_y)$ and $h(pT_x,qT_y)$, then we must ensure that these sampled functions contain sufficient zero values to prevent the end effect or circular convolution. In general, if a nonzero-value function of dimension (N_1,M_1) is convolved with a second nonzero-value function of dimension

(N_2, M_2), the resulting function is of dimension $(N_1 + N_2 - 1, M_1 + M_2 - 1)$. Hence, we should append sufficient zeros to each of the functions, as illustrated in Fig. 11.17, to accommodate this relationship. The appendage of additional zeros may be necessary to make the number of data samples in each row and column compatible with the FFT algorithm being used. After zeros have been appended to functions $r(pT_x, qT_y)$ and $h(pT_x, qT_y)$, then we apply Eqs. (11.31) to (11.33). The result is the desired two-dimensional convolution. All results must be multiplied by the scale factor $T_x T_y$ to approximate the continuous two-dimensional convolution integral.

Example 11.10   Two-Dimensional Convolution

To illustrate the procedure for implementing a two-dimensional discrete convolution by means of the FFT, consider the two data arrays shown in Fig. 11.18(a). We note that the two arrays are of dimension (2,2) and (2,4). Hence, the convolution result is of dimension (3,5) and sufficient zeros must be appended to each array to increase its respective dimensions to this size to prevent the end effect or circular convolution. For a base-2 FFT algorithm, we must add zeros to each data array to obtain data arrays of dimension (4,8). The augmented data arrays are illustrated in Fig. 11.18(b). We input the data arrays of Fig. 11.18(b) to Eqs. (11.31) and (11.32) and then implement Eq. (11.33). The result of this procedure is shown in Fig. 11.18(c).

Example 11.11   Two-Dimensional Convolution When One Function Is Separable

If one of the functions to be convolved is separable, that is,

$$r(pT_x, qT_y) = r_1(pT_x)r_2(qT_y) \tag{11.34}$$

then the two-dimensional discrete convolution can be accomplished by repeated evaluations of one-dimensional discrete convolutions. To show this, substitute Eq. (11.34) into Eq. (11.30):

$$
\begin{aligned}
g(pT_x, qT_y) &= \sum_{j=0}^{M-1} \sum_{i=0}^{N-1} r_1(iT_x)r_2(jT_y)h[(p-i)T_x, (q-j)T_y] \\
&= \sum_{i=0}^{N-1} r_1(iT_x)\left[\sum_{j=0}^{M-1} r_2(jT_y)h[(p-i)T_x, (q-j)T_y] \right]
\end{aligned}
\tag{11.35}
$$

The term inside the brackets is a one-dimensional convolution, which is evaluated for each value of i, where $i = 0, 1, \ldots, N - 1$. That is, we convolve the function $r_2(jT_y)$ with each row of the data matrix $h[iT_x, jT_y]$. This resulting data matrix is then convolved, by column, with the function $r_1(iT_x)$, as described by Eq. (11.35).

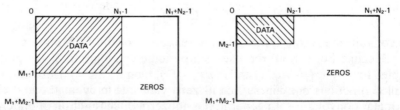

Figure 11.17   Appending zeros in two-dimensional FFT convolution to avoid the end effect.

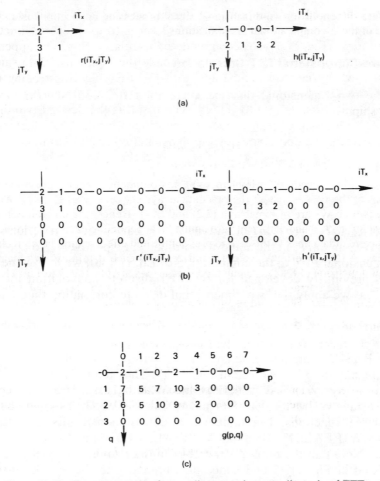

Figure 11.18 Base-2 example of appending zeros in a two-dimensional FFT convolution to avoid the end effect.

To develop the FFT computational approach to this two-dimensional convolution case, we use the following relationship for separable functions (see Prob. 11.3):

$$r_1(iT_x)r_2(jT_y) \iff R_1(n/NT_x)R_2(m/MT_x) \tag{11.36}$$

Application of the FFT convolution theorem procedures described by Eqs. (11.31) to (11.33) requires that we determine the product

$$R(n/NT_x, m/MT_y)H(n/NT_x, m/MT_y)$$

$$= R_1(n/NT_x)R_2(m/MT_x)H(n/NT_x, m/MT_y) \tag{11.37}$$

Recall that $N + M$ one-dimensional FFTs are required to determine $R(n/NT_x, m/MT_y)$. However, because $r(pT_x, qT_y)$ is separable, only 2 one-dimensional FFTs are required. The computational savings is readily apparent and the utilization of separable filter functions in two-dimensional signal-processing applications is common because of these computational simplifications.

Two-Dimensional FFT Correlation

The two-dimensional discrete correlation function is given by the relationship

$$\psi(pT_x,qT_y) = \sum_{j=0}^{M-1} \sum_{i=0}^{N-1} r(iT_x,jT_y)h[(p + i)T_x, (q + j)T_y]$$

$$p = 0, 1, \ldots, N - 1 \qquad\qquad (11.38)$$

$$q = 0, 1, \ldots, M - 1$$

where $\psi(pT_x,qT_y)$, $r(pT_x,qT_y)$, and $h(pT_x,qT_y)$ are periodic functions with periods NT_x and MT_y in the x and y coordinates, respectively. To apply the two-dimensional FFT to the computation of Eq. (11.38), we follow the procedures previously developed for the FFT discrete convolution theorem except that we apply the two-dimensional discrete correlation theorem.

Summary

A BASIC computer program for computing a two-dimensional convolution using the FFT is given in Fig. 11.19. The data to be convolved are stored in arrays W1REAL(II%,JJ%) and W2REAL(II%,JJ%) and the imaginary arrays W1IMAG(II%,JJ%) and W2IMAG(II%,JJ%) should be set to zero for real functions. Two-dimensional convolution results are returned in arrays W1REAL(II%,JJ%) and W2REAL(II%,JJ%) and must be scaled by T_xT_y. Note that the program branches to the two-dimensional FFT program listed in Fig. 11.12 and hence also branches to the one-dimensional FFT program listed in Fig. 8.7. N%, NU%, M%, and MU% must be initialized and XREAL(I%), and XIMAG(I%) must be dimensioned to the larger of N% or M%. The reader is responsible for preventing convolution end effects.

Two-dimensional signal processing often involves large data matrices that can exceed computer memory capacity. For this reason, these types of data are normally stored on magnetic disk or tape units. As a result, one encounters a data-access problem. Recall that the two-dimensional FFT can be computed by first performing a one-dimensional FFT on each row of data. If the data is stored sequentially by rows, then data access for this step is straightforward. However, the next step in computing the two-dimensional transform requires performing the FFT on the columns of the matrix that is generated by row transforms. We now encounter a memory-access problem in extracting column data. One could transpose the matrix,

```
8000 REM:  TWO-DIMENSIONAL FFT CONVOLUTION PROGRAM- THE
8002 REM:  MAIN PROGRAM SHOULD DIMENSION THE DATA
8004 REM:  ARRAYS W1REAL(II%,JJ%),W1IMAG(II%,JJ%),
8006 REM:  W2REAL(II%,JJ%),W2IMAG(II%,JJ%) AND
8008 REM:  DUMMY ARRAYS W3REAL(II%,JJ%), W3IMAG(II%,JJ%).
8010 REM:  N%,NU%,M%, AND MU% MUST BE INITIALIZED.
8012 REM:  XREAL(I%) AND XIMAG(I%) SHOULD BE DIMENSIONED
8014 REM:  THE LARGER OF N% OR M%.THE PROGRAM CALLS
8016 REM:  THE TWO-DIMENSIONAL FFT ROUTINE(FIG. 11.12)
8018 REM:  BEGINNING AT LINE 9000, WHICH IN TURN CALLS
8020 REM:  THE ONE-DIMENSIONAL FFT PROGRAM (FIG. 8.7)
8022 REM:  BEGINNING AT LINE 10000.
8024 REM:  COMPUTE THE TWO-DIMENSIONAL FFT OF W1(N,M)
8030    GOSUB 9000
8040 FOR II%=1 TO N%.
8050    FOR JJ%=1 TO M%
8060        W3REAL(II%,JJ%)=W1REAL(II%,JJ%)
8070        W3IMAG(II%,JJ%)=W1IMAG(II%,JJ%)
8080        W1REAL(II%,JJ%)=W2REAL(II%,JJ%)
8090        W1IMAG(II%,JJ%)=W2IMAG(II%,JJ%)
8100    NEXT JJ%
8110 NEXT II%
8120 REM:  COMPUTE THE TWO-DIMENSIONAL FFT OF W2(N,M)
8130    GOSUB 9000
8140 REM: COMPUTE THE PRODUCT OF W1(II%,JJ%) AND W2(II%,JJ%)
8150 FOR I%=1 TO N%
8160    FOR J%=1 TO M%
8170        W2REAL(I%,J%)=W1REAL(I%,J%)
8180        W1REAL(I%,J%)=W1REAL(I%,J%)*W3REAL(I%,J%)
                        -W1IMAG(I%,J%)*W3IMAG(I%,J%)
8190        W1IMAG(I%,J%)=-W2REAL(I%,J%)*W3IMAG(I%,J%)
                        -W1IMAG(I%,J%)*W3REAL(I%,J%)
8200    NEXT J%
8210 NEXT I%
8220 REM:  COMPUTE THE TWO-DIMENSIONAL FFT OF THE
8222 REM:  PRODUCT CONJUGATE
8230    GOSUB 9000
8240 RETURN
8250 END
```

Figure 11.19   Subroutine in BASIC for two-dimensional FFT convolution.

but this is not possible if the matrix exceeds the computer memory. References [10] to [12] propose various methods for partitioning the matrix so that it will fit into available memory. Alternate transposition procedures are described in Refs. [14] to [17]. Another approach to efficient computation of the two-dimensional discrete transform is to derive a direct two-dimensional FFT. This method is described in Refs. [18] and [19]. An excellent introduction to the field of digital image processing is presented in Oppenheim [5].

PROBLEMS

11.1 Prove each of the following properties of the two-dimensional Fourier transform directly from the defining relationships of Eqs. (11.1) and (11.13):

(a) Addition:

$$h(x,y) + g(x,y) \iff H(u,v) + G(u,v)$$

(b) Shifting:

$$h(x - a, y - b) \iff e^{-j2\pi(au+bv)} H(u,v)$$

(c) Modulation:

$$h(x,y) \cos(2\pi f_0 x) \iff \tfrac{1}{2} H(u + f_0, v) + \tfrac{1}{2} H(u - f_0, v)$$

(d) Scaling:

$$h(ax,by) \iff (1/|ab|)H(u/a, v/b)$$

$$h(x,y)e^{j2\pi(ax+by)} \iff H(u - a, v - b)$$

(e) Convolution:

$$h(x,y) ** g(x,y) \iff H(u,v)G(u,v)$$

11.2 Repeat Prob. 11.1 but derive all results from a one-dimensional viewpoint utilizing the relationships of Eqs. (11.8) and (11.9) and the one-dimensional Fourier transform.

11.3 If a function $h(x,y)$ is separable such that

$$h(x,y) = h_1(x)h_2(y)$$

then the two-dimensional Fourier transform of $h(x,y)$, that is, $H(u,v)$ is separable:

$$H(u,v) = H_1(u)H_2(v)$$

Prove this result.

11.4 Determine the two-dimensional Fourier transform of the following:

(a) $h(x,y) = \cos(2\pi u_0 x)$
(b) $h(x,y) = \sin(2\pi v_0 y)$
(c) $h(x,y) = \cos\{2\pi[x \cos(\theta) + y \sin(\theta)]\}$

(d) $h(x,y) = \sin(2\pi u_0 x)\,\sin(2\pi v_0 y)$

(e) $h(x,y) = \cos(2\pi u_0 x)\,\sin(2\pi v_0 y)$

(f) $h(x,y) = \dfrac{\sin(2\pi u_0 x)}{2\pi u_0 x}\,\dfrac{\sin(2\pi v_0 y)}{2\pi v_0 y}$

(g) $h(x,y) = \delta(y)$

(h) $h(x,y) = \cos(2\pi v_0 y)\delta(x)$

(i) $h(x,y) = 1 \qquad (x^2 + y^2)^{1/2} < \beta$

$\qquad\qquad\quad = 0 \qquad \text{otherwise}$

11.5 Parseval's Theorem for one-dimensional functions is given by

$$\int_{-\infty}^{\infty} h^2(t)\,dt = \int_{-\infty}^{\infty} |\,H(f)\,|^2 df$$

Derive the corresponding relationship for two-dimensional functions.

11.6 Figure 11.2(a) can be redrawn in terms of lines of zero phase, as illustrated in Fig. 11.20. Develop analytic expressions for the following:

(a) A function of the form $y = mx + \eta b$ for the zero-phase lines, where m is the slope, b is the y intercept, and η is an integer.

(b) An expression for θ in terms of the spacial frequency terms u_0 and v_0.

(c) An expression for L, the spatial period, i.e., the distance between zero-phase lines.

11.7 If we introduce the polar coordinates $x = r\cos(\theta)$, $y = r\sin(\theta)$, $u = w\cos(\phi)$, and $v = w\sin(\phi)$, then $h(x,y)$ and $H(u,v)$ become $h_p(r,\theta)$ and $H_p(w,\phi)$, respectively, where

$$h_p(ar,\theta+\theta_0) \;\Longleftrightarrow\; (1/a^2)H_p(w/a,\phi+\theta_0) \qquad (11.39)$$

Note that Eq. (11.39) states that if $h(x,y)$ is rotated through an angle θ_0, then the transform $H(u,v)$ is also rotated through the angle θ_0. Derive Eq. (11.39).

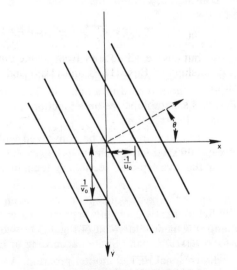

Figure 11.20 Function for Prob. 11.6.

Hint:

$$h(ax+by, cx+dy) \quad \Longleftrightarrow \quad (1/\,|\,ad-cd\,|\,)h(Au+Bv, Cu+Dv)$$

where

$$\begin{bmatrix} A & B \\ C & D \end{bmatrix} = \begin{bmatrix} a & b \\ c & d \end{bmatrix}^{-1}$$

11.8 An interesting special case of two-dimensional discrete Fourier transforms occurs when the sampled function $h(pT_x, qT_y)$ separates as $h_1(pT_x)h_2(qT_y)$, where $p = 0, 1, \ldots, N-1$ and $q = 0, 1, \ldots, M-1$. Show that in this case the discrete Fourier transform $H(n/NT, m/MT)$ can be evaluated by one N-point FFT and one M-point FFT.

11.9 Figure 11.4 shows the two-dimensional FFT as a set of one-dimensional FFTs on the rows of the data matrix followed by a set of one-dimensional FFTs on the columns of the data matrix computed in the first step. Show by sketches, as in Fig. 11.4, that equivalent results can be obtained by FFTing first the columns of the data array followed by the rows of the intermediate computed array.

11.10 Consider the sampled waveform shown in Fig. 11.21. Analytically compute the two-dimensional discrete Fourier transform by using one-dimensional transforms:
   (a) Transform each column and then each row of the result.
   (b) Transform each row and then each column of the result.

11.11 Develop an alternate inversion formula for two-dimensional transforms that allows one to use the forward two-dimensional FFT.

11.12 In Chapter 8, we developed techniques for increasing the efficiency of the one-dimensional FFT for the case of real data. If only two-dimensional real data are being considered, develop the procedures and appropriate relationships for the following:
   (a) Computing the two-dimensional FFT of two real functions simultaneously.
   (b) Computing the two-dimensional FFT of a $(2N, 2M)$ real data array with an (N, M) two-dimensional FFT program.

11.13 It is of value to consider two-dimensional convolution and correlation from an area viewpoint without regard to amplitude. For each of the functions illustrated in Fig. 11.22, determine the area function (i.e., follow Fig. 11.11) for both convolution and correlation.

11.14 If a small two-dimensional area is to be convolved with a much larger two-dimensional area, then sectioning techniques as developed in Chapter 10 must be used. Extend the overlap-add sectioning technique to two-dimensional convolution.

11.15 Figure 11.15 develops the basic concepts of two-dimensional convolution involving impulse functions. By using this approach, extend the two-dimensional Nyquist sampling criteria development of Fig. 5.3 to two dimensions. Assume that the transform $H(u, v)$ is band-limited according to Eq. (11.23).

11.16 Develop a two-dimensional FFT computer program. An input variable should allow the output to be in either a conventional viewing format or standard two-dimensional FFT format.

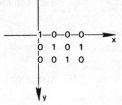

Figure 11.21 Sampled waveform for Prob. 11.10.

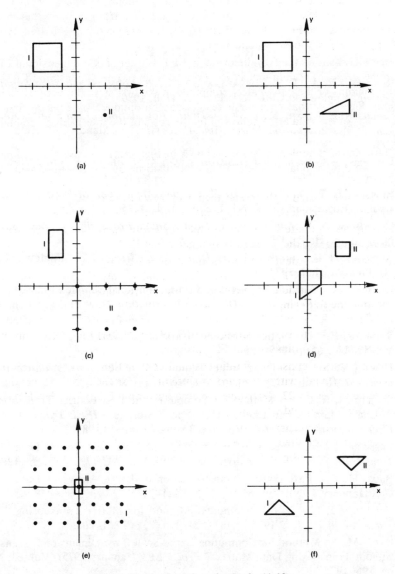

(a)

(b)

(c)

(d)

(e)

(f)

Figure 11.22 Functions for Prob. 11.13.

11.17 Apply the program developed in Prob. 11.16 to the two-dimensional waveform for Fig. 11.1(a). Explain any differences with the theoretical transform results of Fig. 11.1(b).

11.18 Compute the two-dimensional FFT of the function shown in Fig. 11.7.

11.19 Compute the two-dimensional FFT of the waveform shown in Fig. 11.2(a). Explain any differences with the theoretical transform results of Fig. 11.2(b).

11.20 Apply the two-dimensional Hanning weighting function to the results of Prob. 11.18.

REFERENCES

1. ROBINSON, E. A., T. S. DURRANI, AND L. G. PEARDON. *Geophysical Signal Processing*. Englewood Cliffs, NJ: Prentice-Hall, 1986.

2. ROBINSON, E. A., AND M. T. SILVIA. *Digital Foundations of Time Series Analysis, Vol. 2: Wave-Equation Space-Time Processing*. Oakland, CA: Holden Day, 1981.

3. PAPOULIS, A. *Systems and Transforms with Applications in Optics*. New York: McGraw-Hill, 1968.

4. RABINER, L. R., AND B. GOLD. *Theory and Application of Digital Signal Processing*. Englewood Cliffs, NJ: Prentice-Hall, 1975.

5. OPPENHEIM, A. V. *Applications of Digital Signal Processing*. Englewood Cliffs, NJ: Prentice-Hall, 1978.

6. ANDREWS, H. C., AND B. R. HUNT. *Digital Image Restoration*. Englewood Cliffs, NJ: Prentice-Hall, 1977.

7. LEGAULT, R. "Aliasing Problems in Two-Dimensional Sampled Imagery." in L. Biberman, ed., *Perception of Displayed Information*. New York: Plenum, Chap. 7, 1973.

8. SWING, R. E. "The Optics of Microdensitometry." *Opt. Eng.* (November 1973), Vol. 12, No. 6, pp. 185–198.

9. HUANG, T. S. "Two-Dimensional Windows." *IEEE Trans. Audio and Electroacoust.* (March 1972), Vol. AU-20, No. 1, pp. 88–89.

10. SINGLETON, R. C. "A Method for Computing the Fast Fourier Transform with Auxiliary Memory and Limited High Speed Storage." *IEEE Trans. Audio and Electroacoust.* (June 1967), Vol. AU-15, No. 3, pp. 91–98.

11. BRENNER, N. M. "Fast Fourier Transform of Externally Stored Data." *IEEE Trans. Audio and Electroacoust.* (June 1969), Vol. AU-17, No. 3, pp. 128–132.

12. BUIJS, H. L. "Fast Fourier Transformation of Large Arrays of Data." *Appl. Opt.* (January 1969), Vol. 8, No. 1, pp. 211–212.

13. EKLUNDH, J. O. "A Fast Computer Method for Matrix Transposing." *IEEE Trans. Comput.* (July 1972), Vol. C-21, No. 7, pp. 801–803.

14. ONOE, M. "A Method for Computing Large-Scale Two-Dimensional Transforms without Transposing Data Matrix." *Proc. IEEE* (January 1975), Vol. 63, No. 1, pp. 196–197.

15. TWOGOOD, R. E., AND M. P. EKSTROM. "An Extension of Eklundh's Matrix Transposition Algorithm and Its Application in Digital Image Processing." *IEEE Trans. Comput.* (September 1976), Vol. C-25, No. 9, pp. 950–952.

16. DELOTTO, I., AND D. DOTTI. "Two-Dimensional Transform by Minicomputers without Matrix Transposing." *Comput. Graph. and Imag. Proc.* (September 1975), Vol. 4, No. 3, pp. 271–278.

17. ANDERSON, G. L. "A Stepwise Approach to Computing the Multidimensional Fast Fourier Transform of Large Arrays." *IEEE Trans. Acoust. Speech Sig. Proc.* (June 1980), Vol. ASSP-28, No. 3, pp. 280–284.

18. HARRIS, D. B., J. H. MCCLELLAN, D. CHAN, AND H. S. SCHUESSLER. "Vector Radix Fast Fourier Transform." *IEEE Int. Conf. Acoust. Speech Sig. Proc., Rec.* (May 1977), pp. 548–551.

19. HOYER, E. A., AND W. R. BERRY. "An Algorithm for the Two-Dimensional FFT." *IEEE Int. Conf. Acoust. Speech Sig. Proc., Rec.* (May 1977), pp. 552–555.

12

FFT DIGITAL FILTER DESIGN

Digital filtering is the realization of the convolution integral in discrete form. Recall from Ex. 4.4 that the output of a linear system is determined by convolving the system impulse response $h(t)$ with the system input waveform $x(t)$. Common signal-processing terminology characterizes $h(t)$ as a filter, that is, the input signal $x(t)$ is *filtered* by a system with filter impulse response $h(t)$ to produce the output signal $y(t)$.

A straightforward realization of a digital filter can be achieved by sampling the impulse response $h(t)$ and performing the discrete convolution operation with the sampled input waveform $x(kT)$. In the literature, this design approach is termed a *Finite-Impulse Response* (FIR) filter because the sampled impulse response $h(kT)$ is described by N samples. FIR digital filters require considerable computational complexity because each system output value $y(kT)$ requires multiplication of the N samples of the sampled impulse response $h(kt)$ with N sample values of the input signal $x(kT)$ and $N - 1$ additions of these product terms.

Computational complexity can be reduced significantly by implementing *digital recursive filters*. Recursive filtering, as the name implies, is realized by expressing the discrete convolution equation as a summation of weighted input sample values $x(kT)$ and a weighted sum of previously computed output values:

$$y(kT) = \sum_{i=0}^{\alpha} a_i x[(k - i)T] + \sum_{j=0}^{\beta} b_j y[(k - j - 1)T] \qquad (12.1)$$

Recursive filters can reduce the required number of multiplications by an order of magnitude with respect to a FIR filter implementation. However, the design of a recursive digital filter is complicated if the filter function to be realized is not characterized by a well-defined analytical function [1].

In this chapter, we discuss the basic techniques for applying the FFT to the design and implementation of nonrecursive (i.e., FIR) digital filters. FFT digital filter design can be accomplished by two basic techniques. We can begin with the desired time-domain impulse response of the filter (time-domain specification) or with the desired filter frequency-domain response function (frequency-domain specification). In either case, the filter response function can be specified by an analytic function or by experimental samples. FFT digital filter designs are particularly valuable where the impulse or frequency response of the filter has been determined experimentally. Unless the experimental data describes a well-known filter shape, then the design of a suitable recursive digital filter is extremely time consuming (or impossible).

Our design approach is focused on design-time efficiency and the practicality of implementation. Practicing professionals must constantly evaluate the time required to design a *better* digital filter against the savings in data-processing time that results from a more elegant filter design. The FFT design approach presented here is best suited for *quick-solution* laboratory analysis or for signal-processing problems involving *unconventional* filter response functions. Our discussion of system-simulation analysis in Chapter 14 is an example where our techniques are more efficient to apply than more elaborate design techniques.

12.1 FFT TIME-DOMAIN DIGITAL FILTER DESIGN

Assume we are given the time-domain impulse response of a desired filter graphically, analytically, or as numerical values determined from an experiment. It is desired to design a nonrecursive digital filter that produces results equivalent to the specified filter. The digital filter is to be implemented by FFT convolution techniques (Chapter 10) and, therefore, its impulse response must be of finite duration. The design of a FFT digital filter from a time-domain specification is very similar to the development of the discrete Fourier transform that is graphically developed in Fig. 6.2.

Design Procedure

If the time-domain specification is an analytical function, we begin our design by sampling the given impulse-response function. For an experimentally obtained impulse function, we begin with the numerical sampled values. In both cases, we require that the sample interval T is sufficiently small to produce negligible aliasing. Because the FFT is used to implement the designed digital filter, it may be necessary to truncate the sampled impulse response. If truncation is not required, we know from Chapter 6 that the discrete Fourier transform of the digital filter is a good approximation to the continuous Fourier transform of the specified analog filter. The sampled impulse-response function thus satisfies the digital filter design goal in that its frequency function is a good approximation to the specified frequency function. Further, the filter can be implemented by means of FFT convolution techniques.

Recall from Table 10.2 that efficient application of FFT sectioning techniques requires that the fillter impulse response be represented by a small number of samples with respect to N, the number of sample values to be FFTed. If an experimenter's computer capacity accommodates a digital filter designed without impulse-response truncation, then the design is complete. However, it is often the case that the number of nonzero samples that define the impulse response of the digital filter must be minimized.

The number of nonzero samples of the impulse-response function can be modified by multiplication with a truncation (weighting) function. As shown in Figs. 6.2(d) and (e), time-domain truncation can introduce *ripples* in the frequency function unless a weighting function that smoothly tapers to zero is used. To determine an acceptable truncation width, we experimentally decrease the width, or duration, of the weighting function until the resulting digital filter frequency function as computed with the FFT differs unacceptably from the desired analog frequency function. The minimum width of the weighting function that yields acceptable results corresponds to a minimum impulse-response duration and hence a more efficient FFT implementation.

Example 12.1   FFT Digital Filter Design: Time-Domain Specification

To illustrate the FFT filter design procedure, consider the filter impulse function:

$$h(t) = \alpha^2 t e^{-\alpha t} \quad t \geq 0 \tag{12.2}$$
$$= 0 \qquad\qquad t < 0$$

The Fourier transform is given by

$$H(f) = \alpha^2/(\alpha + j2\pi f)^2 \tag{12.3}$$
$$= |H(f)| \, e^{j\theta(f)}$$

where

$$|H(f)| = \alpha^2/\{[\alpha^2 - (2\pi f)^2]^2 + (4\pi f \alpha)^2\}^{1/2} \tag{12.4}$$

$$\theta(f) = \tan^{-1}\{-4\pi f \alpha/[\alpha^2 - (2\pi f)^2]\} \tag{12.5}$$

The parameter α was chosen as 2π to yield a filter frequency-domain 6 dB *cutoff* frequency of 1 Hz. We show the time-domain impulse response function of Eq. (12.2) and the amplitude and phase response functions of Eqs. (12.4) and (12.5) in Figs. 12.1(a) and (b), respectively.

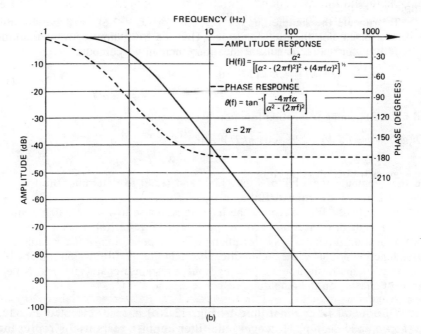

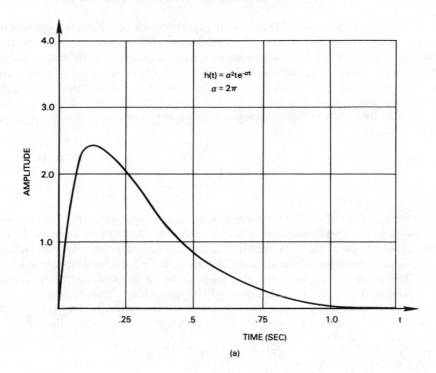

Figure 12.1 Example filter continuous time- and frequency-domain response functions.

To sample the impulse response of Eq. (12.2), we must choose the sample interval T. We know that T must be sufficiently small to minimize the effect of aliasing. Let us assume that our design goal is a digital filter whose magnitude response approximates that shown in Fig. 12.1(b) to approximately 500 Hz. As a result, we must ensure that we choose the sample interval T such that aliasing is negligible up to a frequency of 500 Hz. For a sampling frequency $f_s = 1000$ Hz, the *foldover*, or aliasing, frequency is $f_s/2 = 500$ Hz, and we have ensured that aliasing is reasonably negligible below 500 Hz.

We next choose the maximum number of sample values N that can be conveniently FFTed. We assume for purposes of discussion that $N = 2048$. With $T = 1/f_s = 0.001$ and $N = 2048$, we sample Eq. (12.2) to obtain

$$h(kT) = \alpha^2(kT)e^{\alpha(kT)} \qquad k = 0, 1, \ldots, 2047 \qquad (12.6)$$

The 2048 samples of Eq. (12.6) represent an impulse duration of 2.048 s. We observe from Fig. 12.1(a) that the choice of 2048 samples (or 2.048 s) introduces negligible truncation. Hence, we expect the FFT of the sampled impulse response and the continuous Fourier transform to agree closely. The FFT of the function given by Eq. (12.6) for the chosen parameters is essentially the same as samples of Fig. 12.1(b).

The next step in our time-domain FFT digital filter design procedure is to truncate the impulse response in order to minimize the number of near-zero sample values. From Table 10.2, the number of samples representing the impulse-response function should not exceed 299 for efficient FFT convolution with a 2048-point transform. Hence, we will experimentally sequentially reduce the width of the truncation function with a goal of reducing the digital filter impulse-response duration to 299 sample values (0.299 s).

To truncate the sample impulse response of Eq. (12.6), we have arbitrarily chosen for discussion the rectangular and Hanning truncation (or weighting) functions. Hence, for the rectangular truncation function, we compute

$$
\begin{aligned}
h(kT) &= h(kT) & 0 \leq k \leq W_H/T \\
&= 0 & W_H/T < k \leq N - W_H/T
\end{aligned}
\qquad (12.7)
$$

and for the Hanning truncation function, we compute

$$
\begin{aligned}
h(kT) &= h(kT)[\ \tfrac{1}{2} + \tfrac{1}{2}\ \cos(\pi kT/W_H)] & 0 \leq k \leq W_H/T \\
&= 0 & W_H/T < k \leq N - W_H/T
\end{aligned}
\qquad (12.8)
$$

We then compute the FFT of Eqs. (12.7) and (12.8) to determine the frequency characteristics of the truncated filter.

If we choose W_H to reduce the impulse response to 1 s (i.e., 1000 samples), then the results of truncation become apparent. In Fig. 12.2(a), we see that rectangular truncation produces a small rippling effect. Application of the Hanning truncation function produces the digital filter shown in Fig. 12.2(b), which rather closely approximates the desired amplitude- and phase-response characteristics. Note the effects of aliasing on the amplitude response around 500 Hz.

The digital filter illustrated in Fig. 12.2(b) is characterized as one of *quick and easy* design. However, the filter impulse response is represented by 1000 nonzero sample values. The designer must evaluate the trade-off

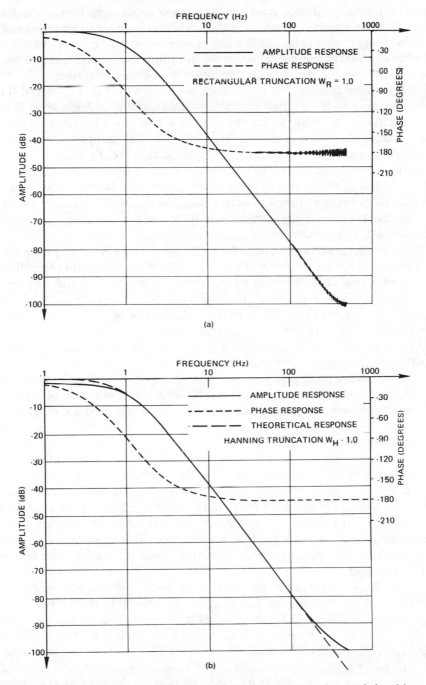

(a)

(b)

Figure 12.2 FFT filter design amplitude- and phase-response characteristics: (a) rectangular truncation at 1.0 s, and (b) Hanning truncation at 1.0 s.

between increased design time and decreased processing time for a filter with a reduced number of nonzero sample values. In general, we recommend that the designer accept the inefficiency of implementation caused by the large number of nonzero sample values of the filter impulse response unless there are compelling reasons to further decrease data-processing time.

To reduce this number of samples, we decrease W_H. The digital filter approximation to the desired characteristics continue to degrade as W_H is decreased. Figure 12.3(a) illustrates amplitude and phase functions resulting from rectangular truncation with $W_R = 0.5$ s (500 samples). As shown, rectangular truncation produces unacceptable rippling. In Fig. 12.3(b), we show the filter design characteristics with Hanning truncation width $W_H = 0.5$ s. Although the design approximates rather closely the theoretical response, the amplitude function differs from the desired frequency response in that the low-frequency components have been attenuated by approximately 4 dB.

Hanning truncation reduces the area under the impulse-response function and, as a result, produces the attenuation that we observe. Recall that the Fourier transform for zero frequency is simply the integral or area of the impulse-response function. If this attenuation is objectionable for the digital filter design of concern, then the effect can be partially compensated. We simply multiply the truncated impulse response by the appropriate constant that yields a windowed impulse response with an area equal to that of the theoretical impulse-response function.

Figure 12.4 illustrates the amplitude-response functions obtained by multiplication by the appropriate constant to increase the area of the truncated impulse response. A comparison of the truncated impulse response of Fig. 12.4 and the truncated impulse response of Fig. 12.3(a) illustrates the effect of multiplication.

The amplitude-response function that results from multiplication agrees more closely with the desired response at the lower frequencies but is shifted to the right at higher frequencies. That is, the amplitude-response characteristic shows the 6 dB cutoff is at approximately 1.5 Hz rather than at the desired 1 Hz. If the exact cutoff frequency is of importance, then we can begin our design procedure with a specified cutoff frequency that is lower (i.e., 0.5 Hz). The phase-response function is not affected by the multiplicative constant.

For $W_H = 0.5$ s, we have developed a digital filter represented by 500 samples values. Although we have not reached our goal of 299, recall from the discussion of Sec. 10.3 that an increase of this optimum number of samples by a factor of 2 increases computing time only slightly. As a result, this digital filter can be implemented efficiently by FFT convolution techniques.

In summary, we have followed the design procedure of successively reducing parameter W_H, the truncation function width. We note that in each reduction of the parameter W_H, we must accept a compromise in the digital

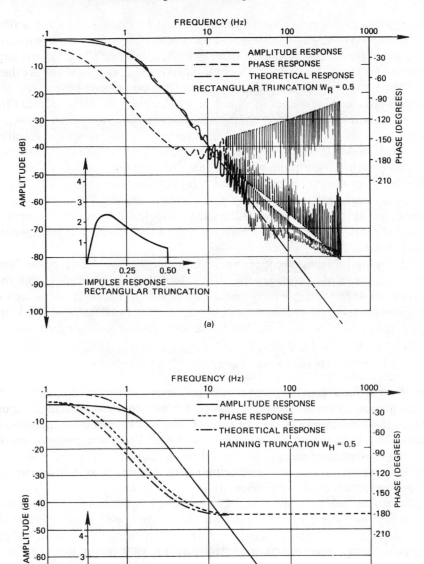

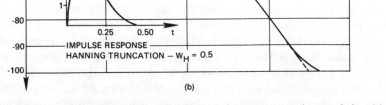

Figure 12.3 FFT filter design amplitude- and phase-response characteristics: (a) rectangular truncation at 0.5 s, and (b) Hanning truncation at 0.5 s.

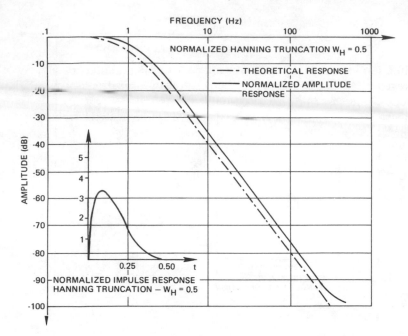

Figure 12.4 FFT filter design amplitude response with Hanning truncation at 0.5 s and normalized impulse-response area.

filter design. As W_H is decreased, we encounter an increased spreading of the frequency function. Hence, a design for maximum implementation efficiency (i.e., minimum number of sample values) deviates from the desired characteristics. However, if we are willing to represent the impulse response by a large number of sample values, then, as shown, we can achieve an excellent approximation to the specified filter frequency response and the design can be achieved very efficiently.

12.2 FFT FREQUENCY-DOMAIN DIGITAL FILTER DESIGN

Frequency-domain specification of a filter implies that the digital filter design begins with an analytical expression for the frequency response of a filter or with numerical values of amplitude and phase obtained from an experiment. As in the time-domain specification case, the goal is to design a digital filter that approximates the given frequency response and that can be implemented by FFT convolution. One could claim that this is identical to the previous discussion; however, there are subtle design differences in the two developments. In this section, we will investigate these differences as well as develop the fundamentals of FFT frequency-domain digital filter design.

Graphical Development

Consider the time- and frequency-response functions of the example filter shown in Fig. 12.5(a). Our approach to digital filter design follows the presentation used to develop the discrete Fourier transform in Chapter 6. Because we assume that the filter characteristics are known to us only in the frequency domain, it is necessary to first sample in the frequency domain. The frequency-sampling function and its inverse Fourier transform are

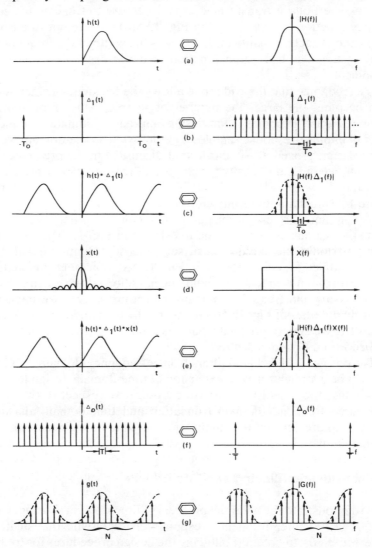

Figure 12.5 Graphical development of FFT digital filter designed from a frequency-domain specification.

shown in Fig. 12.5(b). From the time-convolution theorem, the resulting sampled frequency function and its inverse Fourier transform are shown in Fig. 12.5(c). Because sampling in the frequency domain corresponds to convolution in the time domain, we can have time-domain aliasing. It is necessary that the frequency-domain sampling period $1/T_0$ is sufficiently small to ensure that time-domain aliasing is negligible. As illustrated in Fig. 12.5(c), aliasing is negligible for this example.

Because only a finite number of sample values of the frequency function can be inverse Fourier transformed, it is necessary to truncate the sampled frequency function. As illustrated in Fig. 12.5(d) we attempt to ensure that the truncation function significantly exceeds the width of the frequency function. A rectangular truncation function is used to simplify the graphical development.

We recognize that the wider one allows the frequency-truncation function to become, the larger the number of sample values representing the impulse response becomes. Our design approach ultimately reduces the number of impulse-response samples by employing a time-domain weighting function design procedure similar to that discussed in the previous section. As a result, we incur no penalty for choosing the frequency-truncation function extremely wide. Figure 12.5(e) illustrates the Fourier transform pair obtained by truncating the sampled frequency response.

To complete the description of the digital filter, it is necessary to sample with the time-domain sampling function shown in Fig. 12.5(f). Note that the sampling period T has already been set because NT must equal T_0. The resulting sampled functions shown in Fig. 12.5(g) represent the digital filter as designed from a frequency-domain specification. If we ensure that time-domain aliasing and frequency-domain truncation effects are insignificant, then the digital filter of Fig. 12.5(g) is essentially equivalent to that discussed in Sec. 12.1; we simply proceed from this point using the design techniques described in the previous section.

The design of FFT digital filters from a frequency-domain specification appears to be a straightforward extension of time-domain design techniques. This is in fact the case if we exercise caution with respect to two pertinent assumptions: frequency-domain truncation and time-domain aliasing. We will now explore further the implication of time-domain aliasing in FFT frequency-domain filter design.

Time-Domain Aliasing and End Effects

To demonstrate the potential problem of time-domain aliasing, we show in Fig. 12.6(a) a filter frequency-response characteristic that digital filter designers often try to obtain. Following the design procedures for frequency-domain design of digital filters that were described in Fig. 12.5, we first sample in the frequency domain using the sampling function illustrated in

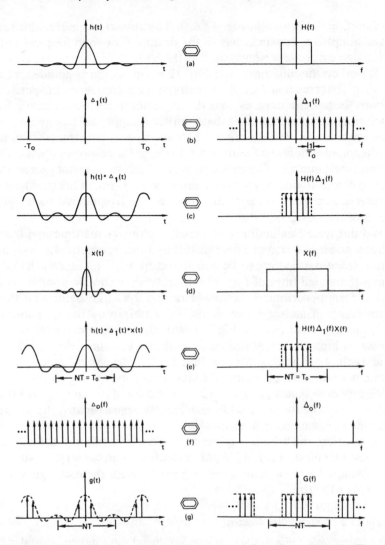

Figure 12.6 Graphical development of an *apparently* perfect rectangular filter.

Fig. 12.6(b). We purposely have chosen a large frequency sample interval so that there is significant time-domain aliasing, as shown in Fig. 12.6(c). Recall that the interval $1/T_0$ is the variable that the FFT filter designer uses to limit time-domain aliasing to an acceptable level.

The filter frequency-response function is of finite duration and, as a result, multiplication by a frequency-truncation or weighting function of greater duration, as shown in Fig. 12.6(d), introduces no distortion but does set the truncation width NT. Time-domain sampling with N samples over the time-function period T_0 is achieved by multiplication with the time-sampling function illustrated in Fig. 12.6(f). The resulting time- and frequency-domain sampled approximations to the desired time- and frequency-domain filter characteristics are shown in Fig. 12.6(g).

Based on the illustrations of Fig. 12.6, one could conclude that a filter has been designed that has a perfectly square frequency response. This, however, is not the case as this design cannot be implemented by FFT convolution techniques. Recall that in order to apply FFT convolution techniques without end effects (Sec. 10.1), it is necessary that the number of nonzero samples defining the digital filter impulse-response function be less than the total number of samples to be FFTed. As a result, zeros must be added to the N points of the first period of $g(t)$, Fig. 12.6(g). The impulse response with appended zeros then defines the frequency-function characteristics of the designed filter.

To illustrate this addition of zeros, we repeat the time and frequency functions of Fig. 12.6(g) in Figs. 12.7(a) and (d), respectively. Assume that the number of data points to be processed by the FFT is $2N$; the N points defining the digital filter of Fig. 12.6(a) must therefore be combined with N zeros to form a periodic function of period $2NT$. To determine the time-domain result of adding these zeros, we multiply by the periodic square-wave function illustrated by Fig. 12.7(b); the corresponding time function is shown in Fig. 12.7(c). Multiplication of the functions of Figs. 12.7(a) and (b) yields the results illustrated in Fig. 12.7(c), a periodic function with $2N$ points per period. The frequency function corresponding to Fig. 12.7(c) is obtained by convolution of Figs. 12.7(d) and (e). This function is illustrated in Fig. 12.7(f); the function no longer closely approximates the desired frequency-domain response function.

As shown, the addition of zeros results in rippling in the digital filter frequency response. To reduce this effect, it is necessary to substitute a suitably shaped time-domain weighting function for the rectangular function used in Fig. 12.7(b).

Example 12.2   Notch Filter Design

As an illustration of FFT frequency-domain digital filter design, consider the frequency function illustrated in Fig. 12.8(a). As shown, the desired filter function has *notch* filter characteristics between 1.5 and 2.5 Hz and a cutoff frequency of 5 Hz. We wish to design a digital filter approximation to the illustrated filter function.

Assume that the data to be filtered has no frequency component above 10 Hz. Hence, our time-domain sample interval T should be less than 0.05 s. For $N = 1024$, choose $T = 0.0391$ s and thus $\Delta f = 0.25$ Hz. When sampling the frequency function illustrated in Fig. 12.8(a), recall that the computation of the inverse FFT requires that we must fold the real frequency function about $n = N/2$. We assume the phase function is zero and hence the imaginary frequency function is zero. Note that because the phase function is zero, then the frequency function is even and hence the impulse-response function is even (noncausal).

The inverse FFT (scaled by T) is shown in Fig. 12.8(b). This impulse-response

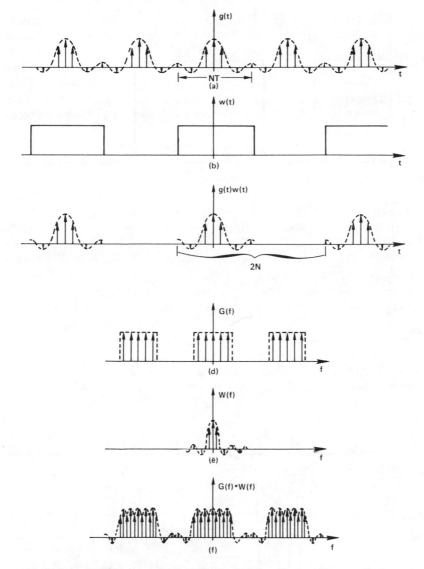

Figure 12.7 Graphical development illustrating the effect on filter amplitude response resulting from appending zeros to the impulse response.

function is also symmetric about $n = N/2$, but we show only values for positive time. Following our design procedure, we apply a Hanning truncation or weighting function to the impulse-response function shown in Fig. 12.8(b). The weighting function is positioned to have unity value at time $t = 0$, zero value at time $t = T_c$, and to be symmetric about $n = N/2$. We compute the FFT of the weighted impulse response to determine the frequency response of the designed filter. The log amplitude response of the FFT designed digital filter is illustrated in Fig. 12.8(c) for T_c

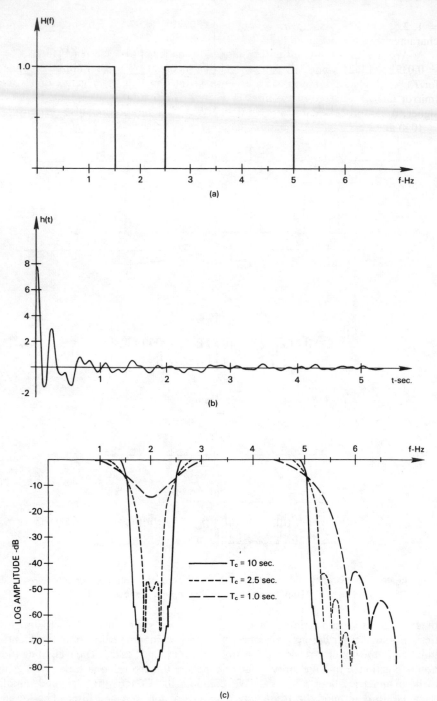

Figure 12.8 FFT notch filter designed from a frequency-domain specification.

= 1, 2.5, and 10 s. As shown, severe truncation ($T_c = 1$ s) results in a filter whose characteristics are not useful.

We use FFT convolution techniques to implement the designed filter. For T = 0.0391 s, a 1024-point FFT can process 40.04 s of data. The filter impulse response for $T_c = 2.5$ s has a duration of 5 s because we must include the negative time values (mirror image) if a zero-phase filter is desired. If this filter is acceptable to the designer, it can process data rather efficiently. A 20-s duration impulse response (T_c = 10 s) gives excellent attenuation characteristics but is less efficient. However, we

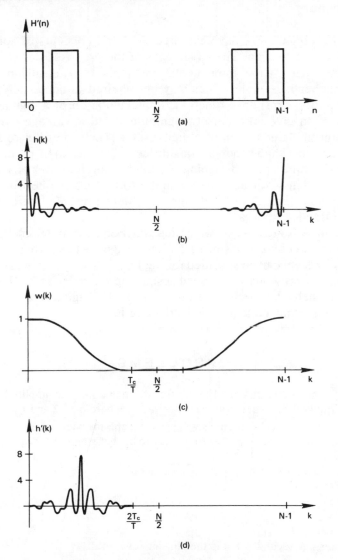

Figure 12.9 Illustrations showing correct formatting of time and frequency functions for the digital filter design of Fig. 12.8.

argue again that the designer should predetermine if design simplicity or data-processing time is of first priority.

For clarity of presentation, Figs. 12.9(a) to (d) illustrate the appropriate frequency function to be sampled, the FFT computed impulse response, the Hanning weighting function, and the zero-phase-shift filter impulse response, respectively. The impulse response of Fig. 12.9(d) is that which must be convolved with the data by means of the FFT if a zero-phase-shift filter is desired.

Summary

In this chapter, we have explored the basic concepts of nonrecursive FFT digital filter design. The advantages of the proposed design procedure are design efficiency and implementation simplicity. A disadvantage is that a large number of sample values may be required to adequately represent the filter impulse-response function. Implementation of a digital recursive filter can often give an order-of-magnitude speed advantage over an FFT convolution implementation of a digital filter [1]. If the filter to be used is of a conventional shape and vast quantities of data are to be processed, then certainly the time spent designing a digital recursive filter is worthwhile. But if the design is for an experimental effort or if the filter function is of *unusual shape*, then the FFT design procedure presented here is a simple and cost-effective approach.

Alternate truncation or weighting functions can be employed to alter the side-lobe characteristics of FFT filter designs. Recall from Sec. 9.2 that the side-lobe level can be specified for the Dolph-Chebyshev weighting function. Using the previously described design approach, Helmns [2] has applied the Dolph-Chebyshev weighting function to FFT digital filter design problems and has achieved a specified side-lobe level.

PROBLEMS

12.1 Analytically determine the Fourier transform and plot the amplitude and phase spectrums of the waveform $h(t) = a^2 t e^{-at}$, where $a = 2$ and $t > 0$.

12.2 If the aliasing level that one is willing to accept has been set at x dB, then why does one select the *crossover* frequency to be that where the frequency function is down $(x + 3)$ dB?

12.3 If the waveform of Prob. 12.1 is sampled at
  (a) 250
  (b) 500
  (c) 1000
  (d) 1500
  samples per second, what is the aliasing level in dB?

12.4 When the Hanning weighting function is applied to the impulse-response function illustrated in Figure 12.1(a), should the weighting function have its maximum amplitude centered on the origin or at the midpoint of the interval $[0, W_H]$? Explain.

12.5 Design a digital filter to be implemented using FFT convolution that approximates the following time-domain specified filters:

(a) $h(t) = e^{-t}$

(b) $h(t) - e^{-t} \cos(2\pi t)$

(c) $h(t) = [\sin^2(t)]/t^2$

12.6 The FFT time-domain specification technique for digital filter design is of considerable importance in those cases where designs for digital recursive filters do not exist. Give examples.

12.7 Refer to the impulse-response function of Fig. 12.5(g). Explain the result if the N sample values shown are used to implement a digital filter.

12.8 Refer to Fig. 12.6. Why can't the problems cited be resolved by selecting the frequency-sampling interval in Fig. 12.6(b) much smaller than that shown?

12.9 Design a digital filter to be implemented using FFT convolution that approximates the following frequency-domain specified filters:

(a) $H(f) = \dfrac{1}{1 + (2\pi f)^2}$

(b) $H(f) = \dfrac{f^3}{f^4 + 1}$

(c) $H(f) = \dfrac{\sin(2\pi f) \cos(2\pi f)}{2\pi f}$

Aliasing should be below -50 dB.

REFERENCES

1. OPPENHEIM, ALAN V., AND R. W. SCHAFER. *Digital Signal Processing*. Englewood Cliffs, NJ: Prentice-Hall, 1975.

2. HELMS, H. D. "Non-recursive Digital Filters: Design Methods for Achieving Specification on Frequency Response." *IEEE Trans. Audio and Electroacoust.* (September 1968), Vol. AU-16, No. 3, pp. 336–342.

3. RABINER, L. R., AND B. GOLD. *Theory and Application of Digital Signal Processing*. Englewood Cliffs, NJ: Prentice-Hall, 1975.

4. RADER, C. M., AND B. GOLD. "Digital Filter Design Techniques in the Frequency Domain." *Proc. IEEE* (February 1967), Vol. 55, No. 2, pp. 149–171.

5. HAMMING, R. W. *Digital Filters*, 2d ed., Signal Processing Series. Englewood Cliffs, NJ: Prentice-Hall, 1983.

6. MCCLELLAN, J. H., T. W. PARKS, AND L. R. RABINER. "A Computer Program for Designing Optimum FIR Digital Filters." *IEEE Trans. Audio and Electroacoust.* (December 1973), Vol. AU-21, No. 6, pp. 506–526.

7. GOLD, B., AND K. L. JORDAN, JR. "A Direct Search Procedure for Designing Finite Duration Impulse Response Filters." *IEEE Trans. Audio and Electroacoust.* (March 1969), Vol. AU-17, No. 1, pp. 33–36.

8. RABINER, L. R., B. GOLD, AND C. A. MCGONEGAL. "An Approach to the Ap-

proximation Problem for Nonrecursive Digital Filters." *IEEE Trans. Audio and Electroacoust.* (June 1970), Vol. AU-18, No. 2, pp. 83–106.

9. RABINER, L. R., AND R. W. SCHAFER. "Recursive and Nonrecursive Realizations of Digital Filters Designed by Frequency Sampling Techniques." *IEEE Trans Audio and Electroacoust.* (September 1971), Vol. AU-19, No. 3, pp. 200–207.

13

FFT MULTICHANNEL
BAND-PASS FILTERING

In radar, sonar, communications, and signal processing systems, the application of the FFT to digital multichannel band-pass filtering is of major importance. These fields of FFT applications are based on the interpretation of each FFT resolution cell as the output of a band-pass filter. In this chapter, we develop graphically and analytically the fundamentals of FFT digital band-pass filtering.

We first explore the analogy of the FFT to a bank of integrate and sample filters. Both graphical and mathematical presentations are developed. We then review the data-weighting function discussed in Sec. 9.2 from a filter-shaping perspective. The relationship of FFT resolution to band-pass filter response characteristics is explored in detail.

The FFT filtering development is then extended to the interpretation of a sequence of FFT outputs as sequential time samples from a bank of band-pass filters. Interpreting FFT outputs as time samples is a concept that is contrary to one's intuition. One normally considers the FFT as a time-to-frequency transform. For these reasons, basic implementation considerations that one employs for FFT multichannel filtering are explored in detail. Numerous examples are presented to solidify the development.

13.1 FFT BAND-PASS INTEGRATE AND SAMPLE FILTERS

An interpretation of the FFT that has been found useful is its realization as a bank of band-pass integrate and sample filters. Such an interpretation requires we show the FFT can be viewed as a time-domain convolution operation. This follows because a filter is a linear system and the output of a linear system is the convolution of the input to the system and the system impulse response (Ex. 4.4). Hence, to show that the FFT can be interpreted as a bank of filters, we must show that the FFT operation is a convolution operation involving a set of impulse-response functions whose Fourier transform (i.e., transfer or system frequency response functions) are those of a band-pass filter bank. In this section, we develop both analytically and graphically this interpretation of the FFT.

Development of Band-Pass Filter Equations

To develop the analytical relationships describing the FFT as a bank of band-pass integrate and sample filters, consider the discrete Fourier transform approximation to the continuous Fourier transform:

$$Y(n/NT) = T \sum_{k=0}^{N-1} y(kT)e^{-j2\pi nk/N} \qquad n = 0, 1, \ldots, N/2 \qquad (13.1)$$

Recall that Eq. (13.1) is simply the rectangular integration approximation to the finite-interval continuous Fourier transform integral. Hence, if T is sufficiently small, then Eq. (13.1) can be written with small error as

$$Y(nf_0) = \int_0^{NT} y(t)e^{-j2\pi nf_0 t}\, dt$$

$$= \int_0^{NT} y(t)\cos(2\pi nf_0 t)\, dt - j\int_0^{NT} y(t)\sin(2\pi nf_0 t)\, dt \qquad (13.2)$$

$$n = 0, 1, \ldots, N/2$$

where $f_0 = 1/NT$. Although Eq. (13.2) is a continuous Fourier transform integral, we are only evaluating the equation for the $(N/2) + 1$ discrete frequencies evaluated in Eq. (13.1), that is, $0, f_0, 2f_0, \ldots, (N/2)f_0$. (Observe that Eq. (13.2) holds also for negative frequencies, but this generalization is omitted for clarity.) From Eq. (13.2), we will proceed to demonstrate the implied convolution operation of the FFT. To do so, we have purposely converted from the discrete (Eq. (13.1)) to the continuous domain (Eq. (13.2)). Although it is not necessary to prove our arguments in the continuous domain, we find that such an approach is generally more easily visualized.

Our approach to the development of the FFT band-pass integrate and sample filter concept is to first consider only the real term of Eq. (13.2) (assume $y(t)$ is a real function):

$$Y_R(nf_0) = \int_0^{NT} y(t)\cos(2\pi nf_0 t)\, dt \qquad n = 0, 1, \ldots, N/2 \qquad (13.3)$$

We will show that Eq. (13.3) can be interpreted as a time-sampled output from a bank of filters with impulse-response functions $u(t) \cos(2\pi n f_0 t)$, where $u(t)$ is unity over the interval $(0, NT)$. To facilitate the development, we first consider the low-pass FFT filter of the filter bank.

Low-Pass Filter Development

Consider Eq. (13.3) for the case $n = 0$:

$$Y_R(0) = \int_0^{NT} y(t)\, dt \tag{13.4}$$

For this case, Eq. (13.3) reduces to a simple integration of $y(t)$ over the interval of 0 to NT. We claim that Eq. (13.4), that is, the FFT output for $n = 0$, is the linear-system convolution equation representing a low-pass filter followed by a sampler. To develop this viewpoint, consider the following arguments.

In Fig. 13.1, we show the procedure for convolving the two waveforms $y(t)$ and $u(t)$:

$$r_0(t) = \int_{-\infty}^{\infty} y(\tau)u(t - \tau)\, d\tau \tag{13.5}$$

If we let $y(t)$ represent the input waveform to a linear system and let $u(t)$ be the impulse response of the system, then Eq. (13.5) determines the system output $r_0(t)$. In Fig. 13.1, we have assumed that the system impulse response $u(t)$ is a unity amplitude function over the interval 0 to NT, as shown in Fig. 13.1(a). The system input $y(t)$ is assumed to be a general waveform, as shown in Fig. 13.1(b). Although Eq. (13.5) describes the linear-system output for all time t, we show the graphical evaluation of Eq. (13.5) for the single point of time $t = t' = NT$.

Now observe that the evaluation of the convolution relationship of Eq. (13.5) for the point in time t' requires only the integration of $y(t)$ over the interval 0 to NT, as illustrated in Figs. 13.1(e) and (f). Also note from Fig. 13.1(e) that multiplication by $u(t)$ required in the convolution procedure of Eq. (13.5) actually determines the integration interval because $u(t)$ is defined to have utility amplitude over the interval 0 to NT and to have zero amplitude elsewhere. Hence, the evaluation of the convolution equation for the point of time $t = t'$, as shown in Fig. 13.1(f), reduces to integration of $y(t)$ over the interval 0 to NT. But this result is exactly Eq. (13.4), the FFT output for the case $n = 0$. We have then shown that Eq. (13.4) can be interpreted as the evaluation of a convolution equation for a single point of time. This operation is called *integrate and sample filtering*.

Because Eq. (13.5) describes the output of a linear system with input $y(t)$ and impulse response $u(t)$, the real FFT output at $n = 0$ can be characterized as the output of a linear system sampled at time $t = t' = NT$. The system-response characteristics are defined by the impulse response $u(t)$,

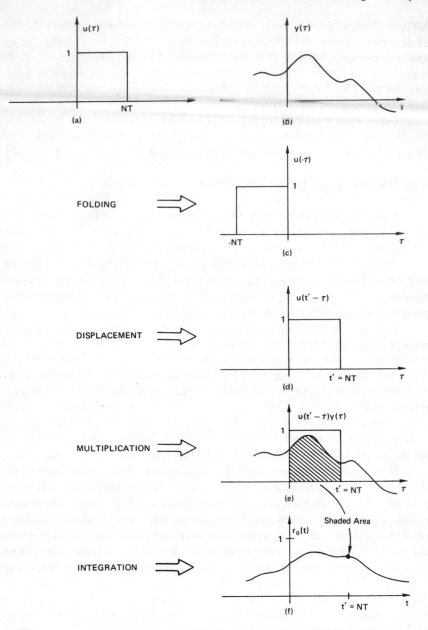

Figure 13.1 Graphical development of the equivalence of the FFT and convolution for the case $n = 0$.

which is illustrated in Fig. 13.2(a). Magnitude of the Fourier transform of this impulse response is shown in Fig. 13.2(b). This frequency function is that of a low-pass filter with a magnitude frequency-response function of

$NT[\sin(\pi f/f_0)]/(\pi f/f_0)$. As a result, the real FFT output for the case $n = 0$, as described by Eq. (13.4), is equivalent to the output of a low-pass integrate and sample filter with a $NT[\sin(\pi f/f_0)]/(\pi f/f_0)$ frequency response. Since Eq. (13.4) evaluates the convolution integral for only one point of time, we interpret the FFT for $n = 0$ as a low-pass filter followed by a sampler.

Note that we have shown that the FFT output can be considered as a sampled value of a time function. This is in direct contrast to the normal interpretation of the FFT as a time-to-frequency-domain transform. We now extend these low-pass filter results to the band-pass filter case.

Band-Pass Filter Development

Let us now consider Eq. (13.3) for the case $n = 1$:

$$Y_R(f_0) = \int_0^{NT} y(t) \cos(2\pi f_0 t) \, dt \tag{13.6}$$

Our objective is to demonstrate that Eq. (13.6) represents the output of a band-pass filter centered at frequency f_0 followed by a sampler. Our approach follows that used for the low-pass filter development.

In Fig. 13.3, we show the basic waveforms obtained in realizing the convolution equation

$$r_1(t) = \int_{-\infty}^{\infty} y(\tau)u'(t - \tau) \, d\tau \tag{13.7}$$

where

$$u'(t) = u(t) \cos(2\pi f_0 t) \tag{13.8}$$

As in the low-pass filter case, Eq. (13.7) characterizes the output of a linear system with input $y(t)$ and impulse response $u'(t)$ defined by Eq. (13.8). Figure 13.3 graphically evaluates Eq. (13.7) for the single point of time $t = t' = NT$. As shown in Figs. 13.3(e) and (f), this single point of the convolution result is determined by multiplying $y(t)$ by $\cos(2\pi f_0 t)$ and integrating over the interval 0 to NT. This sample value of the convolution result is exactly the FFT output for $n = 1$ computed from Eq. (13.6). Hence, Eq. (13.6) can be interpreted as the evaluation of the convolution equation (13.7) for the

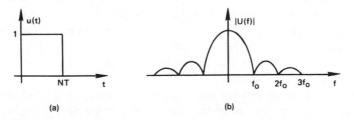

(a)   (b)

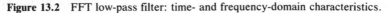

Figure 13.2   FFT low-pass filter: time- and frequency-domain characteristics.

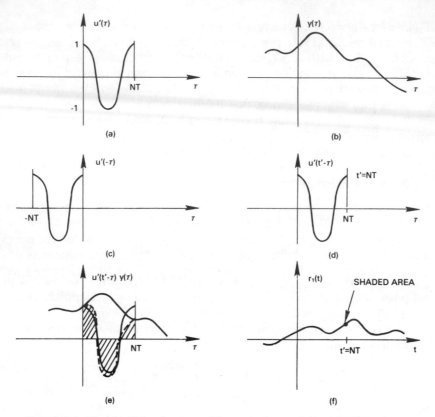

Figure 13.3 Graphical development of the equivalence of the real FFT and convolution for the case $n = 1$.

single point of time $t = t' = NT$, and, as a result, the real FFT output for $n = 1$ is that of a sampled output from a filter with the impulse-response function given by Eq. (13.8).

In Fig. 13.4(a), we show the impulse response of Eq. (13.8). Note that this impulse response is simply the multiplication of the low-pass filter impulse response $u(t)$ and the term $\cos(2\pi f_0 t)$. To determine the corresponding frequency-response function, recall from the frequency-shifting theorem (Eq. (3.23) and Ex. 3.8) that multiplication of the low-pass impulse response $u(t)$ by the $\cos(2\pi f_0 t)$ term translates the low-pass filter frequency-response function illustrated in Fig. 13.2(b) into a band-pass frequency-response function centered at frequency f_0, as shown in Fig. 13.4(b). The band-pass filter frequency characteristic is the $NT[\sin(\pi f/f_0)]/(\pi f/f_0)$ function of the low-pass filter shifted in frequency.

The FFT real output for the case $n = 1$ (Eq. (13.6)) can then be interpreted as a single sample of the output of a band-pass filter centered at frequency f_0 with $NT\{\sin[\pi(f - f_0)/f_0]\}/[\pi(f - f_0)/f_0]$ frequency-response

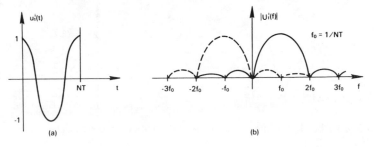

Figure 13.4   FFT band-pass filter centered at frequency $f_0 = 1/NT$: time- and frequency-domain characteristics.

characteristics (magnitude). As in the low-pass filter case, the output of the FFT can be considered as a single value of a time waveform.

If we consider Eq. (13.3) for the general case, we note that the previous arguments apply. For each n, we can draw an illustration analogous to Fig. 13.3. Hence, the FFT real output $Y_R(nf_0, NT)$ can be interpreted as the output of a convolution integral:

$$Y_R(nf_0, NT) = \int_0^{NT} y(t) \cos(2\pi nf_0 t)\, dt$$

$$= \int_{-\infty}^{\infty} y(\tau) u_n^i(NT - \tau)\, d\tau \tag{13.9}$$

$$n = 0, 1, \ldots, N/2$$

where the convolution integral must be interpreted as being evaluated only at $t = NT$ and where $u_n^i(t)$ is the system impulse response given by

$$u_n^i(t) = u(t) \cos(2\pi nf_0 t) \tag{13.10}$$

The superscript i in $u_n^i(t)$ indicates that the filter response is in phase and is determined by multiplication of the impulse response $u(t)$ by a cosine term.

From the frequency-shifting theorem (Ex. 3.8), multiplication of the low-pass filter impulse response $u(t)$ by the function $\cos(2\pi nf_0 t)$ translates the $NT[\sin(\pi f/f_0)]/(\pi f/f_0)$ low-pass filter frequency response to a band-pass filter centered at frequency nf_0. As a result, Eq. (13.3) can be interpreted as the band-pass filter bank illustrated in Fig. 13.5, where it is understood that we sample the outputs of the filter bank at time $t = NT$. All side-lobe characteristics of the $NT\{\sin[\pi(f - nf_0)/f_0]\}/[\pi(f - nf_0)/f_0]$ filters in Fig. 13.5 have been omitted for clarity. The frequency response of each filter is centered at frequency nf_0, where $n = 0, 1, \ldots, N/2$. Note that we can easily extend our arguments to include nf_0, where $n = N/2 + 1, \ldots, N - 1$, to determine the filter response for negative frequencies.

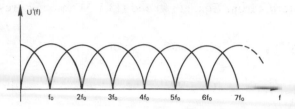

Figure 13.5 FFT band-pass filter bank frequency-domain characteristics (side lobes deleted).

Quadrature Band-Pass Filter Bank

We have neglected the imaginary term in Eq. (13.2). If we treat this term in the same manner as the real term, then we obtain another band-pass filter bank described by

$$Y_I(nf_0, NT) = -\int_0^{NT} y(t)\, \sin(2\pi nf_0 t)\, dt$$

$$= \int_{-\infty}^{\infty} y(t) u^q (NT - \tau)\, d\tau \qquad (13.11)$$

$$n = 0, 1, \ldots, N/2$$

where the convolution integral is evaluated only at $t = NT$ and where $u^q(t)$ is the system impulse response given by

$$u^q(t) = u(t)\, \sin(2\pi nf_0 t) \qquad n = 0, 1, \ldots, N/2 \qquad (13.12)$$

Superscript q in $u^q(t)$ indicates that the impulse response is in quadrature and is determined by multiplication of the impulse response $u(t)$ by a sine term. The negative sign in Eq. (13.11) is absorbed by the folding operation in the convolution process (see Prob. 13.2). A comparison of Eqs. (13.12) and (13.10) reveals that the two impulse-response functions differ only by a phase shift, that is, the difference between $\sin(2\pi nf_0 t)$ and $\cos(2\pi nf_0 t)$. Hence, we use the defining terms *in phase* and *quadrature*. We can then repeat the arguments leading to Fig. 13.3 with the exception that the impulse-response function is now a sine function instead of a cosine function. Completion of this development results in another band-pass filter bank that differs only in phase-response characteristics from those of the previous development. Equation (13.11) can be interpreted as the time-sampled output of a filter bank with impulse response that is in quadrature or 90° out of phase with the filter bank described by Eq. (13.9).

Summary

Based on the previous discussions, FFT results *can be interpreted* as the time-sampled outputs of two band-pass filter banks from which phase and amplitude information can be derived. Summarizing, the real and imaginary FFT terms from Eqs. (13.9) and (13.11) can be expressed as

$$Y_R(nf_0,NT) = \int_0^{NT} y(t) \cos(2\pi nf_0 t)\, dt$$

$$= \int_{-\infty}^{\infty} y(\tau)u(NT - \tau) \cos[2\pi nf_0(NT - \tau)]\, d\tau \qquad (13.13)$$

$$Y_I(nf_0,NT) = -\int_0^{NT} y(t) \sin(2\pi nf_0 t)\, dt$$

$$= \int_{-\infty}^{\infty} y(\tau)u(NT - \tau) \sin[2\pi nf_0(NT - \tau)]\, d\tau \qquad (13.14)$$

$$n = 0, 1, \ldots, N/2$$

Hence, the FFT relationship of Eq. (13.2) can be written as

$$Y(nf_0,NT) = Y_R(nf_0) + jY_I(nf_0)$$

$$= \int_0^{NT} y(t) \cos(2\pi nf_0 t)\, dt -$$

$$j \int_0^{NT} y(t) \sin(2\pi nf_0 t)\, dt$$

$$= \int_0^{NT} y(t)e^{-j2\pi nf_0 t}\, dt \qquad (13.15)$$

$$= \int_{-\infty}^{\infty} y(\tau)u(NT - \tau)e^{j2\pi nf_0(NT - \tau)}\, d\tau$$

$$n = 0, 1, \ldots, N/2$$

where the convolution integral is evaluated only at $t = NT$.

The filter bank described by the real part of Eq. (13.15) is generally termed the *in-phase filter bank* and the filter bank described by the imaginary part of Eq. (13.15) is termed the *quadrature filter bank*. Sampled real and quadrature band-pass filter bank outputs are an alternate way of interpreting the conventional real and imaginary outputs of the FFT.

13.2 FFT BAND-PASS FILTER FREQUENCY-RESPONSE CHARACTERISTICS

Within the context of interpreting the FFT as a bank of filters, it is of value to reexamine some of the terminology associated with the FFT. In particular, we wish to investigate the frequency-response characteristics of FFT filters and reinvestigate FFT resolution and data-weighting functions.

FFT Filter Bank Frequency-Response Characteristics

From Fig. 13.5, we note that there is considerable overlap of the frequency-response functions of adjacent filters in the FFT filter bank. As a result, a single sinusoid input to the FFT filter bank can produce an output at several adjacent filters. To examine this effect further, consider the FFT results shown in Fig. 13.6. The input sinusoid to the FFT is a cosine waveform of frequency 6.5/32 Hz. For $N = 32$ and $T = 1$, the filter frequency responses of the FFT filter bank are centered at integer multiples of the frequency $f_0 = 1/NT = 1/32$. Hence, the frequency of the input sinusoidal is exactly between the center frequencies of the two band-pass filters centered at 6/32 and 7/32 Hz.

As shown, the FFT responds to the input sinusoid with maximum output at the two adjacent filters. The output magnitude is 0.637 of the input. All other filters of the FFT filter bank also respond to the input sinusoid because of the side-lobe frequency response or *side-lobe leakage* of the filter bank. The output value for each FFT filter in the filter bank is determined by the magnitude of the filter main lobe or side lobe at the frequency of the input sinusoid.

We know from Sec. 9.2 that we can reduce side-lobe leakage by use of data-weighting functions. This concept is reexamined in this section in view of interpreting the FFT as a band-pass filter bank.

FFT Resolution

In Chapter 9, we addressed FFT resolution. The interpretation of the FFT as a bank of integrate and sample filters further illustrates the concept of FFT resolution. Recall from previous developments that the frequency

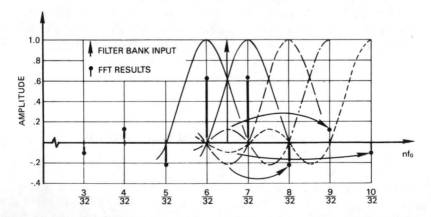

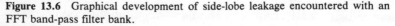

Figure 13.6 Graphical development of side-lobe leakage encountered with an FFT band-pass filter bank.

responses of the filters in the FFT integrate and sample filter bank are centered at integer multiples of the frequency f_0, as illustrated in Fig. 13.7(a). Thus, adjacent frequency responses of filters are separated by $1/NT$, the resolution of the FFT. Note from Fig. 13.7(a) that *crossover* points of adjacent frequency responses are also separated by the resolution term $1/NT$. These crossover points are at the -4 dB values on the filter frequency-response characteristics. This crossover value contrasts to the normally encountered -3 dB crossover definition for a filter bank.

Resolution of a filter is defined as the capability of a filter bank to distinguish between frequencies. Sinusoids of frequencies contained within the bandwidth of any filter of the band-pass filter bank can not be distinguished at the output of that filter. Hence, the term *resolution* is used. When a rectangular data-weighting function is used, the convention is to define the bandwidth of each filter as $f_0 = 1/NT$.

To illustrate the filter bank concept of FFT resolution improvement, consider Fig. 13.7. Because NT is the duration of the time function that is being FFTed, then FFT resolution improvement, that is, a decrease in the bandwidth of each FFT filter, can be achieved by increasing the number of

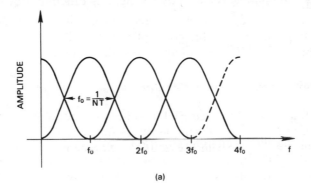

(a)

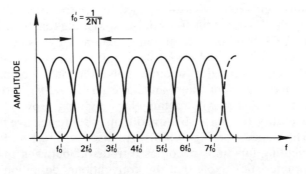

(b)

Figure 13.7   FFT band-pass filter bank for (a) N samples and for (b) $2N$ samples.

data points N (for a given T). In Fig. 13.7(a), we show the FFT band-pass filters with resolution $f_0 = 1/NT$. Figure 13.7(b) illustrates the improved resolution obtained by doubling NT. If we increase the time duration of the data record being FFTed to $2NT$, then the FFT resolution becomes $f'_0 = 1/2NT$, as illustrated in Fig. 13.7(b). As shown, the filter bandwidths decrease by a factor of two.

FFT Data-Weighting Function: A Filtering Viewpoint

As developed in this chapter, the FFT can be considered as a bank of integrate and sample band-pass filters with poor side-lobe characteristics. However, from Sec. 9.2, we know that it is possible to improve side-lobe characteristics by using data-weighting functions. This concept is further clarified by noting in Sec. 13.1 that we show the FFT can be characterized as the output of a linear system with impulse-response function $u'(t)$. If we shape this impulse-response function, we can improve the band-pass filter characteristics.

To modify the impulse-response function of the filter bank, we proceed as in Sec. 9.2 and multiply the data by a weighting function. Equation (13.15) becomes

$$Y(nf_0, NT) = \int_0^{NT} [w(t)y(t)]e^{-j2\pi n f_0 t} \, dt$$

$$= \int_0^{NT} y(t)[w(t)e^{-j2\pi n f_0 t}] \, dt \qquad (13.16)$$

$$n = 0, 1, \ldots, N/2$$

where $w(t)$ is the data-weighting function. Note in Eq. (13.16) that multiplication of the data by a weighting or window function is equivalent to multiplication of the impulse-response function $u(t)e^{-j2\pi n f_0 t}$ by the weighting function $w(t)$.

Analogous to the development in Sec. 13.1, we can show that the frequency-response characteristics of each filter in the FFT filter bank are determined by the Fourier transform of the weighting function. In Fig. 13.8, we show the FFT band-pass filter bank obtained by using the Hanning weighting function. The band-pass filter bank obtained by using the conventional rectangular weighting function is also shown for comparison. Note that the filters in the Hanning filter bank have a bandwidth greater than those of the filter banks obtained with rectangular weighting. However, we accept this loss of resolution (increased bandwidth) in order to achieve the improved side-lobe performance (see Fig. 9.8(b)). As discussed in Sec. 9.2, the utilization of weighting functions becomes a trade-off between resolution and side-lobe characteristics. Note that the conventional definition of FFT res-

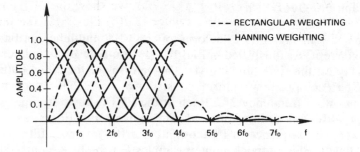

Figure 13.8 Comparison of the FFT band-pass filter bank frequency-response characteristics for rectangular and Hanning weighting functions.

olution ($f_0 = 1/NT$) implies a different bandwidth when weighting functions are used.

Summary

The results of this section do not differ from those of Sec. 9.2. Only the interpretation viewpoint has changed in that we have reexamined some of the basic concepts of the FFT in terms of linear filters. If one's formal education includes linear-system theory, this section should lend additional insight to some of the basic concepts of the FFT.

13.3 MULTICHANNEL BAND-PASS FILTERING BY SHIFTED FFTs

In Sec. 13.1, we developed the interpretation of the FFT as a bank of integrate and sample filters. We now extend that discussion to a series of FFTs where each sequential FFT is hopped or shifted along the time function being transformed. This sequence of FFT outputs can be interpreted as time samples from the outputs of a bank of band-pass filters. That is, we can use a sequence of FFTs to filter a time signal into time samples of individual channels with bandwidth equal to the resolution of the FFT. We will show that the FFT outputs are complex time samples of the outputs of a bank of quadrature filters.

Because one normally considers the FFT as a time-to-frequency transform, the concept of realizing a sequence of FFT outputs as a filtered time sequence is often confusing. For this reason, we find it of value to investigate in detail this application of the FFT. In this section, we develop graphically and analytically the concept of FFT multichannel band-pass filtering.

Graphical Overview

We show in Fig. 13.9 a pictorial of the FFT multichannel band-pass filtering concept. As illustrated in Fig. 13.9(a), multiple FFTs are performed sequentially on the time function $y(t)$. Waveform $y(t)$ of Fig. 13.9(b) is assumed to be a composite waveform, consisting of a constant-value waveform and a sinusoid of frequency $2f_0$. Each of the FFTs shown in Fig. 13.9(a) is equivalent to a bank of integrate and sample band-pass filters, as shown in Fig. 13.9(f). The sequence of FFT outputs for the low-pass filter is shown in Fig. 13.9(c). FFT output complex samples for the filter centered at fre-

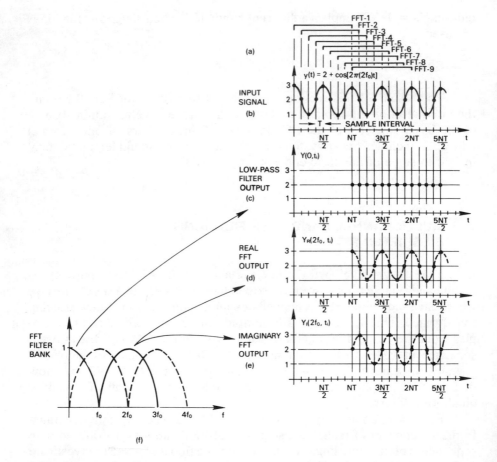

Figure 13.9 Graphical representation of the output of a sequence of FFTs with multiple input signals.

quency $2f_0$ are shown in Figs. 13.9(d) and (e). Each sequence of FFT outputs for a filter is the sampled output-time waveform for that respective filter.

Figure 13.9(a) also illustrates the overlap of sequential FFTs. As shown, the *hop* or *shift* for each FFT is one sample interval T of the input signal. Hence, the sampling rate of the output signal of each filter is $1/T$. In later developments, we increase the FFT shift interval.

Note from Fig. 13.9(c) that the output waveform of the real FFT low-pass filter is the constant term of the input signal. Further, the output wave-form of the real FFT band-pass filter centered at frequency $2f_0$, as shown in Fig. 13.9(d), is the sinusoidal term of the input waveform. The output waveform of the quadrature, or imaginary, FFT band-pass filter centered at frequency $2f_0$, as shown in Fig. 13.9(e), is identical to the output of the real filter with the exception of a time delay (90° phase shift). Because the hop or shift interval of successive FFTs is T, then the sample interval for each FFT filter output is T. We now develop a theoretical basis to support the graphical results of Fig. 13.9.

Theoretical Development

In Fig. 13.9, we show a sampled time function $y(t)$ that we wish to digitally band-pass filter using the FFT. The time domain over which each successive FFT is to be taken is also shown. If we assume a FFT weighting function $w(t)$, then we can write the FFT approximation of Eq. (13.16) for the time interval 0 to NT as

$$Y(nf_0, NT) = \int_0^{NT} y(t)w(t)e^{-j2\pi nf_0t}\, dt \qquad n = 0, 1, \ldots, N/2 \qquad (13.17)$$

Recall that the parameter NT in $Y(nf_0, NT)$ is the end point of the time interval over which the first FFT is to be taken. We can write Eq. (13.17) equivalently as a convolution integral:

$$Y(nf_0, t_1) = \int_{-\infty}^{\infty} y(\tau)w(t_1 - \tau)e^{j2\pi nf_0(t_1 - \tau)}\, d\tau \qquad (13.18)$$

where $t_1 = NT$ and we assume that Eq. (13.18) is evaluated only at $t = t_1$. Recall that the convolution procedure requires that we fold $w(t)$ about the y axis and then shift the folded function. A shift of $t_1 = NT$ is required in Eq. (13.18) to obtain the weighting function with position as illustrated in Fig. 13.9 for FFT-1.

Now consider a second FFT over the interval δ to $\delta + NT$, as shown by the placement for FFT-2 illustrated in Fig. 13.9. For this time interval, the FFT approximation is

$$Y(nf_0, \delta + NT) = \int_{\delta}^{\delta + NT} y(t)w(t - \delta)e^{-j2\pi nf_0(t - \delta)}\, dt \qquad (13.19)$$

Analogous to Eq. (13.18), we can write Eq. (13.19) as a convolution integral:

$$Y(nf_0, t_2) = \int_{-\infty}^{\infty} y(\tau)w(t_2 - \tau)e^{j2\pi nf_0(t_2 - \tau)} \, d\tau \qquad (13.20)$$

where the convolution is evaluated only at $t_2 = \delta + NT$. Note that if we fold $w(\tau)$ according to the rules of convolution, then the shift $t_2 = \delta + NT$ is required to obtain the placement illustrated in Fig. 13.9 for FFT-2.

From Eqs. (13.18) and (13.20), the FFT approximation for any time interval can be written as

$$Y(nf_0, t_i) = \int_{-\infty}^{\infty} y(\tau)w(t_i - \tau)e^{j2\pi nf_0(t_i - \tau)} \, d\tau \qquad (13.21)$$

where t_i is the end point of the interval of duration NT over which the FFT is taken, as illustrated in Fig. 13.9. Note that Eq. (13.21) is simply the convolution of $y(t)$ with the function $w(t)e^{j2\pi nf_0t}$, where we evaluate the convolution at times $t = t_1, t_2, \ldots, t_i, \ldots$.

As discussed previously, the term $w(t)e^{j2\pi nf_0t}$ can be interpreted as the impulse response of a linear system that yields an in-phase and quadrature band-pass filter bank. As a result, for each t_i, Eq. (13.21) and the sequence of FFTs illustrated in Fig. 13.9 can be interpreted as the sampled output (at time t_i) of a bank of analog band-pass filters. The interval $t_i - t_{i-1}$ is the sample interval of the filter-bank outputs.

Consider Eq. (13.21) for the case $n = 0$:

$$Y(0, t_i) = \int_{-\infty}^{\infty} y(\tau)w(t_i - \tau) \, d\tau \qquad (13.22)$$

We observe that Eq. (13.22) is the convolution of the waveform $y(t)$ with the impulse-response function $w(t)$, that is, we have filtered $y(t)$ with a low-pass filter whose frequency-response characteristics are given by the Fourier transform of $w(t)$. Further, Eq. (13.22) is valid only for the points in time $t = t_1, t_2, t_3, \ldots$, which implies that we have sampled the output of the low-pass filter. Hence, the FFT output sequence $Y(0, t_1)$, $Y(0, t_2)$, $Y(0, t_3)$, $\ldots$ is actually a set of time-domain samples of the output of a low-pass filter with input $y(t)$.

Similarly, the FFT output sequence for the band-pass filter centered at frequency f_0, that is, $Y(f_0, t_1)$, $Y(f_0, t_2)$, $Y(f_0, t_3)$, $\ldots$, is the complex sampled time waveform of the output from a band-pass filter (center frequency f_0). The real part of $Y(f_0, t_1)$, $Y(f_0, t_2)$, $Y(f_0, t_3)$, $\ldots$ is the sampled waveform out of the in-phase FFT filter bank and the imaginary part of $Y(f_0, t_1)$, $Y(f_0, t_2)$, $Y(f_0, t_3)$, $\ldots$ is the sampled waveform output of the quadrature FFT filter bank. The imaginary part of $Y(f_0, t_i)$ is shifted in phase $90°$ with respect to the real part of $Y(f_0, t_i)$. Hence, a sequence of FFTs on the time function $y(t)$ yields results equivalent to a bank of quadrature digital band-pass filters.

Example 13.1   FFT Low-Pass Filtering

To further illustrate the special case of FFT low-pass filtering, assume $w(t)$ is the rectangular weighting function. From Eq. (13.22),

$$Y(0,t_i) = \int_{-\infty}^{\infty} y(\tau)w(t_i - \tau) \, d\tau = \int_{t-NT}^{t} y(t) \, dt \qquad (13.23)$$

We show an example of FFT low-pass filtering according to Eq. (13.23) in Fig. 13.10. For the graphical example shown, we have chosen the same input waveform assumed in Fig. 13.9:

$$y(t) = 2 + \cos(2\pi f't) \qquad f' = 2/NT \qquad (13.24)$$

Sample interval T has been set to $NT/8$.

Figure 13.10(a) illustrates the window placement of the first FFT described by Eq. (13.23):

$$Y(0,NT) = \int_{0}^{NT} y(t) \, dt \qquad (13.25)$$

Substitution of Eq. (13.24) into Eq. (13.25) yields a value of $2NT$ because the cosine term integrates to zero. We graphically obtain the same result from Fig. 13.10(a) by determining the area under the waveform over the interval 0 to NT. The output of the low-pass filter (i.e., the FFT) at the time NT is equal to $2NT$, as shown in Fig. 13.10(d). Effectively, we have filtered the signal $y(t)$ with a rectangle impulse-response low-pass filter and sampled the filtered waveform at time $t_1 = NT$, the first sample value shown in Fig. 13.10(d).

To obtain another sample of the low-pass filter output waveform, we shift the FFT window one sample interval, as shown by the weighting-function location FFT-

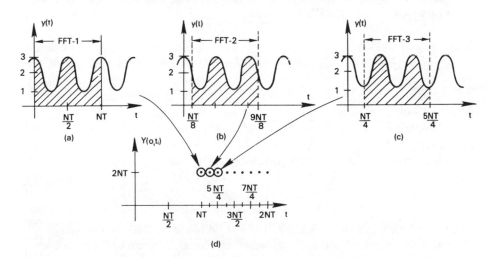

Figure 13.10   Example of FFT low-pass filtering.

2 in Figs. 13.9 and 13.10(b). For this case, Eq. (13.23) becomes

$$Y(0,9NT/8) = \int_{NT/8}^{9NT/8} y(t) \, dt \tag{13.26}$$

Substitution of Eq. (13.24) into Eq. (13.26) yields a value of $2NT$ because the cosine term integrates to zero. From Fig. 13.10(b), we obtain the same result by evaluating the area under the waveform over the interval $NT/8$ to $9NT/8$. This result is equivalent to the output of a low-pass filter (i.e., the FFT output) sampled at time $9NT/8$, as shown in Fig. 13.10(d). Figure 13.10(c) further illustrates the shifting or hopping FFT and the interpretation of the FFT as the sampled output of the low-pass filter. Note that the sample interval of the low-pass filter output waveform is $NT/8$, the FFT shift interval.

Recall that the frequency response of the FFT low-pass filter under the assumption of a rectangular window is a $NT[\sin(\pi f/f_0)]/(\pi f/f_0)$ function, which is zero at frequencies $f_0 = 1/NT, 2/NT, 3/NT, \ldots$ (see Fig. 13.2). Hence, we would expect the FFT low-pass filter to completely eliminate an input sinusoid of frequency $f_0 = 2/NT$. This is the case illustrated in Fig. 13.10.

Note that the zero-frequency input signal to the low-pass filter has an amplitude of 2 and the output amplitude is $2NT$. The amplitude-scaling factor NT should be interpreted as the *gain* (multiplier) of the FFT filter bank.

Example 13.2   FFT Band-Pass Filtering of a Sinusoid

To graphically illustrate the concept of FFT band-pass filtering of a sinusoid, assume $y(t)$ is given by

$$y(t) = 2 + \cos(2\pi f't) \qquad f' = 2/NT \tag{13.27}$$

which is the identical waveform considered in Ex. 13.1. Here we evaluate the sequence of FFT equations (13.21) for the case $n = 2$, that is, the band-pass filter with center frequency $f_0 = 2/NT$ is evaluated.

Substitution of Eq. (13.27) into Eq. (13.21) for $n = 2$ yields

$$Y(2f_0,t_i) = \int_{-\infty}^{\infty} [2 + \cos(2\pi f'\tau)]w(t_i - \tau)e^{j4\pi f_0(t_i - \tau)} \, d\tau \tag{13.28}$$

For simplicity, we let $w(t)$ equal the rectangular function. Equation (13.28) becomes

$$Y(2f_0,t_i) = 2\int_{t_i - NT}^{t_i} [e^{j4\pi f_0 t_i}]e^{-j4\pi f_0 \tau} \, dt \tag{13.29}$$

$$+ \int_{t_i - NT}^{t_i} [e^{j4\pi f_0 t_i}\cos(2\pi f'\tau)]e^{-j4\pi f_0 \tau} \, dt$$

$$= e^{j4\pi f_0 t_i} \int_{t_i - NT}^{t_i} \cos(2\pi f't)[\cos(4\pi f_0 t) - j\sin(4\pi f_0 t)] \, dt \tag{13.30}$$

Since $f_0 = 1/NT$, the first integral of Eq. (13.29) is always over an integer number of periods and is zero. That is, the constant term in $y(t)$ is removed from the output of the filter centered at frequency $2f_0$. Evaluation of Eq. (13.30) for $t_i = NT$, the first of the sequence of FFTs, yields

$$Y(2f_0,NT) = e^{j4\pi f_0(NT)} \int_0^{NT} \cos(4\pi f_0 t)[\cos(4\pi f_0 t) - j\sin(4\pi f_0 t)]\, dt$$

$$= \int_0^{NT} \cos^2(4\pi f_0 t)\, dt - j\int_0^{NT} \cos(4\pi f_0 t)\sin(4\pi f_0 t)\, dt$$

$$= \int_0^{NT} \tfrac{1}{2}\, dt + \int_0^{NT} \tfrac{1}{2}\cos(8\pi f_0 t)\, dt \tag{13.31}$$

$$-j\int_0^{NT}[\,\tfrac{1}{2}\sin(0) + \tfrac{1}{2}\sin(8\pi f_0 t)]\, dt$$

$$= NT/2$$

Computation of Eq. (13.31) is graphically illustrated in Fig. 13.11(a). We show the window placement for the first FFT and the cosine impulse-response function of the FFT filter centered at frequency $2f_0$. The product of this cosine term and the

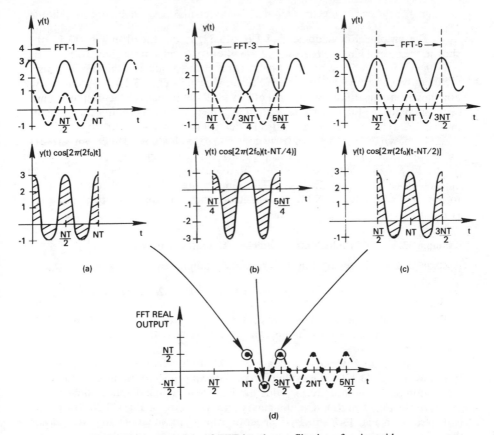

Figure 13.11 Example of FFT band-pass filtering of a sinusoid.

waveform $y(t)$ is also shown. Integration of the product term yields the result of Eq. (13.31), which is shown in Fig. 13.11(d). The corresponding product involving the sine term is not shown because it integrates to zero. The output value $NT/2$ is the sample output of the FFT band-pass filter (center frequency $2f_0$) at time $t_1 = NT$.

We evaluate the output of the same FFT band-pass filter at time $t_i = 9NT/8$ (FFT-2) in a later example (Ex. 13.3). To obtain the FFT output at time $t_i = 5NT/4$, which is FFT-3 in Figs. 13.9 and 13.11(b), we write Eq. (13.30) as

$$Y(2f_0, 5NT/2) = e^{j2\pi f_0(5NT/4)} \int_{NT/4}^{5NT/4} \cos[2\pi(2f_0)t] \, [\cos(4\pi f_0 t)$$

$$- j \sin(4\pi f_0 t)] \, dt$$

$$= - \int_{NT/4}^{5NT/4} \cos^2(4\pi f_0 t) \, dt \qquad (13.32)$$

$$+ j \int_{NT/4}^{5NT/4} \cos(4\pi f_0 t) \sin(4\pi f_0 t) \, dt$$

$$= - NT/2$$

The graphical development of Eq. (13.32) is illustrated in Fig. 13.11(b). As shown, the cosine impulse response of the filter centered at frequency $2f_0$ now ranges over the interval $NT/4$ to $5NT/4$. The integral of the product of this cosine term and the input waveform yields the $- NT/2$ result shown in Fig. 13.11(d). Again, we have not shown the quadrature sine term of the FFT filter because the output is zero.

Evaluation of the FFT filter output at time $t_i = 3NT/2$ is graphically illustrated in Fig. 13.11(c) and can be analytically determined analogous to Eq. (13.28). The result is shown in Fig. 13.11(d). If we continue to evaluate Eq. (13.28) for successive values of t_i, the results depicted in Fig. 13.11(d) result. That is, successive outputs of the FFT band-pass filter centered at frequency $2f_0$ are samples of the input sinusoidal waveform $\cos[2\pi(2f_0)t]$. The sample interval of the filter output is equal to the FFT shift interval $NT/8$. As expected, the constant term of the input signal is filtered by the FFT band-pass filter.

Example 13.3   FFT Multichannel Filtering: Complex Samples

Assume that $y(t)$ is given as in the previous examples by

$$y(t) = 2 + \cos(2\pi f' t) \qquad f' = 2/NT \qquad (13.33)$$

In Ex. 13.2, we purposely chose the FFT shift interval to ensure that all imaginary FFT output samples are zero. In this example, we remove this restriction and address complex samples.

From Ex. 13.2, we first evaluate Eq. (13.30) for the case $n = 2$ and $t_i = NT$. We show in Fig. 13.12(a) the evaluation of the real product term in Eq. (13.30). The integration of this product term yields the FFT real sample output at time $t_i = NT$, as shown in Fig. 13.12(c). The imaginary product term of Eq. (13.30) for $t_i = NT$ is illustrated in Fig. 13.12(b). By inspection, this term integrates to zero and hence the FFT imaginary output sample is zero, as shown in Fig. 13.12(d).

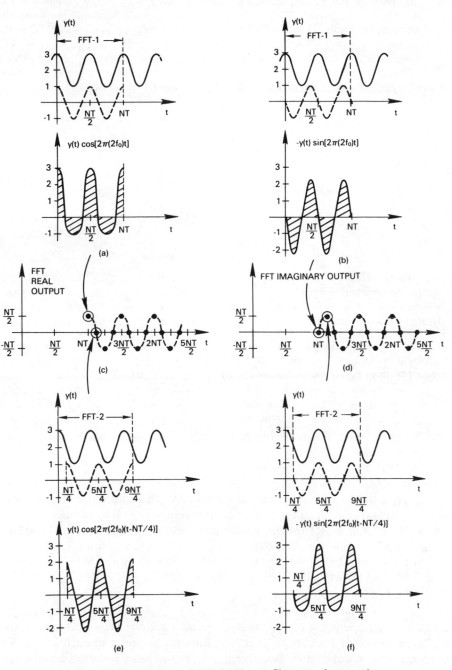

Figure 13.12 Example of FFT band-pass filter complex samples.

To evaluate Eq. (13.30) at time $t_i = 9NT/4$ and $f' = 2/NT$, we write

$$Y(2f_0, 9NT/4) = e^{j2\pi f_0(9NT/4)} \int_{NT/4}^{9NT/4} \cos[2\pi(2f_0)t] \, [\cos(4\pi f_0 t)$$

$$- j \sin(4\pi f_0 t)] \, dt$$

$$= e^{j9\pi/2} \left[\int_{NT/4}^{9NT/4} \cos^2(4\pi f_0 t) \, dt \right.$$

$$\left. - j \int_{NT/4}^{9NT/4} \cos(4\pi f_0 t) \sin(4\pi f_0 t) \, dt \right]$$

$$= e^{j9\pi/2} \left[\int_{NT/4}^{9NT/4} \tfrac{1}{2} \, dt \right. \tag{13.34}$$

$$+ \int_{NT/4}^{9NT} \tfrac{1}{2} \, \cos(8\pi f_0 t) \, dt$$

$$\left. - j \int_{NT/4}^{9NT/4} \cos(4\pi f_0 t) \sin(4\pi f_0 t) \, dt \right]$$

$$= e^{j9\pi/2} \, (NT/2)$$

$$= jNT/2$$

Evaluation of the real term of Eq. (13.34) is graphically illustrated in Fig. 13.12(e). As shown in Fig. 13.12(c), the real product term integrates to zero for $t_2 = 9NT/4$. Graphical evaluation of the imaginary term is shown in Fig. 13.12(f). The imaginary product term integrates to $NT/2$, as shown in Fig. 13.12(d).

If we continue to evaluate the FFTs, Eq. (13.30) for subsequent values of t_i, we obtain complex samples represented by the waveforms illustrated in Figs. 13.12(c) and (d). Note that the real and quadrature filter output sampled waveforms are identical with the exception of a time delay or phase shift.

Summary

Recall from Chapter 9 that the FFT halves the amplitude of sinusoids between the positive and negative frequency outputs. As a result, if we repeat Examples 13.2 and 13.3 for the frequency $-f_0$, we obtain identical results. Hence, if we add both positive and negative frequency results, we obtain a sinusoid with maximum amplitude NT, which is the amplitude (unity) of the input sinusoid multiplied by the FFT filter bank gain NT. The sample rate of the FFT filter outputs is equal to 1/(FFT shift interval).

We have shown that each FFT output is equivalent to a sample from a convolution operation of the input waveform and the impulse response of the respective FFT band-pass filter. The impulse response of each FFT filter is complex or in quadrature because the FFT weighting function (and hence the input signal) is multiplied by a cosine term to obtain the real output and multiplied by a sine term to obtain the imaginary output. Because the impulse responses for each filter in the filter bank only differ by a 90° phase shift, then the output waveforms of the in-phase and quadrature filters are in quadrature and only differ by a 90° phase shift.

13.4 SAMPLE RATE CONSIDERATIONS IN FFT MULTICHANNEL FILTERING

For clarity of presentation, we purposely oversampled the output of each FFT band-pass filter in the previous examples. Because we set the FFT shift or hop interval equal to the sample interval T of the input waveform, then each FFT band-pass filter output waveform is also sampled with interval T. However, a band-pass waveform can be sampled at a rate that is determined by the bandwidth of the waveform and not the maximum frequency component of the signal. The procedures used are termed *down sampling* and *quadrature sampling*. Both sampling techniques are described in detail in Secs. 14.1 and 14.2.

We show in Sec. 14.1 that a bandpass signal with transmission bandwidth B_T can be down sampled (with constraints) if the sampling frequency $f_s \geq 2B_T$. However, we show in Sec. 14.2 that further sampling efficiencies can be obtained by representing the band-pass signal in complex or quadrature form before sampling. A band-pass signal that has been translated or down sampled to zero center frequency in quadrature form (i.e., complex) can be sampled at a rate $f_s \geq B_T$. Both down-sampling and quadrature-sampling techniques can be applied to band-pass waveforms if a band-pass filter is used to control aliasing.

The aliasing level for each FFT band-pass filter is determined by the characteristics of the window or weighting function. In most practical applications of FFT band-pass filtering, weighting-function selection is a trade-off compromise between the time duration of the filter and the desired filter performance in the passband (minimum ripple) and stop band (low side lobes). The weighting-function impulse response of a filter with low side lobes generally has a duration considerably longer than that of a rectangular weighting function with similar bandwidth (see Figs. 9.8(a) and (b)). A longer-duration weighting function increases the number of sample values N for each FFT and hence determines the practicality of FFT band-pass filtering. The normal procedure is to adopt a weighting function that provides reasonable band-pass filtering characteristics and then translate each FFT band-pass filter output to a baseband where a digital recursive filter is applied to improve the filtering characteristics.

In Fig. 13.13, we illustrate the correct procedure for defining the bandwidth of a FFT band-pass filter. We show the frequency-domain characteristics of an arbitrarily selected weighting function. The level of aliasing is determined by the filter side lobes. If an aliasing level considered to be

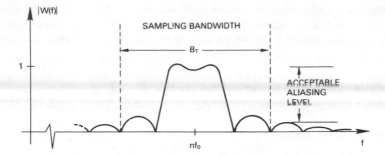

Figure 13.13 Graphical definition of the sampling bandwidth B_T of a band-pass filter.

acceptable is as illustrated, then the output waveform of this band-pass filter has a *sampling bandwidth* B_T as shown. Bandwidth B_T is also termed the *transmission bandwidth* and is used in applying the sampling techniques developed in Secs. 14.1 and 14.2 to the special case of FFT band-pass filtering.

From Sec. 14.1 we know, that a band-pass waveform can be down sampled to its low-pass equivalent. If only the FFT real band-pass filter outputs are used, then Eq. (14.1) determines the appropriate sample rate. We can improve the sampling efficiency if we apply both down-sampling and quadrature-sampling techniques. Because FFT band-pass filtering results in complex or quadrature sampled waveforms, we can down sample to zero center frequency both the real and imaginary outputs of each FFT band-pass filter and recover the original signal even though spectrum overlap (aliasing) occurs. The down-sampling frequency f_s' must satisfy $f_s' > B_T$, Eq. (14.5). Note that each FFT filter is centered at integer multiples of the frequency f_0, where $f_0 = 1/NT$. Hence, from Sec. 14.1, translation to zero center frequency requires we set an integer multiple of the sampling frequency f_s' equal to the center frequency of each band-pass waveform that is to be down sampled. In most FFT band-pass filtering applications, the lowest frequency FFT filter used has a center frequency nf_0, where n is integer valued. We then set the sampling frequency f_s' equal to an integer multiple of f_0, that is, $T' = NT/n$. The sample interval T' for each FFT *complex* band-pass filter must satisfy the following relationships:

$$T' \leq 1/B_T \tag{13.35}$$

$$T' = NT/n \qquad n \text{ is integer valued} \tag{13.36}$$

$$T' = pT \qquad p \text{ is integer valued} \tag{13.37}$$

Note that Eq. (13.37) ensures that the output sample interval T' is selected as an integer multiple of T, the sample interval of the input waveform to the FFT. Sample interval T' is implemented by setting the FFT shift or hop interval to T'.

Quadrature sampling requires we reconstruct a *real* signal from the complex samples representing the aliased zero center frequency down-sampled spectrum. From Eq. (14.2), we multiply these complex time-domain samples by the exponential $e^{-j2\pi f't}$, where f' is the center frequency of the desired baseband signal. The real part of the frequency-shifted complex waveform is the desired signal. Interpolation may be required to increase the sampling rate consistent with the bandwidth of the resulting baseband signal (see Sec. 14.2).

13.5 FFT MULTICHANNEL DEMULTIPLEXING

A practical application of the fundamentals developed in this chapter is the digital demultiplexing of frequency multiplexed signals. We use the example multiplexed signal shown in Fig. 13.14(a) to outline the principles of FFT band-pass filtering for multichannel signals. Each 4-kHz spectrum shown is assumed to be the result of single side-band modulation of a voice signal. Our objective is to FFT band-pass filter each channel of the multiplexed signal and reconstruct the signal so that the voice is audible.

We choose our FFT parameters to center a band-pass filter at 2 kHz. Therefore, let

$$f_0 = 1/NT = 2 \text{ kHz} \tag{13.38}$$

and the corresponding FFT filter bank for a Hanning weighting function (see Fig. 9.8(b)) is shown in Fig. 13.14(b). As illustrated, this selection of f_0 results in a filter whose passband frequency response is broader than required to filter the single side-band voice spectrum. However, if we let $f_0 = 1$ kHz, then the resulting band-pass filter severely attenuates the voice spectrum in each channel. Also note that we have redundant FFT band-pass filters. We use only those FFT outputs that are centered on the frequency of each channel of the multiplexed signal. The selected filters are shown by solid lines.

Assume that the multiplexed signal has been filtered with a low-pass aliasing filter with a cutoff frequency of 20 kHz. The sample interval of the input waveform to the FFT must satisfy the Nyquist sampling criteria:

$$T \leq 1/(2 \times 20 \times 10^3) \tag{13.39}$$

If we use a radix-2 FFT algorithm, the parameter N is chosen equal to an integer power of 2. Equations (13.38) and (13.39) are satisfied for the following selected parameters:

$$f_0 = 2 \text{ kHz}$$

$$N = 32 \tag{13.40}$$

$$T = 1/(64 \times 10^3)$$

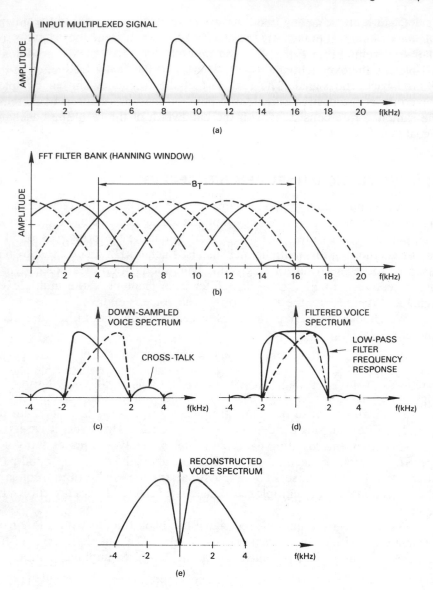

Figure 13.14 (a) Example frequency-multiplexed signal spectrum, (b) FFT filter bank characteristics selected to demultiplex part (a), (c) frequency spectrum of a complex down-sampled FFT band-pass filter output, (d) low-pass filter frequency-response characteristics and the resulting filtered voice spectrum, and (e) the reconstructed voice spectrum.

Assume that an acceptable aliasing level for each FFT band-pass filter is -40 dB. From Fig. 9.8(b), the second side lobe of the Hanning function is down 41 dB. Hence, the sampling bandwidth B_T of each band-pass filter output is 12 kHz, as shown in Fig. 13.14(b). Note that the main lobe and first side lobe of the Hanning function *passes through* the signal in the adjacent channel. We eliminate this *crosstalk* with additional filtering to be discussed later.

The FFT band-pass filter output sample interval T' must satisfy Eqs. (13.35) to (13.37):

$$T' \leq 1/(12 \times 10^3)$$

$$T' = 32T/n \qquad n \text{ is integer valued} \qquad (13.41)$$

$$T' = pT \qquad p \text{ is integer valued}$$

The equation set of Eq. (13.41) is satisfied for $T' = 4T = 1/(16 \times 10^3)$. Each successive FFT is then shifted or hopped over four samples of the input waveform. This down sampling, or decimation, translates the output of each of the selected FFT band-pass filters to zero center frequency.

The frequency spectrum of a complex down-sampled band-pass filter output is shown in Fig. 13.14(c). Note that the desired signal spectrum is overlapped and has a baseband bandwidth of 2 kHz. However, the Hanning band-pass filter characteristics allows frequency components from adjacent channels into the quadrature baseband spectrum. To remove this crosstalk, we apply a digital low-pass recursive filter to both the real and imaginary sample sequences. An assumed low-pass filter characteristic is shown in Fig. 13.14(d) as well as the filtered overlapped spectrum.

To recover the voice waveform so that it is audible, we multiply the complex samples represented by the low-pass filtered spectrum of Fig. 13.14(d) by the complex exponential $e^{-j2\pi f' t}$, where $f' = 2$ kHz. The real part of the complex product is the desired voice waveform with spectrum, as shown in Fig. 13.14(e). Because the sampling rate of the down-sampled signal is 16 kHz, then interpolation is not necessary.

In practical applications, channel selectivity is a key issue in the application of the FFT to demultiplexing. Weighting functions more sophisticated than the Hanning function are generally required. Further, our simplistic example is computationally very inefficient in that our choice of a base-2 FFT algorithm computed the output of all filters in the filter bank. Application of the FFT to the general problem of time-division and frequency-division (TDM and FDM) transmultiplexing is discussed in detail in Refs. [2] to [6]. Our analysis is also applicable to single-sideband frequency-division multiplex (SSB-FDM) modulation and demodulation.

PROBLEMS

13.1 Equations (13.6) to (13.8) and Fig. 13.3 develop the FFT band-pass filter argument for the case $n = 1$ and real functions. Repeat this analytical and graphical development for $n = 2$ and $n = 3$ and only real terms.

13.2 Develop analytically and graphically the quadrature filter bank for the cases n = 0, 1, and 2. Show graphically why the negative sign in Eq. (13.11) does not appear in the convolution form of the equation.

13.3 Develop the interpretation of the FFT as a bank of band-pass integrate and sample filters by proceeding along the lines in Sec. 13.1 but using only discrete arguments. Use the relationship

$$H(n/NT) = \sum_{k=0}^{N-1} h(kT)e^{-j2\pi nk/N} = h(0)e^0 + h(T)e^{-j2\pi n/N} + \cdots +$$

which is equal to the discrete convolution of $h(kT)$ with the sequence 1, $e^{-j2\pi n/N}, \ldots$.

13.4 Repeat the graphical development of Fig. 13.6 for the case of an input cosine waveform of frequency 6.75/32 Hz.

13.5 Repeat the analytical developments of Eqs. (13.3) through (13.15) and the graphical developments of Figs. 13.1 through 13.5 for the case of a Hanning weighting function.

13.6 Repeat Ex. 13.3 for the waveform:

$$y(t) = 1 + \cos(2\pi f_0 t) \qquad f_0 = 1/NT$$

13.7 Let

$$y(t) = 1 + \cos(2\pi f_0 t) + \sin[2\pi(3f_0)t] \qquad f_0 = 1/NT$$

Develop the band-pass filtering equations and FFT output sample results for $t_i = NT$.

13.8 Let

$$y(t) = \cos(2\pi f_1 t) + \sin(2\pi f_2 t)$$

where $f_1 = 3f_0 + f_0/4$ and $f_2 = 3f_0 - f_0/4$. For $f_0 = 1/NT$, use the FFT to band-pass filter $y(t)$.

(a) Discuss and graph the spectrum overlap that results from FFT band-pass filtering and down sampling.

(b) Discuss and show graphically why complex sampling avoids loss of information due to spectrum overlap.

(c) Show how to reconstruct the band-pass filter output signal at a center frequency of $f_0/2$.

13.9 Consider the multiplexed waveform:

$$y(t) = \sum_{n=1}^{4} \cos[2\pi(f_n + f_n/4)] + \sin[2\pi(f_n - f_n/4)] \qquad f_n = n \text{ Hz}$$

Our objective is to use the FFT to demultiplex $y(t)$.

(a) Graphically sketch the spectrum of $y(t)$.

(b) Determine all appropriate parameters for FFT band-pass filtering. Discuss your assumptions.

(c) Analytically reconstruct each demultiplexed signal at a new center frequency of $f_0/2$. Explain how the frequency chosen for reconstruction affects the requirement for interpolation.

REFERENCES

1. HARRIS, F. "The Discrete Fourier Transform Applied to Time Domain Signal Processing." *IEEE Commun. Mag.* (May 1982), Vol. 20, No. 3, pp. 13–22.

2. GREENSPAN, RICHARD L., AND PETER H. ANDERSON. "Channel Demultiplexing by Fourier Transform Processing." *EASCON '74 Proc.* (1974), pp. 360–372.

3. BELLANGER, M. G., AND J. L. DAQUET. "TDM-FDM Transmultiplexer: Digital Polyphase and FFT," *IEEE Trans. Commun.* (September 1974), Vol. Com-22, No. 9, pp. 1199–1205.

4. Special Issue on Transmultiplexers. *IEEE Trans. Commun.* (July 1982), Vol. Com-30, No. 7, pp. 1457–1656.

5. Special Issue on TDM-FDM Conversions. *IEEE Trans. Commun.* (May 1978), Vol. Com-26, No. 5, pp. 489–741.

6. SCHEUERMANN, H., AND H. GOCKLER. "A Comprehensive Survey of Digital Transmultiplexing Methods." *Proc. IEEE* (November 1981), Vol. 69, No. 11, pp. 1419–1450.

14

FFT SIGNAL PROCESSING AND SYSTEM APPLICATIONS

The computing features identified in the previous chapters have resulted in a multitude of signal-processing applications of the FFT. Many commercial and military systems utilize the FFT as an integral processing component. As the price and performance of special-purpose FFT hardware continues to improve, we can expect further growth in FFT signal-processing and system applications. Although it is impossible to enumerate every applications area, the fundamentals of FFT signal-processing techniques are applicable across a broad range of scientific pursuits.

Because every application of the FFT is to sampled waveforms, the basics of signal sampling is of considerable importance to FFT users. For this reason, we first present the details of band-pass- and quadrature-sampling procedures. Then, a broad range of FFT signal-processing and system concepts is presented. An extensive introduction to each field of application is not possible; however, sufficient detail is presented to establish a basic foundation on which the reader can easily build.

14.1 SAMPLING BAND-PASS SIGNALS

The FFT is often used in digital signal-processing applications of band-pass signals. Efficient sampling of band-pass signals is of paramount importance when using the FFT. For this reason, we develop the *band-pass sampling theorem*, a special case of the Nyquist criteria for sampling baseband waveforms, which was developed in Sec. 5.4.

To illustrate the concept of sampling a band-pass signal, consider the time-domain waveform shown in Fig. 14.1. The waveform shown by the solid line in Fig. 14.1(a) is an amplitude-modulated band-pass signal. The dotted line represents the *modulation*, or information content, of the signal. Note that the modulation waveform is sampled at two times per period, but the *carrier frequency* is sampled only once per period. As shown, the samples completely characterize the modulation, or information, waveform even though the sample rate results in aliasing of the band-pass signal. The waveform of Fig. 14.1(a) was sampled in synchronism with the peak of the carrier waveform for clarity of presentation. This is not a requirement for band-pass sampling, as illustrated in Fig. 14.1(b). Here we show the same sample

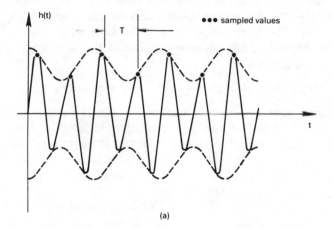

(a)

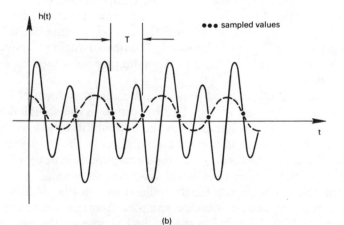

(b)

Figure 14.1   Example of bandpass signal sampling: (a) synchronous sampling with the peak of the carrier waveform, and (b) the general case.

rate as before but with a slight time delay. The dashed-line waveform represented by the samples is the modulation signal.

Band-pass waveforms are assumed to have a nonzero spectrum only over the frequency interval $f_l < |f| < f_h$, where f_h and f_l are the highest and lowest frequencies that bracket the band-pass signal spectrum, respectively. The *transmission* bandwidth of a band-pass signal is defined as $B_T = f_h - f_l$. Using Nyquist criteria, one would sample the band-pass signal at a rate of $2f_h$ samples per second to ensure that overlap aliasing does not occur during sampling. However, recall from Sec. 5.3 that the sampling process produces spectrum images (aliasing) spaced at harmonics of the sampling frequency. We show that aliasing can be used advantageously when sampling band-pass signals and that a sampling rate less than $2f_h$ can be determined ($B_T \ll f_l$) if we associate the band-pass signal with one of the aliasing images. The band-pass-sampling theorem states that a band-pass signal can be reproduced from sample values if the sampling frequency f_s satisfies the relationship

$$2f_h/n \leq f_s \leq 2f_l/(n - 1) \qquad 2 \leq n \leq f_h/(f_h - f_l) \qquad (14.1)$$

and n is integer valued. The condition of Eq. (14.1) ensures that spectrum overlap does not occur and only yields acceptable sampling frequencies for $f_s < 2f_h$. Note that if we let n' equal the largest integer that does not exceed $f_h/(f_h - f_l)$, then the critical (lowest) sampling frequency for a band-pass signal is given by Eq. (14.1) as $f_s' = 2f_h/n'$. Also observe that if we choose $n = f_h/(f_h - f_l)$, then Eq. (14.1) requires $f_s \geq 2(f_h - f_l) = 2B_T$.

We illustrate the concept of efficient sampling of band-pass signals in Fig. 14.2 by means of the convolution theorem. A band-pass time-domain waveform and the corresponding band-pass-frequency spectrum are shown in Figs. 14.2(a) and (c), respectively. Note from Fig. 14.2(c) that the center frequency of the band-pass spectrum is $8f_0$ and the transmission bandwidth B_T is $2f_0$. Choose $f_s = 6f_0$, which satisfies the constraints of the band-pass sampling theorem of Eq. (14.1) for $n = 3$. The time-domain sampling function is shown in Fig. 14.2(b) and the corresponding frequency-domain sampling function is shown in Fig. 14.2(d).

Multiplication of the band-pass time-domain waveform of Fig. 14.2(a) and the sampling function of Fig. 14.2(b) results in the sampled waveform illustrated in Fig. 14.2(e). Recall from the convolution theorem that multiplication in the time domain implies convolution in the frequency domain. Hence, the Fourier transform of the time-sampling function of Fig. 14.2(d) is convolved with the band-pass signal spectrum shown in Fig. 14.2(c). The result is the aliased frequency function illustrated in Fig. 14.2(f).

Note from Fig. 14.2(f) that the sampled frequency function centered at frequency $\pm 2f_0$ is identical to the original band-pass frequency function centered at frequency $\pm 8f_0$. Although the function centered at $\pm 2f_0$ results from aliasing, we have not lost information due to spectrum overlap. The

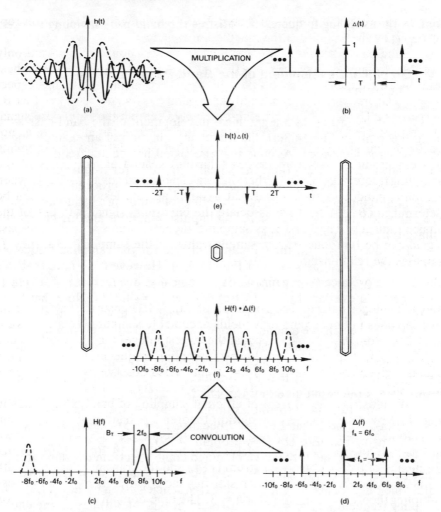

Figure 14.2 Nonoverlapped aliased Fourier transform of a band-pass waveform that is sampled at less than twice the highest frequency component.

sampled frequency functions centered at $\pm 4f_0$ and $\pm 10f_0$ in Fig. 14.2(f) are also the results of aliasing. These terms can be ignored because it can be shown that a low-pass filter with bandwidth $3f_0$ reconstructs the original signal $h(t)$ with only a shift of the center frequency from $8f_0$ to $2f_0$.

The highest frequency component of the band-pass waveform of Fig. 14.2(a) is $9f_0$. Hence, application of the baseband Nyquist sampling theorem requires a sampling frequency of $18f_0$. Because we sampled at a rate of only $6f_0$ with no loss of information, the waveform is said to have been *down sampled* or *decimated*. We can down sample with no spectrum overlap as long as the sampling frequency f_s satisfies the band-pass sampling theorem of Eq. (14.1).

Graphical Development of the Band-Pass Sampling Theorem

We show in Fig. 14.3 a graphical development of the band-pass sampling theorem. In Fig. 14.3(a), the frequency function of an example band-pass signal is illustrated. Assume that the signal has center frequency $14f_0$ and bandwidth $B_T < 2f_0$ (i.e., the signal amplitude for frequencies f_h and f_l equals zero). The graphical frequency-convolution procedure is used in Figs. 14.3(b) through (l) to illustrate the effect of sampling a band-pass signal. We only show the frequency-sampling impulse functions and the convolved (aliased) frequency-domain functions.

Because $B_T < 2f_0$, then a natural choice for the sampling frequency is $f_s = 2B_T = 4f_0$, as shown in Fig. 14.3(b). However, we note that this choice of f_s produces spectrum overlap. Logically, one increases f_s to eliminate spectrum overlap. In Fig. 14.3(c), we set $f_s = 4.25f_0$. Note that there is still some spectrum overlap, but if we increase f_s to $4.33f_0$, as shown in Fig. 14.3(d), we achieve a nonoverlapped sampled frequency spectrum. But if we set $f_s = 4.5f_0$, as illustrated in Fig. 14.3(e), spectrum overlap is again encountered. Using the graphical convolution theorem, we can *tediously* determine the range of f_s that will produce a nonoverlapped sampled spectrum. This is the result given by Eq. (14.1).

Consider Eq. (14.1) for the example band-pass spectrum illustrated in Fig. 14.3(a). We note that $2 \le n \le 7$ because $f_h = 15f_0$ and $f_l = 13f_0$. Let $n = 7$. Then, from Eq. (14.1), we obtain $4.29f_0 \le f_s \le 4.33f_0$. Observe from Figs. 14.3(c) and (e) that we obtained some spectrum overlap for sampling frequencies $4.25f_0$ and $4.5f_0$. By a careful graphical analysis, we can obtain the range of acceptable sampling frequencies given by Eq. (14.1). The sampling frequency $f_s = 4.33f_0$ illustrated in Fig. 14.3(d) lies at one end of this range and, as shown, spectrum overlap does not occur.

Now let $n = 6$ in Eq. (14.1); we obtain a range of acceptable sampling frequencies given by $5f_0 \le f_s \le 5.2f_0$. A graphical illustration of the range of these sampling frequencies is illustrated in Figs. 14.3(l) through (g). For $f_s = 4.5f_0$, as shown in Fig. 14.3(e), we obtain an overlapped spectrum; but for $f_s = 5f_0$, as shown in Fig. 14.3(f), we note that the sampled spectrum is not overlapped. Unacceptable results are obtained for $f_s = 5.5f_0$, as shown in Fig. 14.3(g). As discussed previously, we can carefully adjust f_s to graphically obtain the identical range given by Eq. (14.1) for $n = 6$.

From the results illustrated in Figs. 14.3(h) and (i), we conclude that $6f_0 \le f_s \le 6.5f_0$ is an acceptable range for f_s. Equation (14.1) yields this result for the choice $n = 5$. For $n = 4, 3,$ and 2 in Eq. (14.1), we obtain the results $7.5f_0 \le f_s \le 8.67f_0$, $10f_0 \le f_s \le 13f_0$, and $15f_0 \le f_s \le 26f_0$,

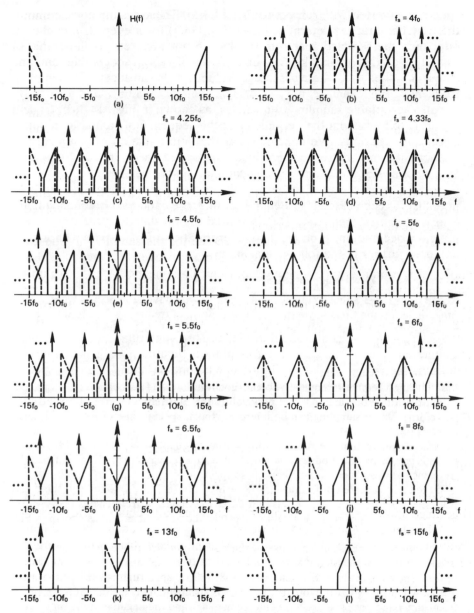

Figure 14.3   Aliased Fourier transform of a band-pass waveform that is sampled at various frequencies.

respectively. Figures 14.3(j) to (l) show acceptable choices of f_s within each of these ranges. As before, we can refine our graphical analysis to produce the ranges defined by Eq. (14.1).

Note that the low-pass spectrum results of Figs. 14.3(f), (j), and (l) are spectrum inverted with respect to Fig. 14.3(a). These results correspond to the frequency ranges determined from Eq. (14.1) for n even. If n is chosen odd, then the sampled spectrum results are not inverted, as illustrated in Figs. 14.3(d), (h), (i), and (k). Also observe that if f_s is an acceptable sampling frequency, then pf_s, where p is integer valued, is also an acceptable sampling frequency. For example, if $f_s = 4.33f_0$, then $f_s = 8.66f_0$ and $f_s = 13f_0$ are also acceptable sampling frequencies, as shown in Figs. 14.3(d), (j), and (k). This follows from the periodicity of the sequence of sampling impulse-frequency functions. We conclude that the determination of sampling-frequency intervals that do not produce overlapped spectrum results is non-trivial. Both Eq. (14.1) and a graphical analysis are helpful.

An alternate way of examining the band-pass sampling theorem is to note that for each case illustrated in Fig. 14.3, the band-pass-frequency function with central frequency $14f_0$ is shifted or translated by the down-sampling, or decimation, process. This interpretation of band-pass sampling is explored further in the following example.

Example 14.1   Down Sampling: A Special Case of Frequency Down Conversion

Recall from Ex. 3.8 that frequency shifting or down conversion occurs when a time function $h(t)$ is multiplied by a sinusoidal waveform of frequency f_0. From the Fourier transform frequency-shifting property, the result of sinusoidal multiplication is to shift $H(f)$, the Fourier transform of $h(t)$, such that the original spectrum is now centered at $f_0 \pm f_c$, where f_c is the center frequency of the spectrum $H(f)$. As shown in the development of Ex. 3.8, spectrum shifting occurs because time-domain multiplication requires frequency-domain convolution. $H(f)$ is convolved with a pair of impulse functions located at $\pm f_0$ which is the Fourier transform of a sinusoidal waveform. Down sampling can be interpreted as a special case of frequency down conversion.

The pair of impulse functions located at frequency $\pm 6f_0$ in Fig. 14.2(d) can be interpreted as the Fourier transform of a cosine waveform. Hence, from the frequency-shifting theorem, the band-pass spectrum centered at $+8f_0$ in Fig. 14.2(c) is shifted and centered at frequencies $(8 - 6)f_0$ and $(8 + 6)f_0$. Similarly, the band-pass spectrum centered at $-8f_0$ is shifted and centered at frequencies $(-8 - 6)f_0$ and $(-8 + 6)f_0$. The result is then the spectrum centered at $\pm 2f_0$, as shown in Fig. 14.2(f), and $\pm 14f_0$.

Similar to the arguments above, the pair of impulse functions located at frequency $\pm 12f_0$, which is not shown in Fig. 14.2(d), result in spectrum being shifted to frequencies of $\pm(12 - 8)f_0$ and $\pm(12 + 8)f_0$. The spectrum pair located at $\pm 4f_0$ are shown in Fig. 14.2(d). Note that this pair is spectrum inverted in that the positive frequency band-pass spectrum is centered at $-4f_0$ and the negative frequency band-pass spectrum is centered at $+4f_0$.

Summary

Because down sampling results in frequency translation of the band-pass signal, it is possible to position the translated spectrum by an appropriate choice of sampling frequency f_s. Note from Fig. 14.3 that the selection of f_s satisfying Eq. (14.1) such that $nf_s = f_l$, where n is integer valued, translates the band-pass signal spectrum interval, $13f_0 < f < 15f_0$, to the frequency interval, $0 < f < 2f_0$, which is generally referred to as the low-pass signal equivalent of the band-pass signal. Sampling frequencies $f_s = 4.33f_0$ for $n = 3$, $f_s = 6.5f_0$ for $n = 2$, and $f_s = 13f_0$ for $n = 1$ satisfy this condition and the graphical results are shown in Figs. 14.3(d), (i), and (k), respectively. For many signal-processing applications, down sampling to the low-pass signal equivalent is the preferred approach. Also note that selection of the sampling frequency such that $nf_s = f_h$ also translates the band-pass signal to a low-pass equivalent, as shown in Figs. 14.3(f) and (i), but the spectrum is inverted.

It is also possible to select a sampling frequency that results in a frequency translation to zero center frequency. Note from the graphical development in Fig. 14.3 that if $nf_s = (f_h - f_l)/2$, that is, the center frequency of the band-pass spectrum, then translation to zero center frequency results. This selection of f_s always produces spectrum overlap and does not satisfy Eq. (14.1). In most cases, spectrum overlap is an irreversible operation (see Prob. 14.3). However, waveforms that are down sampled to zero center frequency are always recoverable if the band-pass signal is quadrature sampled, as is discussed in the following section.

14.2 QUADRATURE SAMPLING

Applications of the FFT are sometimes limited by the sampling rates achievable by analog-to-digital converters. For these cases, it is possible to achieve a lower sampling rate by separating the signal into two waveforms, or channels, and sampling each channel. This concept is based on the principle that a signal can be expressed in terms of two waveforms called *quadrature functions*. Each of the two quadrature functions occupies only one-half the bandwidth of the original signal. Hence, it is possible to sample each quadrature function at one-half the sample rate required to sample the original signal. We now develop the concept of quadrature functions and quadrature sampling.

Quadrature Functions

To demonstrate the concept of quadrature functions, consider Fig. 14.4. We show in Fig. 14.4(a) an example waveform that is assumed to be band-limited with bandwidth f_h, as illustrated in Fig. 14.4(c). To derive the quadrature functions for this waveform, it is necessary to multiply Fig. 14.4(a) by both cosine and sine waveforms.

We show the required cosine waveform $y_i(t)$ in Fig. 14.4(b). We use

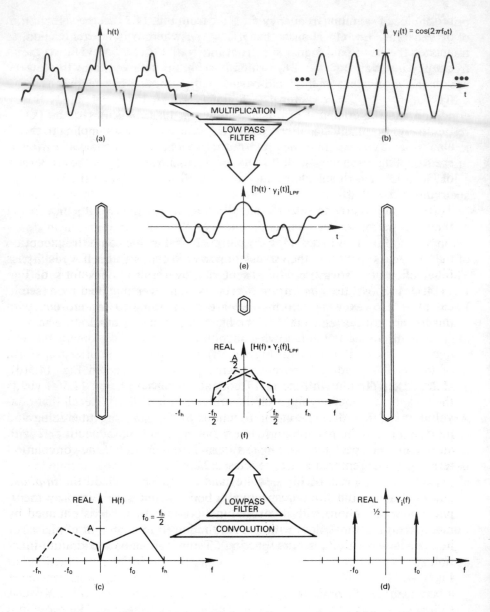

Figure 14.4 Fourier transform of the *in-phase* waveform used in quadrature sampling.

the subscript *i* to indicate that the cosine waveform has been chosen as the reference or *in-phase* sinusoid. Note that the frequency of this sinusoid is $f_0 = f_h/2$, the center frequency of the positive frequency range of the band-limited signal, as shown in Fig. 14.4(c). Multiplication in the time domain requires convolution in the frequency domain and the overlapped spectrum of Fig. 14.4(f) results. For ease of discussion, we have constructed $h(t)$ as an even function and hence $H(f)$ is real. Therefore, the convolution result shown is a real frequency function. Note that we have low-pass filtered the convolution result and have eliminated the convolution terms that are centered at frequencies $\pm 2f_0$.

The modulation process just described shifts or translates to zero center frequency the spectrum shown in Fig. 14.4(c). An inspection of Fig. 14.4(f) also reveals that the translated waveform has a bandwidth of $f_h/2$. However, spectrum foldover has occurred. As a result, if one samples the waveform of Fig. 14.4(e) with sample frequency f_h satisfying the Nyquist criteria, the original time waveform cannot be recovered because of the folded spectrum. To recover the original waveform, it is necessary to define the second of the two quadrature functions.

In Fig. 14.5, we repeat the development of Fig. 14.4 with the exception that we multiply by the sine waveform shown in Fig. 14.5(b). If we low-pass filter the product of Figs. 14.5(a) and (b), the resulting waveform of Fig. 14.5(e) is termed the *quadrature function* in that it is obtained by multiplication with a sine waveform that is 90° out of phase or *in quadrature* with the cosine waveform of Fig. 14.4(b).

By applying the frequency-convolution theorem, we obtain the frequency function illustrated in Fig. 14.5(f). Note that the Fourier transform of the sine waveform is purely imaginary, as illustrated in Fig. 14.5(d). Hence, convolution with the real frequency function of Fig. 14.5(c) yields the imaginary frequency function illustrated in Fig. 14.5(f). Recall that convolution requires that one of the functions be *flipped* prior to shifting and multiplication. The resulting frequency function has a bandwidth $f_h/2$ with an overlapped spectrum. Low-pass filtering has eliminated the convolution terms that are centered at frequencies $\pm 2f_0$.

The waveforms of Figs. 14.4(e) and 14.5(e) are termed the *in-phase* and *quadrature* functions, respectively, because one is obtained by multiplication (translation) with a cosine function and the other is obtained by multiplication (translation) with a sine function (that is 90° out of phase, or in quadrature, with the cosine function). The advantage of quadrature-function representation can be seen by comparing the frequency functions of Figs. 14.4(f) and 14.5(f). Both the in-phase and quadrature waveforms have a bandwidth of $f_h/2$. Hence, each can be sampled according to the Nyquist sampling criteria at a sample rate of f_h samples per second. We show in a later development that these sampled results can be appropriately combined to eliminate the spectrum overlap that occurs.

Note that the *total* number of samples per unit of time that result from sampling the quadrature functions is exactly the same as obtained by sampling a function with bandwidth f_h. However, with quadrature functions, we have separated the original waveforms into two channels, an in-phase

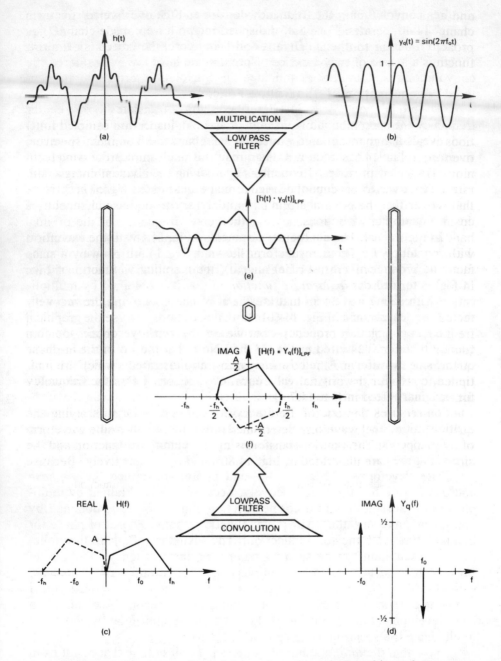

Figure 14.5 Fourier transform of the *quadrature* waveform used in quadrature sampling.

and a quadrature channel, where the analog-to-digital converter for each channel can operate at one-half the speed required in a single-channel approach. If analog-to-digital-converter or digital-processor speed is a limiting function, a factor of two could be of primary concern.

Recombination of Quadrature Functions

A careful recombination of the in-phase and quadrature sampled functions is necessary to obtain the original waveform and eliminate spectrum overlap. In Fig. 14.6, we show a diagram of the quadrature-processing technique described previously as well as a procedure for recombining quadrature functions to produce the original real band-limited signal $h(t)$. Note that we recover the original signal by multiplying the in-phase channel by a cosine waveform with frequency f_0, the center frequency of the original band-limited signal. The quadrature channel is multiplied by a sine waveform with frequency f_0. These results are then added and multiplied by a scale factor of two to recover the original signal. The interpolation function shown in Fig. 14.6 is discussed later.

Figures 14.7 and 14.8 illustrate the rationale underlying the recovery technique diagrammed in Fig. 14.6. In both illustrations, we use the graphical frequency-convolution procedure, but we only show the frequency-domain functions. Figures 14.7(b) and (c) are the Fourier transforms of the in-phase quadrature function and the cosine waveform, respectively, which are multiplied to recover the original waveform. Convolution yields the frequency function illustrated in Fig. 14.7(a).

Figure 14.8 depicts the frequency-domain results of multiplying the quadrature channel waveform determined in Fig. 14.5(e) by a sine waveform of frequency f_0. The Fourier transforms of the quadrature function and the sine waveform are illustrated in Fig. 14.8(b) and (c), respectively. Because

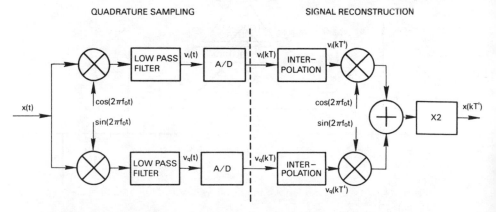

Figure 14.6 Block diagram of the quadrature-sampling and signal-recombination processes.

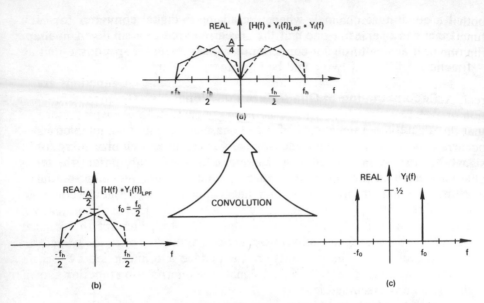

Figure 14.7 Frequency function resulting from the cosine modulation of the in-phase channel waveform.

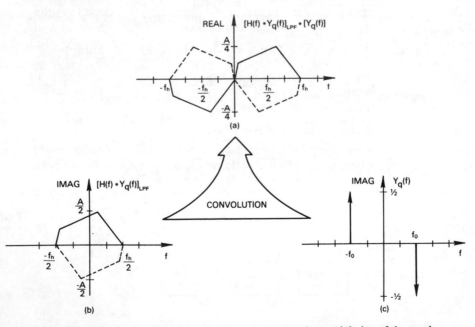

Figure 14.8 Frequency function resulting from the sine modulation of the quadrature channel waveform.

both frequency functions are imaginary, then their convolution is a real function, as shown in Fig. 14.8(a). To produce the results of Fig. 14.8(a), flip one of the functions prior to convolution and take into account the $j^2 = -1$ term.

Now consider Figs. 14.7(a) and 14.8(a); both frequency functions are real. Addition of the two functions gives the original signal-frequency function of Fig. 14.4(c) (except for a scale factor of two). Because we assumed that the band-limited signal spectrum $H(f)$ was real, then no complex terms occurred in the example signal-recovery process. The general procedure for signal reconstruction is to multiply the complex sampled signal (i.e., the in-phase and quadrature samples, $a + jb$) by the sampled complex exponential $e^{-j2\pi f_0 t}$. The desired waveform is then the real part of this complex product:

$$\text{Real } \{(a + jb)[\cos(2\pi f_0 t) - j \sin(2\pi f_0 t)]\}$$

$$= \text{Real } \{[a \cos(2\pi f_0 t) + b \sin(2\pi f_0 t)] \tag{14.2}$$

$$+ j[-a \sin(2\pi f_0 t) + b \cos(2\pi f_0 t)]\}$$

$$= a \cos(2\pi f_0 t) + b \sin(2\pi f_0)t$$

The signal-reconstruction process requires that we *translate* (frequency shift) the quadrature functions *up* in frequency. As a result, the required sampling rate must be increased and interpolation is required because we have sampled at the lower rate. As discussed previously in the development of the Nyquist sampling theorem, $[\sin(t)]/t$ interpolation yields exact results. *We interpolate both the in-phase and quadrature waveforms prior to multiplication by the complex exponential.* For the example shown, we need interpolate only one sample between each output sample of the analog-to-digital converter. The bandwidth (highest frequency) of the signal after multiplication by the complex exponential (i.e., after translation) determines the number of interpolated samples that are required.

Example 14.2   Quadrature Sampling of Band-Pass Signals

We show in Fig. 14.9 an example band-pass waveform with transmission bandwidth $B_T = f_0$ and center frequency $5f_0$. To derive the quadrature waveforms for the band-

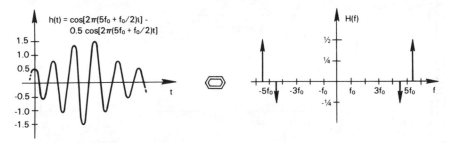

Figure 14.9   Time- and frequency-domain representations of the band-pass signal for Ex. 14.2.

pass signal, we multiply by cosine and sine waveforms with frequency $5f_0$. The in-phase component after low-pass filtering is given by

$$h(t)y_i(t) = \{\cos[2\pi(5f_0 + f_0/2)t] - \tfrac{1}{2} \cos[2\pi(5f_0 - f_0/2)t]\} \cos[2\pi(5f_0)t]$$

$$= \tfrac{1}{2} \cos[2\pi(f_0/2)t] - \tfrac{1}{4} \cos[2\pi(f_0/2)t]$$

$$(14.3)$$

The quadrature component after low-pass filtering is given by

$$h(t)y_q(t) = \{\cos[2\pi(5f_0 + f_0/2)t] - \tfrac{1}{2} \cos[2\pi(5f_0 - f_0/2)t]\} \sin[2\pi(5f_0)t]$$

$$= - \tfrac{1}{2} \sin[2\pi(f_0/2)t] - \tfrac{1}{4} \sin[2\pi(f_0/2)t]$$

$$(14.4)$$

In Figs. 14.10 and 14.11, we use the graphical frequency-convolution theorem to develop the frequency functions corresponding to Eqs. (14.3) and (14.4). Figures 14.10(b) and (c) are the Fourier transforms of the band-pass signal $h(t)$ and the cosine waveform with frequency $5f_0$, respectively. Convolution and low-pass filtering yield the frequency function illustrated in Fig. 14.10(a), the Fourier transform of Eq. (14.3). Note that the frequency function has a bandwidth $f_0/2$ and spectrum foldover has occurred.

In Fig. 14.11, we develop the frequency function corresponding to Eq. (14.4). Convolution of the frequency functions of Figs. 14.11(b) and (c) yields the quadrature frequency function illustrated in Fig. 14.11(a). This frequency function has bandwidth $f_0/2$ with an overlapped spectrum. Note that both Eqs. (14.3) and (14.4) could be

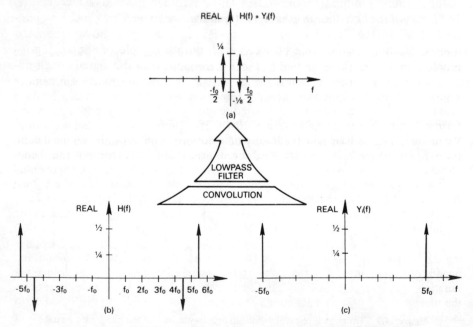

Figure 14.10   Fourier transform of the quadrature waveform for Ex. 14.2.

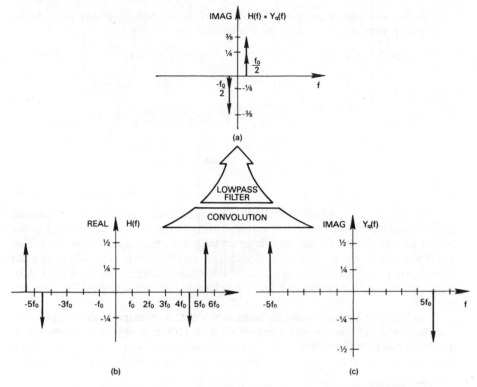

Figure 14.11   Fourier transform of the in-phase waveform for Ex. 14.2.

reduced to a single term through addition. This is the mathematical evidence of the problem of spectrum overlap that has been graphically addressed.

Both the in-phase spectrum of Fig. 14.10(a) and the quadrature spectrum of Fig. 14.11(a) have a bandwidth of $f_0/2$. Hence, each can be sampled at a Nyquist rate of f_0 samples per second as compared to sampling the original band-pass waveform with the Nyquist sampling rate of $2B_T = 2f_0$. To reconstruct a real waveform, we multiply the in-phase and quadrature time-domain samples by the sampled complex exponential $e^{-j2\pi f't}$, where f' is the desired center frequency of the reconstructed waveforms. Assume $f' = f_0$. Multiplication by the complex exponential $e^{-j2\pi f_0 t}$ translates each quadrature spectrum to a center frequency of f_0. The highest frequency of the translated waveform is then $3f_0/2$ and both the in-phase and quadrature functions of Eqs. (14.3) and (14.4) must be interpolated before multiplication to obtain a sampling rate three times the original sample rate.

Figure 14.12(a) illustrates the frequency-domain results obtained by multiplying the in-phase waveform of Eq. (14.3) by $r_i(t) = \cos(2\pi f_0 t)$. Multiplication of the quadrature function of Eq. (14.4) by $r_q(t) = \sin(2\pi f_0 t)$ results in the frequency function shown in Fig. 14.12(b). Addition of the two frequency functions cancels the unwanted overlapped frequency components. The result $H'(f)$ in Fig. 14.12(c) is the original signal-frequency spectrum except the spectrum is now centered at frequency f_0 and must be multiplied by a scale factor of two.

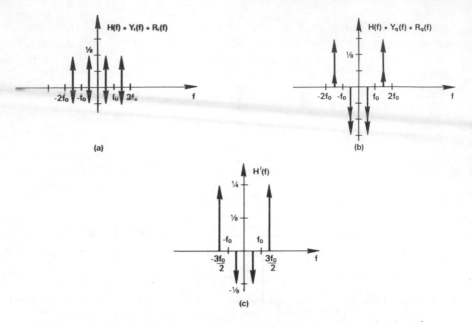

Figure 14.12 Example frequency function for Ex. 14.2: (a) cosine modulation of the in-phase channel, (b) sine modulation of the quadrature channel, and (c) summation of parts (a) and (b).

Summary

As shown, quadrature sampling can be applied to baseband and bandpass signals. If a baseband or band-pass signal with bandwidth B_T is translated in quadrature to zero center frequency, then each quadrature function can be sampled according to the relationship:

$$f_s \geq B_T \tag{14.5}$$

without loss of information. The quadrature sampled waveforms must be reconstructed by the technique diagrammed in Fig. 14.6 to recover the original signal. From Eq. (14.5), quadrature sampling allows the analog-to-digital converter to operate at one-half speed. As experimenters continue to press the state of the art in digital signal processing, analog-to-digital-converter speed is often the limiting constraint.

Recall from Chapter 13 that the real and imaginary FFT outputs are in quadrature (i.e., complex). Hence, FFT band-pass filter applications are a special case of quadrature sampling in that the output waveforms of each FFT band-pass filter are in quadrature. It then follows that these quadrature waveforms can be translated, or down sampled, to zero center frequency without incurring irrecoverable spectrum overlap. This similarity follows in that the processes of quadrature frequency translation to zero center frequency and sampling can be interchanged. In quadrature sampling, we develop zero-center-frequency waveforms in quadrature prior to sampling. Conversely, in FFT band-pass filtering, we first obtain samples of the quadrature band-pass waveform and then translate, or down sample, to zero center frequency.

14.3 FFT SIGNAL DETECTION

An important application of the FFT is in signal detection. The detection of a narrow-band signal buried in noise is a common signal-processing problem in communications, radar, and sonar systems. We describe in this section example experimental results of applying this basic signal-analysis property of the FFT. Application of the FFT to digital matched filtering is also explored.

Signal Extraction Through FFT Resolution Improvement

In Fig. 14.13(a), we show a sampled time-domain sinusoid buried in white noise. As illustrated, there is no appearance of a sinusoid, only noise. The signal-to-noise ratio is -12 dB. Obviously, one could never *detect* the presence of the sinusoid by examination of the time-domain samples shown.

In the frequency domain, we know that the periodic sinusoid has its energy concentrated in a very narrow frequency band, whereas the noise power is spread throughout the frequency domain. Hence, if we take the FFT of the waveform illustrated in Fig. 14.13(a), we expect all the sinusoidal signal energy to be concentrated in a few contiguous samples of the FFT output. Recall from Chapter 13 that the N-output FFT samples can be interpreted as the output of $N/2$ contiguous band-pass filters.

Figure 14.13(b) illustrates the FFT of the waveform of Fig. 14.13(a). For this example, $N = 64$; the 32 sample points shown in Fig. 14.13(b) represent the output power of each FFT filter. Power is computed as the sum of the square of the real and imaginary component of each filter output. This result is doubled to account for negative frequency results. Although the sample value representing the sinusoidal signal has a larger amplitude than other samples, an experimenter could not be certain that the sample represents a periodic signal.

To firmly establish the presence of a signal, it is necessary to spread the noise over more data points. Hence, we increase N to 512. In Fig. 14.13(c), we show the resulting 256 FFT sample outputs. The presence of a periodic signal in the noisy spectrum is very identifiable.

To compute the signal-to-noise ratio improvement that is achieved in Fig. 14.13(c), note that the noise power has been evenly spread throughout

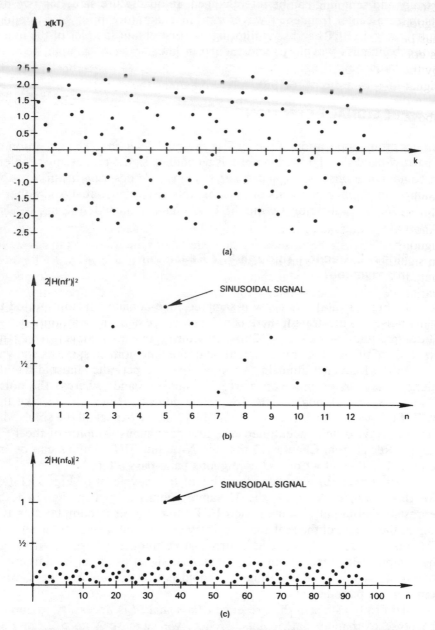

Figure 14.13 Example signal buried in noise, where $S/N = -12$ dB: (a) time-domain presentation, (b) FFT spectrum results for $N = 64$, and (c) FFT spectrum results for $N = 512$.

256 samples. Equivalently, the wideband noise has been filtered by 256 contiguous filters and the noise power output in the filter that contains the signal has been reduced by $10 \log_{10} (1/256) = -24$ dB. Because the signal power is concentrated in a single FFT filter, then the signal power is not reduced by the FFT. If the original signal-to-noise level was -18 dB, then the output signal-to-noise ratio of the FFT filter containing the signal is $-18 - (-24) = 6$ dB. The sample indicating the signal in Fig. 14.13(c) is clearly visible above the noise.

FFT Averaging

Signal-to-noise enchancement as previously discussed cannot be extended indefinitely. Sometimes the size of the FFT (i.e., the number of filters) cannot be increased further because of computer memory limitations. Another limiting factor is that the signal itself can spread over several contiguous FFT filters because of its bandwidth. For these cases, improvement in signal detectability can be achieved by averaging successive FFT power outputs. The effect of averaging is to smooth and reduce wild amplitude variations that could be interpreted as sinusoidal signal components.

In Fig. 14.14(a), we show the FFT spectrum ($N = 512$) of a periodic signal buried more deeply in noise than that previously considered. The periodic component is not detectable. Assume that constraining factors limit

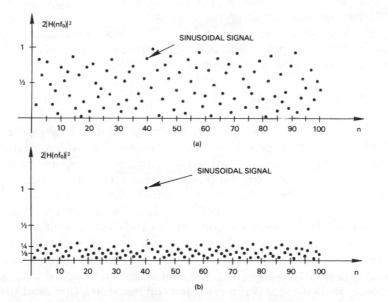

Figure 14.14   Example signal buried in noise, where $S/N = -24$ dB: (a) FFT spectrum results for $N = 512$, and (b) averaged spectrum for 64 successive FFTs with $N = 512$.

the experimenter to an FFT of size $N = 512$. Averaging successive FFTs of this size is the appropriate signal-analysis procedure.

The resulting averaged spectrum is shown in Fig. 14.14(b) for 64 averaged spectrums. The noise is now smoothed significantly and the signal is clearly visible above the smoothed noise level. It is to be noted that the power outputs of each FFT filter are averaged for the successive FFTs. Hence, the phase of the input sinusoid is considered unknown and is not taken into consideration.

The mathematics involved in analyzing the signal-to-noise enhancement illustrated in Fig. 14.4(b) is extremely complicated (Refs. [9] and [21]). However, as a summary, in those cases where the original signal-to-noise ratio is -30 dB or less, the described averaging enhances signal detectability by approximately $1.5 \log_2 Q$ dB, where Q is the number of successive FFTs that are averaged. For original signal-to-noise ratios well above -30 dB, the detectability gain approaches $3.0 \log_2 Q$ dB.

The signal illustrated in Fig. 14.14(a) is 24 dB below the noise level. The number of successive FFTs averaged together is $64 = 2^6$ and the FFT size is $N = 512$. Averaging yields an enhancement of $1.5 \log_2 Q = 1.5 \log_2 2^6 = 9$ dB. From the results leading to Fig. 14.13(c), we know that the 512-point FFT results in a 24 dB gain in signal-to-noise ratio. Hence, the processed signal-to-noise ratio is $24 + 9 - 24 = 9$ dB and the signal is clearly visible, as shown in Fig. 14.14(b).

The examples presented may appear as a simplistic application of the FFT to the signal-detection problem. However, it can be shown that the optimum signal-detection receiver for narrow-band signals with random phase, unknown frequency, and constant amplitude is a bank of band-pass filters followed by a decision threshold (Ref. [23]).

FFT Matched Filtering

A matched filter is the signal processor design that optimizes the peak received signal-to-noise power ratio in the presence of additive white Gaussian noise. Mathematically, a matched filter frequency response is given by $S^*(f)$, where * indicates conjugation if the received signal $s(t)$ has a Fourier transform $S(f)$. Practical high-speed matched-filter realizations are easily achieved because of the ease of FFT frequency-domain processing.

Figure 14.15 illustrates the concept of a FFT matched-filter signal processor. The input signal is transformed to the frequency domain using the FFT and is multiplied by a stored frequency-domain conjugate replica of the received signal. At inverse FFT yields the desired matched-filter output waveform. This output waveform is then compared to a threshold to determine the presence or absence of the desired signal. The optimum signal detector of a phase-modulated sinusoidal pulse in white noise is a set of

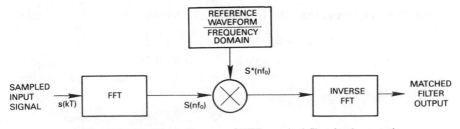

Figure 14.15   Block diagram of FFT matched-filter implementation.

matched filters to the in-phase and quadrature-phase components of the signal. For this reason, the FFT can be used in radar signal processors.

Possibly the most important aspect of FFT matched filtering is the flexibility allowed the signal designer. Waveform variations are easily processed by simply storing the appropriate FFT coefficients. One can envision a system where the signal and hence the matched filter can be changed rapidly.

14.4 FFT CEPSTRUM ANALYSIS: ECHO AND MULTIPATH REMOVAL

Cepstrum signal-processing techniques (Refs. [2] and [3]) are of considerable utility. Specifically, these procedures are based on the premise that when one examines the frequency transform of the logarithm of the Fourier transform, certain contaminating components, such as noise, unwanted signals, etc., can be isolated. Cepstrum analysis is applied to many problem areas, including noise reduction in speech, sonar echo removal, radio-frequency multipath interference rejection, image processing, and removal of multiple reflections in seismology. Because the analysis approach is generally an *art-science*, we find it more meaningful to examine FFT cepstrum analysis techniques by means of specific examples. In this section, we describe FFT cepstrum analysis as applied to the detection and removal of multipath interference or echos from a desired waveform (Ref. [12]).

Multipath or Echo-Removal Problem Definition

Assume that a received signal $s_r(t)$ is given by

$$s_r(t) = s(t) + a_0 s(t + \tau_l) \tag{14.6}$$

where $s(t)$ is the transmitted or desired signal and $s(t + \tau_l)$ is a multipath or echo component. The constant a_0 is the relative attenuation between the direct and multipath components of the signal. If we take the Fourier transform of Eq. (14.6) and then the logarithm, we obtain

$$\begin{aligned}
\log S_r(f) &= \log[S(f) + a_0 S(f)e^{j2\pi f\tau_l}] \\
&= \log[S(f)(1 + a_0 e^{j2\pi f\tau_l})] \\
&= \log S(f) + \log(1 + a_0 e^{j2\pi f\tau_l}) \\
&= \log S(f) - \sum_{n=1}^{\infty} (-1)^n (a_0^n/n)e^{j2\pi f\tau_l}
\end{aligned}$$

(14.7)

where the last term in Eq. (14.7) was obtained by a series expansion. The Fourier transform of Eq. (14.7) is given by

$$C[\log S_r(f)] = C[\log S(f)] - \sum_{n=1}^{\infty} (-1)^n (a_0^n/n)\delta(\tau - n\tau_l) \quad (14.8)$$

where $C[\ ]$ is taken as the Fourier transform. The Fourier transform of the log of the spectrum frequency function is called the *cepstrum*. Note that the first term on the right-hand side of Eq. (14.8) is the cepstrum of the transmitted or desired signal. The remaining term is a sequence of impulse functions.

Cepstrum analysis then transforms the unwanted echo or multipath signal component into a series of evenly spaced impulse functions. In Fig. 14.16(a), we show the cepstrum of the signal only, and in Fig. 14.16(b), we show the cepstrum of the signal plus the unwanted multipath signals. Theoretically, the echo can be removed by removing the impulse functions and then inverting the whole process to recover the transmitted signal.

FFT Echo and Multipath Removal Implementation

We implement the cepstral analysis procedure by means of the block diagram illustrated in Fig. 14.17. The log operation implies an exponential

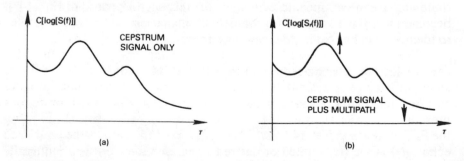

Figure 14.16 Example depicting cepstrum analysis that identifies multipath signal components as impulse functions.

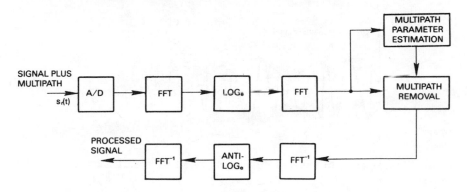

Figure 14.17 Block diagram of cepstrum signal-processing procedure for removing unwanted echo or multipath signals.

logarithm of both the real and imaginary parts of the FFT output. In particular, because the output of the FFT is of the form $[R(nf_0) + jI(nf_0)]$, then

$$\log_e[R(nf_0) + jI(f_0)] = \log_e\{[R^2(nf_0) + I^2(nf_0)]^{1/2}e^{j\theta_n}\}$$

$$= \log_e[R^2(nf_0) + I^2(nf_0)]^{1/2} + \log_e e^{j\theta_n} \quad (14.9)$$

$$= \log_e[R^2(nf_0) + I^2(nf_0)]^{1/2} + j\theta_n$$

where

$$\theta_n = \tan^{-1}[I(nf_0)/R(nf_0)] \quad (14.10)$$

The real part of the $\log_e$ operation is placed in the real part of the FFT and the imaginary part is placed in the imaginary part of the FFT. The resulting FFT yields the *quefrequency* series of Eq. (14.8). Because FFT use implies finite-duration waveforms, then the impulse-response functions of Eq. (14.4) become $[\sin(\tau)]/\tau$ functions. Manual or automated procedures can be used to identify and supress or filter the unwanted $[\sin(\tau)]/\tau$ functions.

We show in Fig. 14.18 the results of the implementation of the FFT cepstrum processing procedure defined in Fig. 14.17. In Fig. 14.18(a), we show an example transmitted signal with no multipath or echoes. Figure 14.17(b) illustrates the example signal with interfering multipath or echoes present. Cepstrum analysis of the contaminated signal of Eq. (14.8) using the FFT is shown in Fig. 14.18(c). The $[\sin(\tau)]/\tau$ function at τ_l is the undesired echo signal. In Fig. 14.18(d), we set the echo functions to zero. Sequential inverse FFT, antilogarithm, and inverse FFT operations result in the recovered signal shown in Fig. 14.18(e).

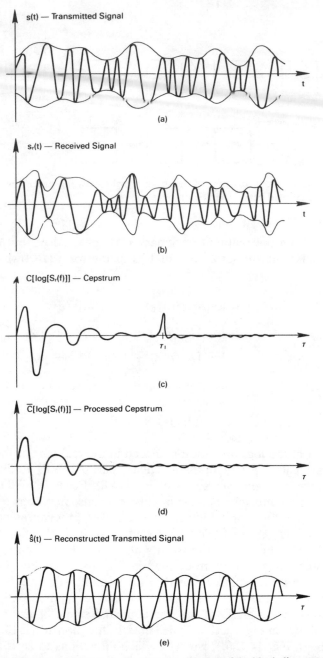

Figure 14.18   Example FFT results of the application of the block diagram of Fig. 14.17.

14.5 FFT DECONVOLUTION

In the discussion of the FFT digital filter design in Chapter 12, we assumed that we could always compute a finite-duration filter impulse response. Deconvolution filter design problems do not normally satisfy this requirement and modifications to the techniques described in Chapter 12 are necessary. We develop in this section an FFT procedure for designing digital deconvolution filters. The technique is applicable to a wide variety of problems: spectral broadening in spectography, well logging in oil exploration, seismic exploration in geophysics, contrast enhancement in optics, and restoration of the output waveform of band-limiting filters (Refs. [2], [10], and [22]).

Deconvolution Problem Definition

To define the deconvolution problem, consider Fig. 14.19. If a signal is passed through a filter whose bandwidth is less than that of the signal, the result is a smearing or broadening of the input waveform. Often compensation of the filter itself can be employed to remove this unwanted distortion. An alternate or sequential approach is to apply appropriate mathematical procedures to the output waveform and thereby *restore* the input waveform. Because the output of a filter can be written as the convolution of the input waveform and the impulse response of the filter, then the mathematical operation for attempting to remove this convolution operation is termed *deconvolution*.

Mathematically, the deconvolution problem is stated as follows. Recall from Ex. 4.4 that a linear system is characterized by the convolution integral:

$$y(t) = \int_{-\infty}^{\infty} x(\tau)h(t - \tau) \, d\tau \tag{14.11}$$

where $x(t)$ is the input signal, $h(t)$ is the system impulse response, and $y(t)$ is the output signal. In this discussion, we assume that the impulse response is known. Given $h(t)$ and the output $y(t)$, it is desired to determine the input signal $x(t)$. From the convolution theorem, we can write Eq. (14.11) equivalently in the transform domain as

$$Y(f) = X(f)H(f) \tag{14.12}$$

The theoretical inverse filter $R(f)$ can be determined by solving for $X(f)$ in

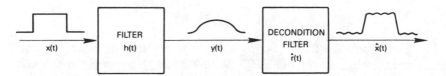

Figure 14.19   Graphical definition of the deconvolution filtering problem.

Eq. (14.12):

$$X(f) = [1/H(f)]Y(f) = R(f)Y(f) \qquad R(f) = 1/H(f) \qquad (14.13)$$

or, equivalently,

$$x(t) = r(t) * y(t) \tag{14.14}$$

The meaning of the term *deconvolution filter* is now apparent. Theoretically, we can recover the signal $x(t)$ perfectly, but as will be seen in the following discussion, practical considerations force us to determine $\hat{x}(t)$, an estimate of $x(t)$.

FFT Deconvolution Filter Design

From Eq. (14.13), the inverse filter is defined in the frequency domain as $R(f) = 1/H(f)$; thus, we have a frequency-domain specification FFT digital filter-design problem. However, in general, $H(f)$ tends to zero as frequency increases and, as a result, $R(f)$ tends to infinity as frequency increases. For this reason, it is normally impossible to sample this frequency function and compute an inverse filter impulse response $r(t)$ of finite duration.

A logical way to approach this issue is to multiply $1/H(f)$ by a truncation function $W(f)$. The resulting frequency function is then zero for all frequencies greater than the truncation frequency f_c. This apparently solves the problem in that it is now feasible to evaluate the inverse discrete transform of $W(f)/H(f)$. However, we know from previous discussions that truncation in the frequency domain yields ripples in the time domain. As a result, $W(f)$ must be chosen to be a function that gently tapers to zero for some frequency f_c and is zero for $f > f_c$. A good compromise is the Hanning function. Figure 14.20 illustrates the proposed frequency-domain modification concept.

The required frequency-domain approximation is obtained by modification of Eq. (14.13):

$$\hat{X}(f) = [W(f)/H(f)]Y(f) = R(f)Y(f) \tag{14.15}$$

where $W(f)$ is the truncation function and

$$R(f) = W(f)/H(f) \tag{14.16}$$

The approximation of Eq. (14.15) is the inverse filtering equation that we implement by means of the FFT. Note that $R(f)$ is simply a frequency-domain specification of a filter, as discussed in Chapter 12.

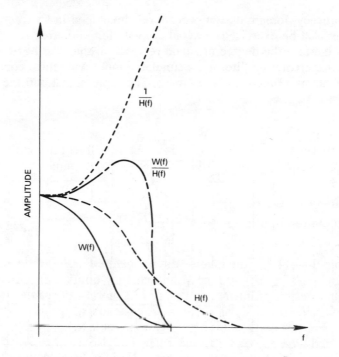

Figure 14.20 Weighting-function modification of the inverse filter frequency response.

FFT Deconvolution Implementation

To illustrate deconvolution filter design, assume that the impulse response of a nonphysically realizable system is given by the function:

$$h(t) = \tfrac{1}{2}\alpha e^{-\alpha t} \qquad (14.17)$$

This impulse-response function is representative of system responses found in many practical signal-restoration problem areas. The Fourier transform of Eq. (14.17) is

$$H(f) = 1/[1 + (2\pi f/\alpha)^2] \qquad (14.18)$$

The analytical expression for the inverse filter frequency response for a Hanning truncation function is

$$R(f) = \frac{\tfrac{1}{2} + \tfrac{1}{2}\cos(\pi f/f_c)}{\alpha^2/[\alpha^2 + (2\pi f)^2)]} \qquad -f_c \le f \le f_c$$
$$= 0 \qquad\qquad\qquad f > f_c \qquad (14.19)$$

This frequency-response function is sampled and a filter is designed by the

FFT frequency-domain design procedures developed in Chapter 12. Recall that care must be exercised to avoid convolution end effects.

To indicate the degree of signal restoration that can be accomplished by FFT deconvolution filtering, a simulated input waveform consisting of a sum of Gaussian functions is assumed. The input signal and the waveform

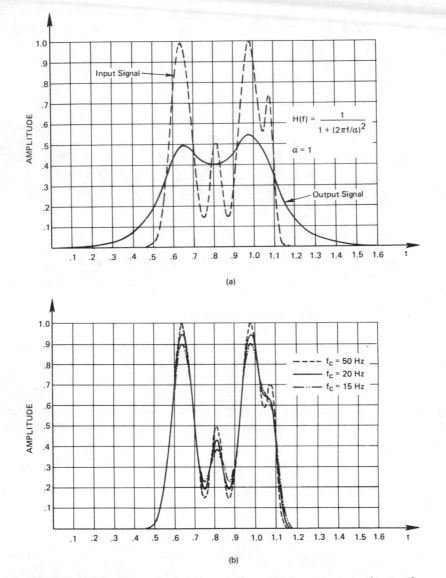

Figure 14.21   Example deconvolution waveforms: (a) low-pass system input and output waveforms, and (b) deconvolution results as a function of the truncation frequency f_c.

resulting from its convolution and the exponential impulse response of Eq. (14.17) are shown in Fig. 14.21(a). It is this output signal to which the inverse digital filter of Eq. (14.19) is applied.

Figure 14.21(b) illustrates deconvolved waveforms as a function of the parameter f_c. Because parameter f_c determines the width of the frequency-domain truncation function, we observe that as f_c is increased, the deconvolved waveform more closely approximates the input signal. Note that for all practical purposes, the input signal is completely restored; the degree of deconvolution that is possible is limited principally by the presence of noise.

If we assume that the signal and noise cannot be identified with respect to the statistics required for the application of sophisticated statistical deconvolution techniques, then the procedure developed here is experimentally applied. We simply decrease the value of the parameter f_c until satisfactory results are achieved. In general, if a high level of noise is added to either the impulse response or the output, then reasonably accurate deconvolution is not possible. The deconvolution approach proposed here must be modified if the filter function is zero-valued for $f < f_c$, (see Prob. 14.17). Silverman [22] describes a theoretically more correct although more complicated FFT deconvolution procedure.

14.6 FFT ANTENNA DESIGN ANALYSIS

The Fourier transform has long been recognized as a useful tool in the solution of antenna design problems. However, these analyses were largely limited to those cases for which the Fourier integrals could be evaluated by classical methods. With the FFT, Fourier transform analysis is considerably more effective.

In this section we develop the fundamentals for applying the FFT to antenna design analysis. Our approach is limited to a consideration of one-dimensional apertures. This may appear inadequate in that antennas are generally considered in two dimensions. However, the treatment is adequate for a great many antennas whose directivity is separable into a product of directivities of one-dimensional apertures and where spacial patterns are surfaces of revolution of the two-dimensional pattern that is produced by the one-dimensional aperture. Further, the one-dimensional case develops the analogy of antenna patterns and the Fourier transform. Our results are readily extendable to two dimensions.

Fourier Transform Relationship Between Antenna Aperture Distribution and Far-Field Pattern

Consider the electric field distribution over the aperture of length a, as shown in Fig. 14.22. This electric field aperture distribution model rep-

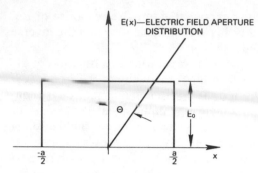

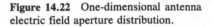

Figure 14.22 One-dimensional antenna electric field aperture distribution.

resents a conventional electromagnetic *horn* antenna or a simple *dipole* antenna. As shown, the electric field is zero over the part of the plane occupied by the conductor but has a uniform electric field distribution over the horn opening (or dipole dimension).

The far-field pattern, as a function of θ, where θ is measured from the perpendicular to the aperture distribution, is given by (Refs. [1] and [13]):

$$E(\theta) = \int_{-\infty}^{\infty} E(x)e^{-j2\pi x[\sin(\theta)]/\lambda} \, dx \qquad (14.20)$$

where $E(x)$ = electric field aperture distribution, volts/meter;
$E(\theta)$ = far-field radiation pattern, volts;
θ = direction of antenna field pattern measured from perpendicular to aperture dimension, degrees.

Equation (14.20) is a Fourier transform, where the aperture dimension x is analogous to time t and the direction function $[\sin(\theta)]/\lambda$ is analogous to frequency f in the conventional Fourier transform relationship.

Note that the analogous relationship between frequency f and $[\sin(\theta)]/\lambda$ must be interpreted correctly in that the variable f is defined from $-\infty$ to $+\infty$, whereas θ is periodic over the interval 0 to 2π. As a result, the Fourier transform relation of Eq. (14.20) is uniquely defined over a finite range of the variable θ. We further explore this antenna pattern Fourier transform interpretation problem in the following example.

Example 14.3   Antenna Far-Field Pattern Fourier Transform Computation

Assume that the electric field aperture distribution $E(x)$ is as shown in Fig. 14.22. Determine the far-field pattern from the Fourier transform relationship of Eq. (14.20) and compare with the conventional Fourier transforms results if Fig. 14.22 is considered a *time*-domain waveform, that is, if the length dimension x is interpreted as a time dimension t.

First, let us compute the conventional Fourier transform of the waveform illustrated in Fig. 14.22:

$$E(f) = \int_{-\infty}^{\infty} E(t)e^{-j2\pi ft}\, dt = \int_{-\infty}^{\infty} E_0 e^{-j2\pi ft}\, dt \tag{14.21}$$

$$= E_0\,[\sin(\pi af)]/\pi af \tag{14.22}$$

As expected, the pulse waveform yields the $[\sin(f)]/f$ function of Eq. (14.22). This result is plotted in Fig. 14.23(a) for parameter $a = 1$.

To determine the far-field antenna pattern, we use Eq. (14.20) and Fig. 14.22:

$$E(\theta) = \int_{-\infty}^{\infty} E_0 e^{-j2\pi x[\sin(\theta)]/\lambda}\, dx \tag{14.23}$$

$$= E_0\,\frac{\sin\{\pi a[\sin(\theta)]/\lambda\}}{\pi a[\sin(\theta)]/\lambda} \tag{14.24}$$

To plot the antenna pattern of Eq. (14.24), we must relate the antenna aperture dimension a to the frequency at which the antenna is to be used. Assume that $a = 2\lambda$. For this case, Eq. (14.24) yields the antenna pattern shown in Fig. 14.23(b).

Now let us compare the results of Eqs. (14.22) and (14.24), that is, Figs. 14.22(a) and (b). As shown in Fig. 14.23(a), the frequency function is defined for all frequency values from $+\infty$ to $-\infty$. (The negative frequency function is a mirror image of the positive frequency function of Fig. 14.23(a) and is not shown for clarity.) In contrast, the antenna pattern in Fig. 14.23(b) is periodic over the interval $-90°$ to $+90°$. Hence, when one attempts to compare the two results, it is readily apparent that the conventional Fourier transform results of Fig. 14.23(a) must be *truncated* if we are to convert these results to those of Fig. 14.23(b).

To determine the appropriate conversion factor and the truncation value, compare the defining relationships of Eqs. (14.21) and (14.23). We note the following equalities:

$$x = t \tag{14.25}$$
$$[\sin(\theta)]/\lambda = f$$

Hence, we convert Fig. 14.23(a) to Fig. 14.23(b). We determine θ from the relationship

$$\theta = \sin^{-1}(f\lambda) \tag{14.26}$$

Because the maximum nonperiodic value of θ is $90°$, then the maximum value of f (i.e., the truncation value) occurs for $f\lambda = 1$. Recall that λ was chosen as $a/2$ in Fig. 14.23(b) and a was chosen as 1 in Fig. 14.23(a). Hence, $\lambda = \frac{1}{2}$ and the truncation value of f is 2 Hz. As a result, to convert Fig. 14.23(a) to Fig. 14.23(b), we use only the main lobe and first side lobe of Fig. 14.23(a) and determine the abscissa axis θ from Eq. (14.26). Note that we have illustrated by means of symbols on Figs. 12.22(a) and (b) several conversion values. As shown, we truncate the conventional Fourier transform results of Fig. 14.23(a) at $f = 2$ Hz.

The results of Fig. 14.23(b) are plotted in conventional polar-coordinate form in Fig. 14.23(c). Observe that the results are symmetrical for angles greater than $\pm90°$. This follows from our electric field aperture distribution assumption in that

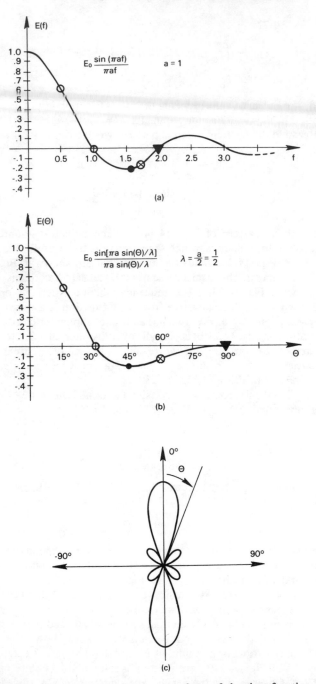

Figure 14.23 (a) Conventional Fourier transform of the time function of Fig. 14.22, (b) far-field antenna pattern for the aperture distribution of Fig. 14.22, and (c) polar coordinate graphical presentation of part (b).

Fig. 14.22 can be interpreted as the aperture distribution in any plane revolved around the abscissa axis. Hence, the antenna pattern is expected to be symmetrical.

FFT Antenna-Pattern Computation

To apply the FFT to the computation of antenna patterns, we simply implement the basic principles previously established. That is, we consider the aperture electric field distribution as a time-domain waveform; compute the FFT of this waveform and then apply the appropriate abscissa scale-conversion factor of Eq. (14.26).

In Fig. 14.24(a), we show an example electric field aperture distribution that alternates in phase and has constant amplitude. Note that the aperture distribution function is symmetrical about the origin. We must be careful to preserve this relationship when applying the FFT. This is accomplished by sampling the aperture distribution function, as shown in Fig. 14.24(b). We use the fact that the sampled function to which the FFT is to be applied must be periodic. The number of zeros that one introduces is strictly a matter of choice as to the desired FFT frequency spacing to allow one to easily trace the side-lobe structure of the antenna pattern.

Figure 14.24(c) illustrates the FFT of the sampled aperture distribution of Fig. 14.12(b). This result must be converted or transformed, as is described in Ex. 14.3. Let us assume that the distances shown in Fig. 14.24(a) are in meters and that we wish to determine the antenna pattern for a wavelength $\lambda = \frac{1}{2}$ m. Then, from Eq. (14.26), the truncation frequency value is 2 Hz. As a result, we transform or convert the FFT results of Fig. 14.12(c) to those of Fig. 14.24(d) by means of Eq. (14.26). Only the results for the frequencies $0 \leq f \leq 2$ Hz are converted. As before, the antenna pattern for angles greater than $\pm 90°$ is a replica of the pattern for angles less than $\pm 90°$. The corresponding polar plot is shown in Fig. 14.24(e).

Recall that as the wavelength of the antenna becomes small with respect to the aperture dimension, then the main lobe of the antenna becomes narrow and the number of side lobes is increased. To see this effect, let us convert or transform the FFT results of Fig. 14.24(c) for a wavelength $\lambda = \frac{1}{5}$ m. From Eq. (14.26), the truncation frequency is now 5 Hz. We illustrate the resulting converted polar plot in Fig. 14.24(f).

We have developed a simplified application of the FFT to antenna-pattern analysis. Our approach requires a conversion of the far-field radiation integral of Eq. (14.20) to a Fourier Integral. A more detailed application of our approach is given in Ref. [25]. Results presented here can be extended to the two-dimensional analysis of antenna apertures. The radiation pattern of reflector antennas is determined in Refs. [5] and [14] by the FFT and a $[\sin(u)]/u$ sampling approach. Incorporation of the FFT with the conjugate gradient method is used to solve for the aperture fields and the induced

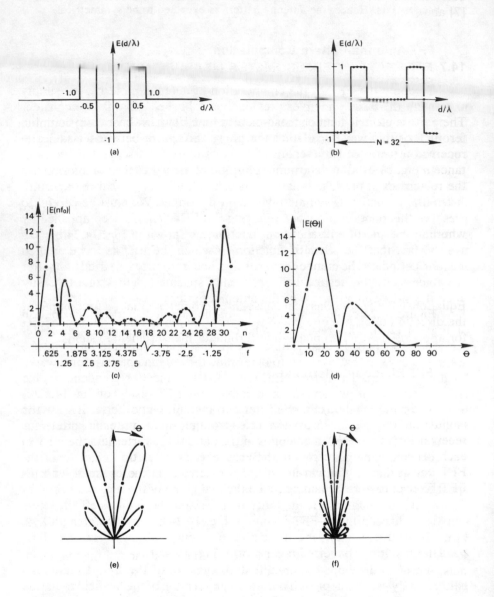

Figure 14.24 (a) Example one-dimensional antenna aperture distribution, (b) sampled aperture distribution for FFT computation, (c) FFT of the sampled aperture distribution of part (b), (d) angle transformation of the FFT results of part (c) for λ = 0.5 meters, (e) polar plot of part (d) for λ = 0.5 meters, and (f) polar plot of part (d) for λ = 0.2 meters.

current densities for wire, wire mesh, and rectangular plate antennas in Refs. [7] and [20].

14.7 FFT PHASE-INTERFEROMETER MEASUREMENT SYSTEM

The FFT can implement a phase-measurement system based on the interferometer principle. Recall that the phase difference between waveforms received at two spatially separated sensors (antennas) separated by a distance d can be used to determine the angle of arrival of the waveform from the relationship (see Fig. 14.25):

$$\theta = \sin^{-1}(\lambda\phi/2\pi d) \tag{14.27}$$

where θ = angle of arrival
λ = signal wavelength
ϕ = phase difference
d = antenna separation

Equation (14.27) is the classical phase-interferometer equation for computing the direction of arrival of a plane wavefront. We now show the procedures for applying the FFT to phase-interferometer measurement systems.

FFT Phase Interferometer

The block diagram of an FFT interferometer direction-of-arrival system is illustrated in Fig. 14.25. As shown, the output of each sensor or antenna/receiver is sampled by an analog-to-digital (A/D) converter and the FFT of each sensor output is computed. Because each resolution element of the FFT consists of a real and an imaginary term, then the phase θ_n of each FFT filter output can be computed as

$$\theta_n = \tan^{-1}\left[\frac{\text{Real Output }(R_n)}{\text{Imag Output }(I_n)}\right] \quad n = 0, 1, \ldots, N/2 \tag{14.28}$$

Equation (14.28) is computed for the FFT outputs for each of the two channels. Phase difference ϕ_n is then determined by simple subtraction for each FFT resolution cell (filter output).

The next process step, system phase correction, is the single most practical consideration in considering the application of the FFT to interferometer signal processing. A limiting factor in the accuracy of a direction finding system is the differential phase error between the two channels. System designers attempt to perfectly match the two channels from sensor

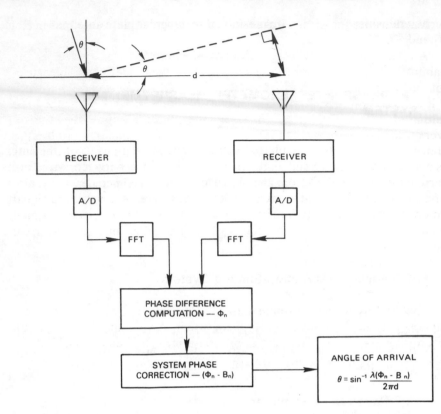

Figure 14.25 Block diagram of an FFT phase-interferometer measurement system.

(or antenna) to receiver output, but in practice phase error exists. Calibration is often necessary to achieve sufficient accuracy.

With an FFT implementation, it is possible to calibrate the system for all frequencies within the passband of the receiver. For example, a broadband signal can be *injected* perpendicular to the sensor array. The phase difference between the two channels should be zero. If the phase difference between corresponding FFT cells differs from zero, this differential is due to system inaccuracies for that frequency cell and the error can be stored (by cell) as a system phase correction. System calibration can be repeated as often as required.

The angle of arrival can be determined for each frequency cell of the FFT. If the signal bandwidth is greater than the FFT bandwidth, adjacent FFT cells should give near-identical results. Note that the proposed implementation concept also yields the angle of arrival of multiple signals if their frequencies do not overlap.

Measurements in the Presence of Interference

Another distinct advantage of an FFT interferometer system is the capability to cope with interfering signals. A conventional phase-measurement system normally is reasonably well-matched to the signal bandwidth and if an interfering signal overlaps any portion of this bandwidth, then the resulting phase measurement is in error. With the FFT approach, the receiver output is divided by the FFT into a band of narrow-band filter outputs. The phase difference is computed simultaneously for each of the filter outputs across the receiver bandwidth. In most instances, the interfering signal differs in angle of arrival from the signal of interest. Hence, on an angle-vs.-frequency plot, one will see two straight-line segments, one for the signal of interest and one for the interfering signal. Unless the interfering signal completely overlaps in frequency the desired signal, then an angle-of-arrival measurement can be made.

FFT Monopulse Direction-Finding System

The FFT can also be applied to the development of an amplitude-comparison monopulse direction-finding system. An amplitude comparison of each filter output provides the appropriate measurement. Note that, as in the interferometer case, it is straightforward to develop a calibration procedure for each FFT resolution cell.

14.8 FFT TIME-DIFFERENCE-OF-ARRIVAL MEASUREMENT SYSTEM

The accurate measurement of the time difference of arrival for narrow-band signals arriving at spacially separated sensors is an excellent application of the FFT. Analog correlators can be used, but system inaccuracies severely limit the fields of application. In this section, we address the basics in applying the FFT to time-difference-of-arrival measurements.

Problem Definition

The FFT is applied to time-difference-of-arrival measurements by implementing classical correlation techniques. From Chapter 4, the correlation function for a waveform $s_1(t)$ that arrives at a sensor at some time t_0 and a replica of that same waveform, $s_2(t)$, arriving at a different sensor at some later time $t_0 + \tau$ is given by

$$z(\tau) = \int_{-\infty}^{\infty} s_1(t)s_2(t + \tau) \, dt \qquad (14.29)$$

By definition, the correlation function measures the degree of match or correlation between a waveform and a shifted replica of the waveform. Hence, Eq. (14.29) reaches a maximum for that shift τ that corresponds to the time difference of arrival of the waveforms at the sensors. We determine the value of τ_{max} at the correlation peak by using the FFT to compute the discrete correlation theorem discussed in Sec. 7.4

FFT Time-Difference-of-Arrival Measurement

The basic computational procedure for FFT application to the time-difference-of-arrival measurement is illustrated in Fig. 14.26. As shown, waveforms $s_1(t)$ and its replica $s_2(t)$ arrive at spacially separated sensors at time difference τ. Each sensor output is sampled by an analog-to-digital converter and input to an FFT. Recall from Chapter 7 that zeros must be appended to the sampled waveforms to avoid end effects. From the FFT outputs, we form the correlation theorem product $S_1(f)S_2^*(f)$, where $S_2^*(f)$ is the complex conjugate of $S_2(f)$. The resulting complex function is termed the *cross spectrum*. The cross spectrum can be viewed as an amplitude and phase spectrum, as shown in Fig. 14.26. The inverse FFT of the cross spectrum is the desired *cross-correlation function*. The cross-correlation function peaks at the value of τ_{max} corresponding to the desired time-difference-of-arrival measurement.

Although the procedure described appears straightforward, there is one shortcoming. Examine the output cross-correlation in terms of an accurate determination of the peak value. The time resolution of the cross-correlation function is determined by the sampling interval of the waveforms $s_1(t)$ and $s_2(t)$. If the sampling interval is T s, then the cross-correlation time resolution is T s, which is not sufficiently accurate for practical applications. As a result, we must interpolate between the samples of the cross-correlation function to determine τ_{max}. As long as the Nyquist sampling rate for the input signals is observed, then, theoretically, the continuous cross-correlation waveform can be reconstructed.

FFT Phase-Domain Time-Difference-of-Arrival Measurement

An alternate approach to measure the time difference of arrival is to compute the slope of the phase-domain function. Note from Fig. 14.26 that the phase slope is equal to $2\pi\tau_{max}$. This follows from the time-shifting theorem (Sec. 3.4). Hence, rather than implementing an interpolation procedure to accurately estimate τ_{max}, we simply estimate the slope of the phase spectrum.

Several alternatives for slope estimation should be explored based on the specifics of the problem. A weighted least-square approach based on

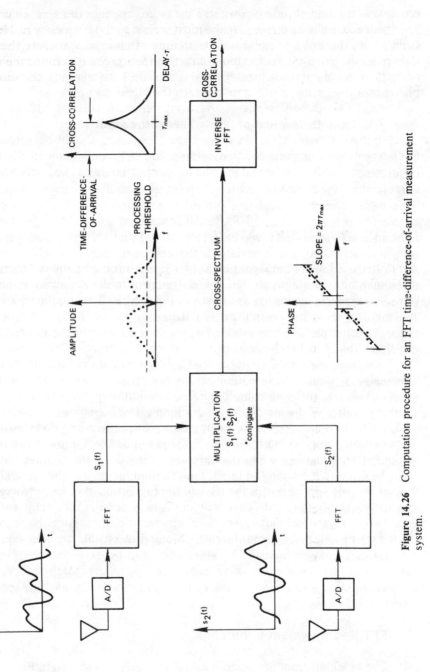

Figure 14.26 Computation procedure for an FFT time-difference-of-arrival measurement system.

cross-spectrum amplitude (shown in Fig. 14.26) appears the most practical. This procedure allows one to give the most weight to the frequency resolution cells where the signal-to-noise ratio is greatest. In fact, a procedure to eliminate from the slope estimation those data from low cross spectrum amplitude cells is probably most advantageous. One can also average consecutive phase-spectrum data in order to increase the signal-to-noise ratio.

An FFT time-difference-of-arrival system can be calibrated, as discussed in the previous section. Note that if the input waveforms have no delay between them, then the cross-phase spectrum should be zero. Any deviations from zero are due to system phase-differential errors. This calibration data can be stored and the phase-spectrum data can be appropriately corrected for each measurement.

14.9 FFT SYSTEM SIMULATION

Accurate prediction of system performance often requires the development of simulation techniques to verify the design criteria. Radar, communications, sonar, and imaging system designers use the FFT for digital simulation analysis to reduce hardware design cost.

The general class of systems for which FFT simulation is applicable are those that can be characterized by open-loop transfer functions. This follows because the FFT requires that a *block* of data is processed simultaneously. Systems whose nonlinearities are characterized in the time domain can also be easily simulated. FFT simulation is appealing to a system's analyst because of the simplicity of designing the simulation. As we saw in Chapter 12, either a time- or frequency-domain specification of system functions can be implemented by the FFT. This implies that equations familiar to the system's hardware engineer are used directly in the simulation.

To explore the potential of FFT simulation techniques, we describe in this section the application of the FFT to the prediction of radar performance in a specified environment. This problem is in general nontractable by conventional analysis techniques and is characteristic of classical background clutter, electronic countermeasure (ECM) and electronic counter-countermeasure (ECCM) problems. The simulation can be extended to more sophisticated radar signal-processing techniques, including matched receivers, doppler filtering, optimum signal design, chirp waveforms, and phased arrays.

FFT Radar-System Simulation

The block diagram of a simplified radar receiver is shown in Fig. 14.27. The mixer and the local oscillator convert the radio-frequency (RF) signal to an intermediate frequency (IF), where the converted signal is amplified

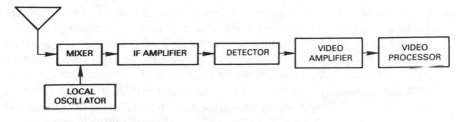

Figure 14.27   Block diagram of a simplified radar receiver.

and filtered. The (RF) pulse modulation is extracted by the detector and amplified by the video amplifier. Target range information is extracted by the video processor. The key to radar performance in a clutter or jamming environment is generally determined by the capability of the video processor.

The classical radar-analysis problem is to determine the degradation of range extraction by the video processor as a function of input-noise characteristics. System degradation is normally measured in terms of probability of detection and probability of false alarm. If the received noise is Gaussian, then closed-form solutions for system performance can be obtained. However, it is necessary to resort to a simulation in order to evaluate system degradation if the noise is an interfering signal with specified modulation characteristics. A digital FFT simulation of the block diagram illustrated in Fig. 14.27 is a cost-effective method for evaluating system performance in these cases.

Figure 14.28 illustrates the radar model chosen for simulation and the corresponding FFT simulation block diagram. The received waveform is simulated by generating samples of the additive combination of the pulsed IF waveform and the interfering signal. Both the IF and video amplifiers (filters) are simulated by sampling their respective transfer functions in the frequency domain, as discussed in Chapter 12.

The IF amplifier is assumed to be a cascade combination of Butterworth filters whose center and cutoff frequencies are chosen to enhance the rolloff

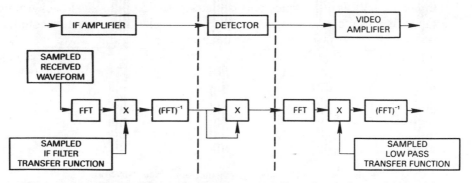

Figure 14.28   FFT simulation model of a radar receiver.

of the skirts of the resulting transfer function. Conventional analog filter design equations are sampled in the frequency domain. Filtering at IF is accomplished by forming the product of the FFT of the sampled input and the sampled filter function; this product is then inversely transformed to obtain the time-domain output of the filter.

Nonlinear square-law detection is simulated by squaring the IF time-domain waveform. Simulation of the video amplifier (filter) is accomplished in the same manner as the IF amplifier. The video amplifier is assumed to be a three-pole Butterworth low-pass filter that is simulated by frequency-domain sampling. The resulting output video waveform is representative of the receiver performance in the presence of the simulated interference.

FFT Radar-System Simulation Results

To illustrate the waveforms that can be obtained by this simulation method, Gaussian noise is added to the input waveform shown in Fig. 14.29. An estimate of the power spectrum of the input waveform obtained by computation of a Hanning weighted FFT is shown in Fig. 14.30(a). A similarly computed IF output power spectrum is shown in Fig. 14.30(b). The detected video output is illustrated in Fig. 14.30(c). If a sequence of system video outputs is generated by the simulation, each with independent noise samples, statistical parameters such as probability of detection, probability of false alarm, error rate, etc. can be evaluated as a function of the characteristics of the interfering signal and the parameters of the video processor.

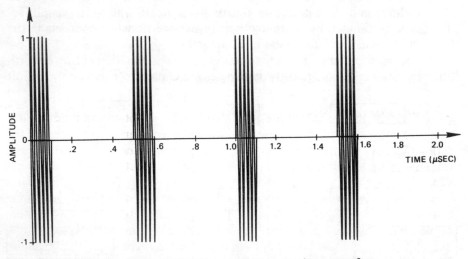

Figure 14.29   FFT radar-system simulation input waveform.

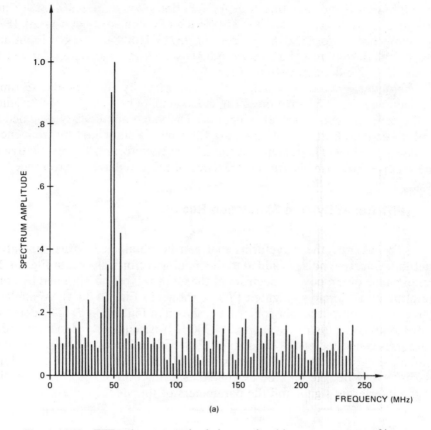

Figure 14.30 FFT radar-system simulation results: (a) power spectrum of input signal plus noise, (b) IF output power spectrum, and (c) detected video output signal.

Communication-System Simulation

FFT simulation techniques are also readily adaptable to communication systems. In digital data systems, a common problem encountered is the estimation of intersymbol intererence as a function of noise level, data rate, transmitter bandwidth, transmitter filter rolloff characteristics, and system synchronization parameters. Analogous to the radar problem, an FFT simulation can be implemented to evaluate the probability of correctly decoding a transmitted message as a function of each parameter degrading system performance. An FFT simulation approach to communication-system analysis allows one to include real-world constraints that are normally unwieldly in closed-form analysis.

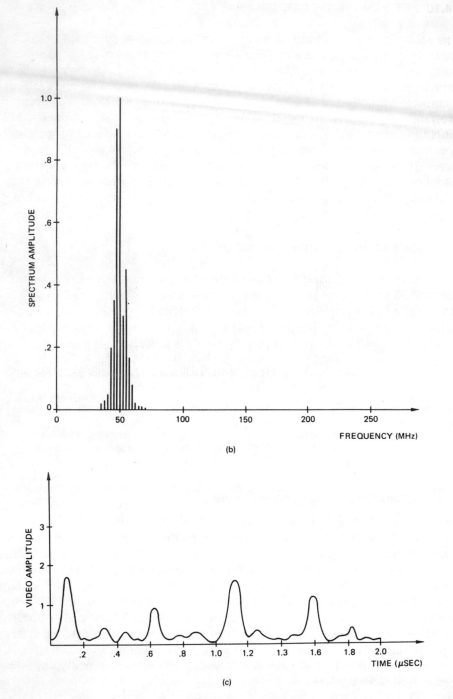

(b)

(c)

Figure 14.30 (*continued*)

14.10 FFT POWER-SPECTRUM ANALYSIS

The measurement of power spectra is a difficult and often misunderstood topic. Because the FFT readily yields frequency and amplitude information, many investigators proceed to *estimate* the magnitude of a sampled waveform by applying the FFT. If the waveform is periodic or deterministic, then a correct interpretation of FFT results is likely. However, when waveforms are random processes, it is necessary to develop a statistical approach to amplitude estimation. We describe in this section the fundamentals of power-spectrum estimation, introduce the terminology, and provide FFT procedures for computing the power spectrum. As is shown, the FFT computational procedures are straightforward; however, the statistical interpretation of the results is difficult. A detailed development of statistical estimation is beyond the scope of this discussion.

Correlation Spectrum Estimation

Let $x(t)$ be a random function of time. In constrast to a deterministic function, future values of a random function cannot be predicted exactly. However, it is possible that the value of the random function at time t_1 influences the value at time t_2. We express this statistical characteristic by means of the *autocorrelation* function, which is given by

$$\phi(\tau) = \lim_{L \to \infty} 1/L \int_{-L/2}^{L/2} x(t)[x(t + \tau)] \, dt \qquad (14.30)$$

The power-spectral-density function $\Phi(f)$ and the autocorrelation function $\phi(\tau)$ are defined as a Fourier transform pair:

$$\phi(\tau) = \int_{-\infty}^{\infty} \Phi(f)e^{j2\pi f\tau} \, df \quad \Longleftrightarrow \quad \Phi(f) = \int_{-\infty}^{\infty} \phi(\tau)e^{-j2\pi f\tau} \, d\tau \qquad (14.31)$$

Function $\Phi(f)$ is called by many terms including the power-spectrum, the power-density, the spectral-density, and the power-spectral-density function. We use these terms interchangeably, as does the literature. Note that if we set $\tau = 0$ in Eqs. (14.30) and (14.31), we obtain

$$\int_{-\infty}^{\infty} \Phi(f) \, df = \phi(0) = \int_{-\infty}^{\infty} x^2(t) \, dt \qquad (14.32)$$

The right-hand of Eq. (14.32) is the total energy or power of the random function (see Sec. 2.4). Because the integral of $\Phi(f)$ is equal to the total signal power, then the terminology *power*, or *spectral density*, has been adopted.

If the autocorrelation function is known, then the calculation of the power spectrum is determined directly from the Fourier transform. However, the general case is that we must determine $\phi(\tau)$. Equation (14.30) is appealing but the relationship requires a knowledge of $x(t)$ for $-\infty < t < \infty$. In practice, $x(t)$ is known only over a finite interval, and we must *estimate* $\phi(\tau)$ based on only this finite duration of data. The estimator for $\phi(\tau)$ that is generally used is

$$\hat{\phi}(\tau) = \frac{1}{L - |\tau|} \int_0^{L-|\tau|} x(t)x[(t + |\tau|)] \, dt \qquad |\tau| < L \qquad (14.33)$$

where $x(t)$ is assumed to be known only over the finite duration L.

Because $\hat{\phi}(\tau)$ is not defined for $\tau > L$, then, as shown in Fig. 14.31, we multiply Eq. (14.33) by a window function that is nonzero where Eq. (14.33) is defined and is zero elsewhere. Function $w(\tau)$ is termed a *lagged window* because we can visually describe our observation of $\hat{\phi}(\tau)$ as looking

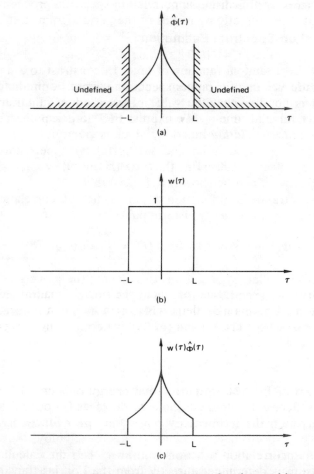

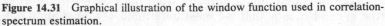

Figure 14.31 Graphical illustration of the window function used in correlation-spectrum estimation.

through the window $w(\tau)$. The modified autocorrelation function $w(\tau)\hat{\phi}(\tau)$ exists for all τ and hence its Fourier transform exists. We can then obtain an estimate of the power spectrum using the relationship of Eq. (14.31):

$$\hat{\Phi}_c(f) = \int_{-\infty}^{\infty} w(\tau)\hat{\phi}(\tau)e^{-j2\pi f\tau} \, d\tau \tag{14.34}$$

where $w(\tau) = 1$ for $|\tau| < L$ and is zero elsewhere. $\hat{\Phi}_c(f)$ is normally defined as the correlation or lagged-product estimator for the power spectrum. This approach to spectral analysis is commonly referred to in the literature as the Blackman-Tukey procedure [27].

Periodogram Power-Spectrum Estimation

An alternate approach to the correlation spectrum procedure is to estimate the spectrum directly by means of the periodogram. Let

$$\hat{\Phi}_p(f) = (1/L) \left| \int_0^L x(t)e^{-j2\pi ft} \, dt \right|^2 \tag{14.35}$$

Subscript p indicates that the power-spectrum estimate is obtained by means of the periodogram. Because Eq. (14.35) is in the form of a Fourier transform over a finite interval, we can then use the FFT to compute the spectrum estimate.

Although the periodogram and the correlation spectrum-estimation procedures appear quite different, they are theoretically equivalent under certain conditions. It can be shown (Ref. [28]) that

$$\hat{\Phi}_p(f) = \int_{-L/2}^{L/2} (1 - |\tau|/L)\hat{\phi}(\tau)e^{-j2\pi f\tau} \, d\tau \tag{14.36}$$

The inverse Fourier transform of Eq. (14.36) yields

$$\hat{\phi}_p(\tau) = (1 - |\tau|/L)\hat{\phi}(\tau) \qquad |\tau| < L \tag{14.37}$$

Hence, if we modify the lagged-product spectrum-estimation technique by simply using a triangular (Bartlett) window rather than a rectangular lag window, then the two procedures are equivalent. Using the convolution theorem, we can rewrite Eq. (14.36) as

$$\hat{\Phi}_p(f) = W_B(f) * \hat{\Phi}_c(f) \tag{14.38}$$

where $W_B(f)$ is the Bartlett frequency-domain window function. Hence, the periodogram spectrum estimate is equal to the lagged-product spectrum estimate convolved with the Bartlett window frequency function.

Correlation spectrum estimation theoretically employs the rectangular lag window, and the periodogram spectrum-estimation procedure can be interpreted as employing the triangular lag window. In practice, we employ neither of these two windows, as will now be described.

Spectral Windows

In the previous discussion, we showed that the correlation and periodogram estimation procedures can both be interpreted as using frequency-domain window or weighting functions. In estimation problems, one strives to achieve an estimator whose mean value (the average of multiple estimates) is the parameter being estimated. It can be shown (Ref. [28]) that the mean value of both the correlation and periodogram estimation procedures is the true spectrum $\Phi(f)$ convolved with the frequency-domain window function:

$$\text{Mean}[\hat{\Phi}_c(f)] = \text{Mean}[\hat{\Phi}_p(f)] = W(f) * \Phi(f) \qquad (14.39)$$

Hence, the mean value of the power-spectrum estimate equals the true spectrum only if the frequency-domain window function is an impulse function (i.e., the data record length is infinite in duration). If the mean of the estimate is not equal to the true value, then we say that the estimate is *biased*.

From our previous discussion of FFT data-weighting functions (Sec. 9.2), we know that detail is lost if we smooth (convolve) with a broad spectral (frequency-domain) window. Said differently, amplitude values adjacent to a true peak in the spectrum become biased due to the smoothing that occurs in the convolution operation with the spectral frequency window function. Hence, one could conclude that a narrow spectral window is desirable. However, this is not a valid conclusion because the more narrow the spectral window, the larger the variance of the estimate (Refs. [27] and [28]). This statement follows intuitively because the variance of the estimate of several random variables that are summed has a smaller variance than that of a single random variable. Hence, to reduce the variance of the estimate, we must broaden the spectral window that averages or smoothes adjacent estimates due to the convolution operation of Eq. (14.39).

The normal method for characterizing the width of the frequency-domain window is to define its bandwidth. In spectral analysis, bandwidth is defined as

$$\text{Bandwidth } (BW) = 1 \Big/ \left\{ \int_{-\infty}^{\infty} W^2(f) \, df \right\} \qquad (14.40)$$

Spectral window bandwidth determines the resolution of the spectrum estimate as well as the variance of the estimate. A compromise between small variance and high fidelity (resolution) is the crux of the power-spectrum estimation problem. We follow the conclusion of Jenkins [28] that any a priori optimality criteria that sets too rigid a mathematical formulation for this trade-off is not practical. A more useful and flexible approach is to use an experimental spectrum-estimation approach that allows one to learn the appropriate bandwidth of the spectral window from the data. After defining a FFT procedure for computing the power spectrum, we develop such an experimental technique.

Smoothed Periodogram FFT Spectrum Estimation

The spectral window for the periodogram is of the form $\{[\sin(f)]/f\}^2$. This follows from the developments leading to Eq. (14.38), where it was shown that the periodogram spectral estimate was equivalent to a correlation estimate using the triangular or Bartlett window. Recall from Sec. 9.2 that the Bartlett frequency window has relatively high side lobes with respect to other window functions. However, Jones [29] has shown that very good periodogram spectral estimates can be obtained from the $\{[\sin(f)]/f\}^2$ spectral window by averaging (smoothing) adjacent spectrum estimates. The smoothed periodogram (sp) estimate is given by

$$\hat{\Phi}_{sp}(f) = W_D(f) * \hat{\Phi}_p(f) \qquad (14.41)$$

where $W_D(f)$ is the rectangular frequency window first suggested by Daniel [28];

$$
\begin{aligned}
W_D(f) &= 1/(\beta f_0) \qquad -\beta f_0/2 \le f \le \beta f_0/2 \\
&= 0 \qquad \text{otherwise}
\end{aligned}
\qquad (14.42)
$$

Note that parameter βf_0 specifies the frequency range over which the periodogram is averaged ($f_0 = 1/L$). Hence, the smoothed periodogram window is that obtained by averaging the appropriate number of $\{[\sin(f)]/f\}^2$ spectral windows that are spaced at intervals of $f_0 = 1/L$. We show in Fig. 14.32 the $\{[\sin(f)]/f\}^2$ periodogram spectral window and the smoothed periodogram spectral window for $\beta = 10$. That is, we have averaged 10 adjacent periodogram windows. A comparison of the smoothed periodogram spectral window with the Hanning and Parzen windows is shown in Fig. 14.33 under the constraint of equal bandwidths. Spectral windows with equal bandwidths,

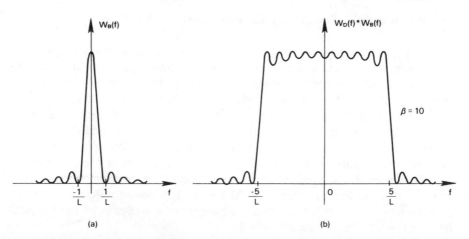

Figure 14.32   (a) Periodogram spectral window, and (b) the smoothed periodogram spectral window for $\beta = 10$.

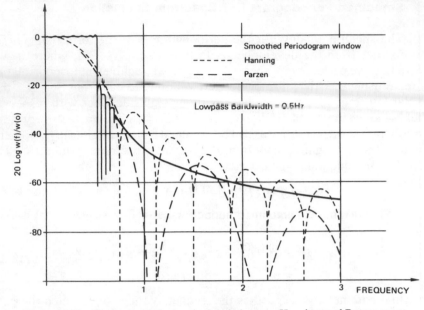

Figure 14.33 Comparison of smoothed periodogram, Hanning, and Parzen spectral windows under the equal-bandwidth constraint.

as determined from Eq. (14.43), produce a spectrum estimate with equal variances. The bandwidth or resolution of the smoothed periodogram is given by β/L. We show in Fig. 14.34 a smoothed periodogram computing procedure using the FFT. As shown, we average the FFT computed estimates in groups of β, except that the first group contains only $\beta/2$ terms.

Experimental Procedure for FFT Spectral Analysis

A practical procedure for power-spectrum estimation is to progressively reduce the spectral analysis bandwidth. This approach allows one to learn significant features of the spectrum during the course of the analysis. The initial choice of a wide bandwidth masks fine detail in the spectrum. However, a wide bandwidth produces a stable (low-variance) estimate. If we allow the analysis bandwidth to become smaller, then additional detail can be explored. The practicality of this approach is limited by interpretation problems that result from the instability (large variance) of the estimates.

To illustrate the concept of spectral bandwidth closing, we generate samples of a random process ($T = 0.1$ s) with a power spectrum that for the present we assume is unknown. Our objective is to deduce from the data the true form of the spectrum.

FFT spectrum estimates according to the procedure of Fig. 14.34 are computed in Fig. 14.35 for $N = 64$ and $BW = 0.8, 0.4,$ and 0.2 Hz. We note

1. Sample $x(t)$ for $0 \leq t \leq L$:

$$x(kT) = x(t)\big|_{kT} \qquad k = 0, 1, \ldots, N - 1$$

2. Compute the FFT of $x(kT)$.

$$X(nf_0) = \sum_{k=0}^{N-1} x(kt)e^{-j2\pi nk/N}$$

$$f_0 = 1/NT$$

3. Compute the periodogram of $X(nf_0)$:

$$\hat{\Phi}_p(nf_0) = (T/N)\{\text{Re}^2[X(nf_0) + \text{Im}^2[X(nf_0)]\}$$

4. Compute the smoothed periodogram:

$$\hat{\Phi}_{sp}(0) = 2/\beta \sum_{n=0}^{\beta/2-1} \hat{\Phi}_p(nf_0)$$

$$\hat{\Phi}_{sp}(\beta f_0/2) = 1/\beta \sum_{n=\beta/2}^{3\beta/2-1} \hat{\Phi}_p(nf_0)$$

$$\hat{\Phi}_{sp}(3\beta f_0/2) = 1/\beta \sum_{n=3\beta/2}^{5\beta/2-1} \hat{\Phi}_p(nf_0)$$

$$\vdots$$

Figure 14.34 FFT computational procedure for smoothed periodogram spectrum estimation.

in Fig. 14.35 that as we close the bandwidth from 0.8 to 0.4 Hz, the estimated spectrum contains several spectrum peaks. As we further close the bandwidth to 0.2 Hz, these peaks become even more pronounced. Before reaching a conclusion that these peaks are real, it is necessary to establish that the peaks are not the result of variability or instability of our estimate. We use the concept of *confidence intervals* to make this assessment.

We also show in Fig. 14.35 the 90-percent confidence limits (amplitude range) for the estimate produced for each bandwidth selection. Because a log amplitude scale is used, then the confidence interval is valid for any frequency estimate of the power spectrum. The confidence limit, or amplitude range, is interpreted in that the true power spectrum for any frequency falls within the noted interval with probability 0.9. Hence, the confidence limit is a measure of the statistical variance of the estimate *if we assume that there is no bias in the spectral estimate*. For wide spectral bandwidth, we know that bias is possible. To determine the confidence limit for each

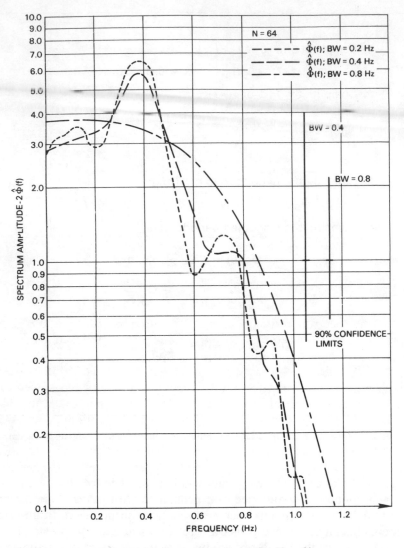

Figure 14.35 Spectrum estimate for $N = 64$.

bandwidth, we use the graphs shown in Fig. 14.36. To use the graph, we must compute the parameter $\eta = 2L(BW)$, where L is the data record length, $L = NT$. Parameter η is referred to as the *number of degrees of freedom* and can be interpreted as the number of squared random variables that have been summed. Intuitively, we expect the variance of summed random variables to decrease as the number of variables summed is increased. Hence, the larger the number of degrees of freedom, the smaller the variance of the spectrum estimate.

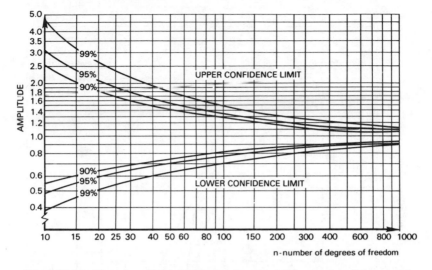

Figure 14.36 Plot of confidence limits as a function of the number of degrees of freedom η.

For the spectrum estimate determined with a bandwidth of 0.8 Hz, we compute $\eta = (2)(0.1)(64)(0.8) = 10.24$. From Fig. 14.36, we obtain the values 2.2 and 0.58 from the upper-limit and lower-limit graphs, respectively. These limits are plotted as a vertical line, as shown in Fig. 14.35. Because the confidence interval is valid for any frequency estimate of the power spectrum, we slide this vertical line along our estimate to the peak value of 0.4 Hz in Fig. 14.36. We conclude that the 90-percent confidence amplitude range is so large with respect to the peak variation in our spectrum estimate that we cannot conclude if the results are statistically significant. To ensure that the peak is real, we must reduce the amplitude range of the confidence interval.

To improve the confidence of our estimate, we increase the number of data points N, which increases η, the number of degrees of freedom of the estimate. In Fig. 14.37, we show spectrum estimates for $N = 512$ and $BW = 0.8, 0.4$, and 0.2 Hz. Note that the spectrum peaks that were observed previously are significantly different. This gives evidence that the previously defined peaks were due to statistical instability. Observe in the lower-frequency region that the spectral estimate decreases in magnitude as the spectral bandwidth is decreased. This observation leads one to the conclusion that there is estimation bias for the wider spectral windows. This same effect is observed in the upper-frequency region (0.8 to 1.2 Hz). In the frequency region of 0.3 to 0.6 Hz, the opposite effect occurs and the amplitude increases as the spectral window bandwidth is reduced. This trend gives evidence that there may be a peak in this region. However, we note that the range of the

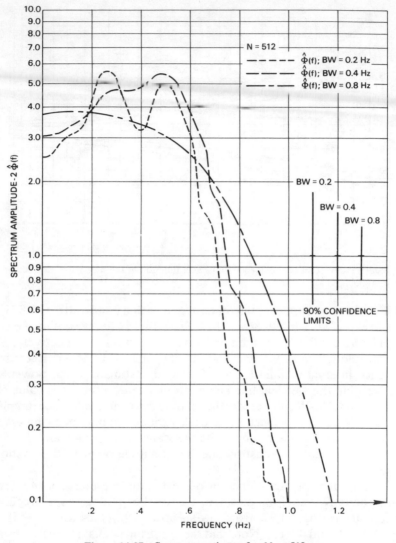

Figure 14.37 Spectrum estimate for $N = 512$.

90-percent confidence limit is still larger than the excursion of the peak we are trying to validate.

In Fig. 14.38, we repeat the spectrum estimates for the case $N = 2048$. We observe a definitive trend toward a peak in the spectrum at approximately 0.5 Hz. The spectrum estimate for $BW = 0.4$ is relatively smooth, which gives credibility to the estimated peak. The estimate for $BW = 0.2$ Hz is questionable because of noticeable variability. Note that for $BW =$

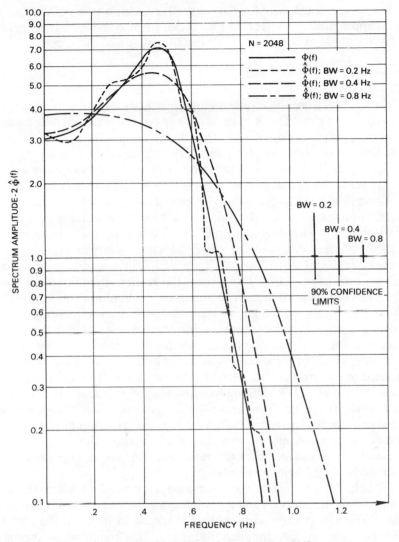

Figure 14.38   Spectrum estimate for $N = 2048$.

0.4 and 0.2 Hz, there is essentially no change in the estimate in the range 0.0 to 0.2 Hz and only a small change in the range 0.3 to 0.6 Hz. We conclude that the spectral estimate ($BW = 0.4$ Hz) has minimum bias in the lower-frequency range. The same arguments and conclusion can be reached for the upper-frequency region. This estimate also has a reasonably small confidence interval. Recall that the confidence interval assumes that there is no bias in the estimate. Therefore, we cannot conclude that the estimate for $BW = 0.8$ Hz is the *best* estimate. Based on these observations, we conclude that the spectrum estimate for $BW = 0.4$ Hz yields a true peak in the spectrum at approximately 0.5 Hz. In Fig. 14.38, we also show the true spectrum. We have estimated reasonably well the true shape of the spectrum.

Summary

In the literature, there are multiple methods described for computing the power spectrum. The Blackman-Tukey procedure is the most popular, but there is no particular reason other than tradition for its use. Spectrum-estimation procedures based on the periodogram yield results that are as good or better than other methods and is computationally more efficient. As discussed, the application of the FFT to spectrum analysis is complex and is primarily a statistical estimation problem. As long as we can continue to increase the data record length, then estimates with reduced variability can be obtained. However the practical problem normally encountered is one of insufficient data and the power-spectrum analysis problem quickly enters the realm of *art-science*. With the FFT, it is quite easy to produce spectrum estimates and for this reason the reader is cautioned to use this section as only an introduction to spectrum estimation. The interpretation of these FFT results is the key to the power-spectrum estimation problem. Readers should beware that there is considerable discussion in the literature concerning the selection of optimal window functions. In most practical spectrum-analysis applications, window selection is of minor importance compared to the problem of spectrum interpretation. The literature also describes the use of data-weighting functions in the periodogram computation of the power spectrum. This approach is used to reduce the side lobes of the Bartlett spectral window that, as discussed, is inherent in the computation of the periodogram. From a statistical viewpoint, this approach is not sound unless the random background noise is of minor importance to the deterministic signal being evaluated.

Welch [32] describes a spectrum-analysis procedure where the data is sectioned into subintervals and a periodogram is computed for each section of data. These periodograms are then averaged to improve the variability of the spectrum estimate for each frequency. This computation approach is limited by the leakage properties of the Bartlett spectral window and data windows are generally used to improve the window characteristics. Additional spectrum analysis applications of the FFT are given in Refs. [30], [31], and [33].

14.11 FFT BEAMFORMING

The conventional delay-and-sum technique for array beamforming in radar, communications, sonar, and seismic applications is illustrated in Fig. 14.39.

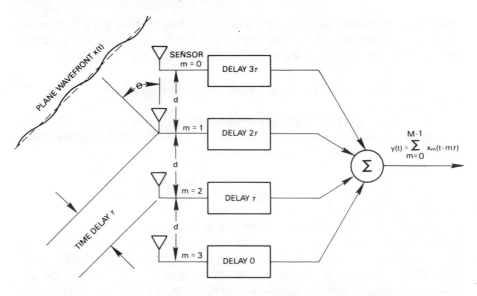

Figure 14.39 Conventional delay-and-sum technique for array beamforming.

As shown, a plane wavefront arriving at an angle θ to a sensor array with spacing d between elements is delayed by an amount τ between each adjacent sensor pair. If we are to recombine, in the proper phase, the output of each sensor, then we must compensate for these time delays. The relationship between the sensor spacing d and the delay τ is given by

$$\tau = (d/c)\cos(\theta) \qquad (14.43)$$

where c is the velocity of propagation of the wavefront. To recombine the sensor outputs for an incoming wavefront at an angle θ, we must delay the output of sensor m by

$$m\tau = (md/c)\cos(\theta) \qquad (14.44)$$

where the wavefront angle θ and the sensor m are defined in Fig. 14.39. Signal recombination, or spatial array beamforming, is then achieved by the coherent (in-phase) addition of the delayed sensor outputs:

$$y(t) = \sum_{m=0}^{M-1} x_m(t - m\tau) \qquad (14.45)$$

where we have assumed M sensors.

Beamforming by means of delay lines becomes quite cumbersome from a hardware viewpoint as the number of array sensors increase. If digital delay lines are used, the sensor outputs must be sampled at a rate much higher than the Nyquist rate to minimize side-lobe degradation in narrow-band linear-array beam patterns. However, with the FFT, it is possible to perform the equivalent of a time delay in the frequency domain. We develop in this section the procedures for applying the FFT to spatial-array beam-forming. Our approach follows that of Refs. [19] and [34].

Frequency-Domain Single-Beam Relationships

Recall from the time-shifting property (Sec. 3.5) that a shift in time by an amount τ is equivalent to multiplication by $e^{-j2\pi f\tau}$ in the frequency domain. Equation (14.45) can then be written as

$$y(t) = \sum_{m=0}^{M-1} x_m(t - m\tau) \iff Y(f) = \sum_{m=0}^{M-1} X_m(f)e^{-j2\pi fm\tau} \qquad (14.46)$$

The term on the right-hand side of Eq. (14.46) is the appropriate frequency-domain relationship for combining the M sensor outputs with the appropriate delays. Hence, we take the Fourier transform of the output for each sensor, multiply by the complex exponential $e^{-j2\pi fm\tau}$, and add the results according to the right-hand side of Eq. (14.46). The inverse Fourier transform yields $y(t)$.

Note that Eq. (14.46) is valid only for one specific value of the parameter τ, that is, the array has been *pointed* in the direction θ defined in Eq. (14.43).

Frequency-Domain Multiple-Beam Relationships

Let us assume that we desire to simultaneously implement the correct delays in order to combine the M sensor outputs for various azimuth angles θ_i. In particular, we must compute the frequency-domain summation of Eq. (14.46) for each value τ_i associated with a beam direction θ_i. To form M azimuth beams, we define increments of τ_i as

$$\tau_i = i(d/M) \qquad i = 0, 1, \ldots, M - 1 \qquad (14.47)$$

where d is the distance between sensors and M is the number of sensors. The beam, or azimuth, angle θ_i for each delay value τ_i can be determined from Eq. (14.43):

$$\theta_i = \cos^{-1}(c\tau_i/d) = \cos^{-1}(ic/M) \qquad (14.48)$$

The right-hand side of Eq. (14.46) then becomes

$$Y_i(f) = \sum_{m=0}^{M-1} X_m(f)e^{-j2\pi f(mid/M)} \qquad (14.49)$$

Equation (14.49) requires that we compute the Fourier transforms of each sensor output, multiply the transform for each sensor by the exponential $e^{-j2\pi f(mid/M)}$ for each desired beam direction θ_i, and then perform the indicated summation. The inverse Fourier transform yields the time-domain function associated with a beam direction θ_i. The computation of $X_m(f)$ and $Y_i(f)$ are easily formulated in terms of a two-dimensional FFT.

Two-Dimensional FFT Array Processing

Assume that we sample each sensor output $x_m(t)$ with sample interval T to form $x_m(kt)$, where $k = 0, 1, \ldots, N - 1$. Then we use the FFT to compute the frequency function for each sensor m:

$$X_m(nf_0) = \sum_{k=0}^{N-1} x_m(kt)e^{-j2\pi(kT)(nf_0)} \qquad n = 0, 1, \ldots, N - 1$$

$$f_0 = 1/NT \tag{14.50}$$

If we replace the continuous variable f in Eq. (14.49) with the discrete frequencies nf_0 obtained by substituting Eq. (14.50), then Eq. (14.49) becomes

$$Y(i, nf_0) = \sum_{m=0}^{M-1} \left\{ \sum_{k=0}^{N-1} x_m(kt)e^{-j2\pi(kT)(nf_0)} \right\} e^{-j2\pi f(mid/M)} \tag{14.51}$$

The two-dimensional FFT relationship of Eq. (14.51) is a function of the beam number i and frequency nf_0. The two-dimensional inverse FFT yields the appropriate time-domain waveform associated with each beam pointed in the direction θ_i. If the number of sensors is large, then digital beamforming using the FFT is attractive.

Summary

The preceding derivation has been for a linear array of M elements. Our approach can be extended to cases of circular, cylindrical, and spherical arrays. Beam-pattern side lobes can be minimized by using a weighting function. It is also possible to design adaptive methods in which each array output is weighted from a calculated expression based on actual received data.

PROBLEMS

14.1 Consider the band-pass waveforms illustrated in Fig. 14.40. Analogous to Fig. 14.3, graphically and analytically develop the range of acceptable sampling frequencies that produce nonoverlapped aliased images of the band-pass spectrum. What do you conclude concerning spectrum inversion?

14.2 Assume a band-pass waveform with a frequency spectrum as shown in Fig. 14.41. Show graphically a sampling frequency that results in down sampling

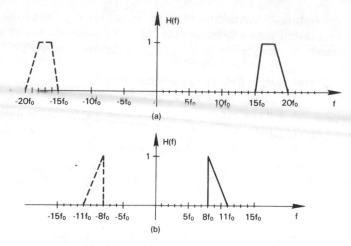

(a)

(b)

Figure 14.40 Band-pass waveforms for Prob. 14.1.

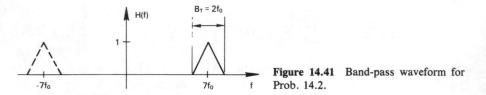

Figure 14.41 Band-pass waveform for Prob. 14.2.

to a center frequency of f_0. Are there additional sampling frequencies that produce the same result?

14.3 For the example frequency function shown in Fig. 14.3(c), choose a sampling frequency that down samples the spectrum to a zero center frequency. Show graphically your results. What are your conclusions? Hint: Consider double side-band amplitude modulation.

14.4 Assume that the frequency function shown in Fig. 14.42 results from a single side-band modulation of a voice signal that inverts the voice spectrum as shown. Develop graphically a down-sampling procedure that simultaneously demodulates the signal and inverts the spectrum. Your results should be identical to those of Fig. 13.14(e).

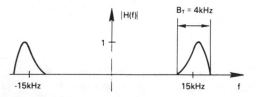

Figure 14.42 Frequency function for Prob. 14.4.

14.5 Repeat the graphical and analytical developments of Ex. 14.2 if $h(t)$ is given as

$$h(t) = \cos[2\pi(5f_0 + f_0/2)t] - \tfrac{1}{2} \sin[2\pi(5f_0 + f_0/2)t]$$

14.6 Assume a bandpass signal has a center frequency of $16f_0$ and a bandwidth B_T = $3f_0$. If quadrature sampling is used, what is the minimum sample rate for the in-phase and quadrature channels? If the signal is to be reconstructed with a center frequency of $5f_0$, determine the sample interpolation requirements.

14.7 Assume that the frequency functions illustrated in Fig. 14.43 are narrow-band signals to be sampled by quadrature-sampling techniques. Use a graphical analysis procedure analogous to Figs. 14.4 and 14.5 to develop the Fourier transform of the in-phase and quadrature functions resulting from quadrature sampling. Discuss for each case the required sampling rate to prevent aliasing. Show by a graphical analysis analogous to Figs. 14.7 and 14.8 that no information is lost even though quadrature sampling results in overlapped frequency functions. Also comment on the increase in sample rate that is required for each case if the sampled waveforms are recombined at a center frequency of $2f_0$.

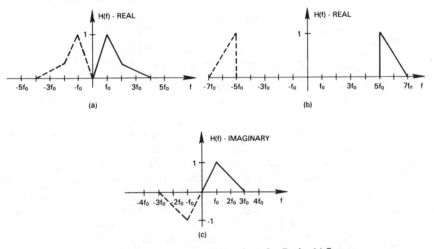

Figure 14.43   Frequency functions for Prob. 14.7.

14.8 Assume that extremely fine resolution is desired in the narrow frequency band B_z shown in Fig. 14.44. Also, assume that the desired resolution cannot be achieved across the total signal bandwidth due to computer memory limitations. Use the concept of frequency translation followed by low-pass filtering to develop a procedure for increased FFT resolution (Zoom FFT, Ref. [24]).

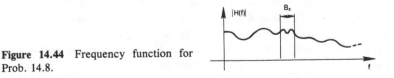

Figure 14.44   Frequency function for Prob. 14.8.

14.9 Assume that a single frequency sinusoid is buried in noise with a $S/N = -20$ dB. If a $S/N = +10$ dB is required to clearly establish the presence of the signal, determine the appropriate FFT parameters.

14.10 A narrow-band signal of bandwidth BW_s is buried in wide-band noise of bandwidth BW_n. Determine the maximum processing gain which can be achieved by FFT signal processing.

14.11 Assume that hardware limits FFT processing to a size $N = 512$. If a single frequency sinusoid is buried in noise with a $S/N = -35\ dB$, determine the number of successive FFTs which must be added to yield a processed signal-to-noise ratio greater than $+6\ dB$.

14.12 Explain how you would process the signal of Prob. 14.11 if $S/N = -55\ dB$ and the frequency of the sinusoid is known.

14.13 Use the FFT to implement the block diagram matched filter processor shown in Fig. 14.15. Experiment with different waveforms and compute the matched filter output. If your waveforms are to be used as a radar, discuss each from the perspective of range resolution and probability of detection.

14.14 Let a received signal $s_r(t)$ be given by

$$s_r(t) = s(t) + a\, s(t + \tau_1)$$

where

$$s(t) = \cos(2\pi f_0 t) \qquad f_0 = 1\ \text{Hz}$$

$$a = -0.9$$

$$\tau_1 = 0.75$$

Our objective is to use cepstrum processing to remove the echo $a[s(t + \tau_1)]$.
  (a) Compute the cepstrum of $s(t)$ using the FFT.
  (b) Compute the cepstrum of $s_r(t)$ using the FFT.
  (c) Implement the block diagram of Fig. 14.17 and compare your results with the waveform $s(t)$ computed in part (a).

14.15 In deconvolution, the inverse filter impulse response is $r(t)$, as determined from Eq. (14.13). What is the theoretical result if $r(t)$ and $h(t)$ are convolved? What is the result if a Hanning weighting function is employed according to Eq. (14.16)?

14.16 Theoretically, deconvolution can be accomplished with no error. Describe several practical limitations to obtaining theoretical results.

14.17 Develop an approach to design a deconvolution filter for the case where $H(f)$ is zero-valued, for example, $H(f) = [\sin(f)]/f$. Hint: Use the Hanning weighting function, where the truncation frequency f_c is the first zero value of $H(f)$. Use a second Hanning weighting function between the first and second zero values of $H(f)$. Repeat as necessary to achieve signal restoration.

14.18 Repeat Ex. 14.3 for the cases $a = 4\lambda$ and $a = \lambda/2$. What do you conclude concerning the relationship between parameters a and λ.

14.19 In Fig. 14.24(b), what is the effect of increasing the number of zero-valued samples?

14.20 Using the procedures developed in Sec. 14.6, compute and plot the far-field antenna pattern for each of the electric field aperture distributions shown in Fig. 14.45.

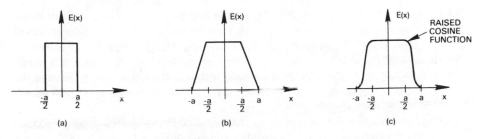

Figure 14.45 Electric field aperture distributions for Prob. 14.20.

14.21 Refer to Eq. (14.27), which determines the angle of arrival of a waveform. For a fixed d, what is the effect of setting $\lambda = \lambda_0, 2\lambda_0, 4\lambda_0, \ldots$? What do you conclude concerning the relationship between parameters d and λ?

14.22 Practical implementation of Eq. (14.27) involves measurement of the parameter ϕ in the presence of noise. If a measurement is made in the presence of noise, what is the effect of:
(a) increasing d for a fixed λ
(b) decreasing λ for a fixed d
What do you conclude is the optimum relationship between the parameters d and λ if noise is considered?

14.23 Assume that an FFT phase-interferometer direction-finding system is to perform measurements over the wavelength range λ_0 to $10\lambda_0$ in the presence of noise. In view of Probs. 14.21 and 14.22, propose a system solution which will insure accurate phase difference measurement over the wavelength range. (Hint: Consider multiple antennas.)

14.24 Modify Fig. 14.25 so that only FFT cells with a signal-to-noise ratio that exceeds a preset threshold enter into the phase difference computation.

14.25 In an FFT time-difference-of-arrival system, how does one determine which signal, $s_1(t)$ or $s_2(t)$, is the first to arrive?

14.26 Prove that the phase slope of the cross-spectrum (Fig. 14.26) is equal to the time-difference-of-arrival multiplied by 2π.

14.27 A weighted least-squares procedure to estimate the phase slope in Fig. 14.26 has been suggested. On what parameters should the weights be based?

14.28 Propose an FFT simulation to evaluate the performance of a radar system employing a specially designed transmitted waveform and a matched filter signal processor. How does your simulation change if the radar is operating in the presence of a known interference (jamming)?

14.29 Let $N = 512$, $BW = 0.1, 0.3$, and 0.9 Hz. If $T = 0.1$ secs, determine the 90% and 95% confidence limits for each case.

14.30 Compute and graph on a log scale the smoothed periodogram spectral window for $\beta = 5, 10, 20$, and 50. Show that the windows are roughly rectangular in shape and that the sidelobes fall off at 6 dB per octave. Observe that the initial fall-off of the sidelobes is a function of β.

REFERENCES

1. BALANIS, A. C. *Antenna Theory, Analysis, and Design.* New York: Harper & Row, 1982.

2. KEMERAIT, R. C., AND D. G. CHILDERS. "Signal Detection and Extraction by Cepstrum Techniques." *IEEE Trans. Info. Theory* (November 1972), Vol. IT-18, No. 6, pp. 745–759.

3. BOGERT, B. P., M. J. HEALY, AND J. W. TUKEY. "The Quefrequency Analysis of Time Series for Echos; Cepstrum, Pseudo Autocovariance, Cross Cepstrum and Saphe Cracking." In M. Rosenblatt, ed., *Time Series Symposium*, pp. 201–243. New York: Wiley, 1963.

4. BROSTE, N. A. "Digital Generation of Random Sequences." *IEEE Trans. Auto. Cont.* (April 1971), Vol. AC-16, No. 2, pp. 213–214.

5. BUCCI, O. M., AND D. M. GUISEPPE. "Exact Sampling Approach for Reflector Antenna Analysis." *IEEE Trans. Ant. Prop.* (November 1984), Vol. AP-32, No. 11, pp. 1259–1262.

6. CARSON, C. T. "The Numerical Solution of Waveguide Problems by Fast Fourier Transform." *IEEE Trans. Micro. Theory Tech.* (November 1968), Vol. 16, No. 11, pp. 955–958.

7. CHRISTODOULOU, C. G., AND J. F. KAUFFMAN. "On the Electromagnetic Scattering from Infinite Rectangular Grids with Finite Conductivity." *IEEE Trans. Ant. Prop.* (February 1986), Vol. AP-34, No. 2, pp. 144–154.

8. CROCHIERE, R. E., AND L. R. RABINER. *Multirate Digital Signal Processing.* Englewood Cliffs, NJ: Prentice-Hall, 1983.

9. HELSTROM, C. W. *Statistical Theory of Signal Detection.* New York: Pergamon, 1960.

10. HUNT, B. R. "Application of Constrained Least Squares Estimation to Image Restoration By Digital Computers." *IEEE Trans. Comput.* (September 1973), Vol. C-22, No. 9, pp. 805–812.

11. JERRI, A. J., "The Shannon Sampling Theorem—Its Various Extensions and Applications: A Tutorial Review." *Proc. IEEE* (November 1977), Vol. 65, No. 11, pp. 1565–1569.

12. JONES, W. R. "Precision FFT Correlation Techniques for Nondeterministic Waveforms." *IEEE EASCON Conv. Rec.* (October 1974), pp. 375–380.

13. KRAUS, J. D. *Antennas.* New York: McGraw-Hill, 1950.

14. LAM, P. T., S. LEE, C. C. HUNG, AND R. ACOSTA. "Stategy for Reflector Pattern Calculation: Let the Computer Do the Work." *IEEE Trans. Ant. Prop.* (April 1986), Vol. AP-34, No. 4, pp. 592–595.

15. LINDEN, D. A. "A Discussion of Sampling Theorems." *Proc. IRE* (July 1959), Vol. 47, No. 7, pp. 1219–1226.

16. NAGAI, K. "Measurement of Time Delay Using the Time Shift Property of the Discrete Fourier Transform (DFT)." *IEEE Trans. Acoust. Speech Sig. Proc.* (August 1986), Vol. ASSP-34, No. 4, pp. 1006–1008.

17. OPPENHEIM, A. V. *Application of Digital Signal Processing.* Englewood Cliffs, NJ: Prentice-Hall, 1978.

18. RABINER, L. R., AND B. GOLD. *Theory and Application of Digital Signal Processing*. Englewood Cliffs, NJ: Prentice-Hall, 1975.

19. RUDNICK, P. "Digital Beamforming in the Frequency Domain." *J. Acoust. Soc. America* (November 1969), Vol. 46, No. 5, pp. 1089–1090.

20. SARKAR, T. K., E. ARVAS, AND S. M. RAO. "Application of FFT and the Conjugate Gradient Method for the Solution of Electromagnetic Radiation from Electrically Large and Small Conducting Bodies." *IEEE Trans. Ant. Prop.* (May 1986), Vol. AP-34, No. 5, pp. 635–640.

21. SCHWARTZ, MISCHA, AND L. SHAW. *Signal Processing*. New York: McGraw-Hill, 1975.

22. SILVERMAN, H. F., AND A. E. PEARSON. "On Deconvolution Using the Discrete Fourier Transform." *IEEE Trans. Audio and Electroacoust.* (April 1973), Vol. AU-21, No. 2, pp. 112–118.

23. WILLIAMS, J. R., AND G. G. RICKER. "Signal Detectability Performance of Optimum Fourier Receivers." *IEEE Trans. Audio and Electroacoust.* (October 1972), Vol. AU-20, No. 4, pp. 264–270.

24. YIP, P. C. Y. "Some Aspects of the Zoom Transform." *IEEE Trans. Comput.* (March 1976), Vol. C-25, No. 3, pp. 287–296.

25. MCDOUGAL, J. R., L. C. SURRATT, AND J. F. STOOPS. "Computer Aided Design of Small Superdirective Antennas Using Fourier Integral and Fast Fourier Transform Techniques." *SWIEECO Rec.* (1970), pp. 421–425.

26. BRAULT, J. W., AND O. R. WHITE. "The Analysis and Restoration of Astronomical Data via the Fast Fourier Transform." *Astronomy and Astrophysics* (July 1971), Vol. 13, No. 2, pp. 169–189.

27. BLACKMAN, R. B., AND J. W. TUKEY. *Measurement of Power Spectra*. New York: Dover, 1959.

28. JENKINS, G. M., AND D. G. WATTS. *Spectral Analysis and Its Applications*. San Francisco: Holden Day, 1968.

29. JONES, R. H. "A Reappraisal of the Periodogram in Spectral Analysis." *Technometrics* (November 1965), Vol. 7, No. 4, pp. 531–542.

30. BINGHAM, C., M. D. GODFREY, AND J. W. TUKEY. "Modern Techniques of Power Spectrum Estimation." *IEEE Trans. Audio Electroacoust.* (June 1967), Vol. AU-15, No. 2, pp. 55–66.

31. HINICH, M. J., AND C. S. CLAY. "The Application of the Discrete Fourier Transform in the Estimation of Power Spectra, Coherence, and Bispectra of Geophysical Data." *Rev. Geophysics* (August 1968), Vol. 6, No. c, pp. 347–362.

32. WELCH, P. D. "The Use of Fast Fourier Transform for the Estimation of Power Spectrum: A Method Based on Time Averaging Over Short, Modified Periodograms." *IEEE Trans. Audio Electroacoust.* (June 1967), Vol. AU-15, No. 2, pp. 70–74.

33. CHILDERS, D. G., (ed.). *Modern Spectrum Analysis*. New York: IEEE Press, 1978.

34. WILLIAMS, J. R. "Fast Beam-Forming Algorithm." *J. Acoust. Soc. America* (1968), Vol. 44, No. 5, pp. 1454–1455.

A

THE IMPULSE FUNCTION: A DISTRIBUTION

The impulse function $\delta(t)$ is a very important mathematical tool in continuous and discrete Fourier transform analysis. Its usage simplifies many derivations that would otherwise require lengthy, complicated arguments. Even though the concept of the impulse function is correctly applied in the solution of many problems, the basis or definition of the impulse is normally mathematically meaningless. To ensure that the impulse function is well-defined, we must interpret the impulse not as a normal function but as a concept in the theory of distributions.

Following the discussions by Papoulis [1, Appendix I] and Gupta [2, Chapter 2], we describe a simple but adequate theory of distributions. Based on this general theory, we develop those specific properties of the impulse function that are necessary to support the developments of Chapter 2.

IMPULSE FUNCTION DEFINITIONS

Normally, the impulse function (δ-function) is defined as

$$\delta(t - t_0) = 0 \qquad t \neq t_0 \tag{A.1}$$

$$\int_{-\infty}^{\infty} \delta(t - t_0) \, dt = 1 \tag{A.2}$$

That is, we define the δ-function as having undefined magnitude at the time of occurrence and zero elsewhere, with the additional property that the area under the function is unity. Obviously, it is very difficult to relate an impulse to a physical signal. However, we can think of an impulse as a pulse waveform of very large magnitude and infinitely small duration such that the area of the pulse is unity.

We note that with this interpretation, we are, in fact, constructing a

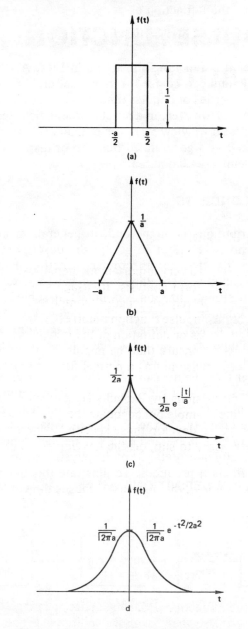

Figure A.1 Representations of the δ-function.

series of functions (i.e., pulses) that progressively increase in amplitude, decrease in duration, and have a constant area of unity. This is simply an alternate method for defining a δ-function. Consider the pulse waveform illustrated in Fig. A.1(a). Note that the area is unity and, hence, we can write mathematically the δ-function as

$$\delta(t) = \lim_{a \to 0} f(t, a) \qquad (A.3)$$

In the same manner, the functions illustrated in Figs. A.1(b) to (d) satisfy Eqs. (A.1) and (A.2) and can be used to represent an impulse function.

The various properties of impulse functions can be determined directly from these definitions. However, in a strict mathematical sense, these definitions are meaningless if we view δ(t) as an ordinary function. If the impulse function is introduced as a generalized function or distribution, then these mathematical problems are eliminated.

DISTRIBUTION CONCEPTS

The theory of distributions is vague and, in general, meaningless to the applied scientist who is reluctant to accept the description of a physical quantity by a concept that is not an ordinary function. However, we can argue that the reliance on representation of physical quantities by ordinary functions is only a useful idealization and, in fact, is subject to question. To be specific, let us consider the example illustrated in Fig. A.2.

As shown, the physical quantity V is a voltage source. We normally assume that the voltage v(t) is a well-defined function of time and that a measurement merely reveals its values. But we know in fact that there does not exist a voltmeter that can measure *exactly* v(t). However, we still insist on defining the physical quantity V by a well-defined function v(t) even though we cannot measure v(t) accurately. The point is that because we cannot measure the quantity V exactly, then on what basis do we require the voltage source to be represented by a well-defined function v(t)?

A more meaningful interpretation of the physical quantity V is to define it in terms of the effects it produces. To illustrate this interpretation, note that in the previous example, the quantity V causes the voltmeter to display

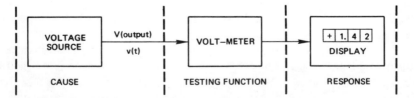

Figure A.2   Physical interpretation of a distribution.

or assign a number as a response. For each change in V, another number is displayed or assigned as a response. We never measure $v(t)$ but only the response; therefore, the source can be specified only by the totality of the responses that it causes. It is conceivable that there is not an ordinary function $v(t)$ that represents the voltage parameter V. But because the responses or numbers are still valid, then we must assume that there is a source V causing them and the only way to characterize the source is by the responses or numbers. We now show that these numbers in fact describe V as a distribution.

A distribution, or generalized function, is a process of assigning to an arbitrary function $\phi(t)$ a response or number

$$R[\phi(t)] \qquad (A.4)$$

Function $\phi(t)$ is termed a *testing function* and is continuous, is zero outside a finite interval, and has continuous derivatives of all orders. The number assigned to the testing function $\phi(t)$ by the distribution $g(t)$ is given by

$$\int_{-\infty}^{\infty} g(t)\phi(t) \, dt = R[\phi(t)] \qquad (A.5)$$

The left-hand side of Eq. (A.5) has no meaning in the conventional sense of integration, but rather is defined by the number $R[\phi(t)]$ assigned by the distribution $g(t)$. Let us now cast these mathematical statements in light of the previous example.

With reference to Fig. A.2, we note that if the voltmeter is modeled as a linear system, then the output at time t_0 is given by the convolution integral

$$\int_{-\infty}^{\infty} v(t)h(t_0 - t) \, dt$$

where $h(t)$ is the time-domain response of the measuring instrument. If we consider $h(t)$ as a testing function (that is, each particular voltmeter has different internal characteristics and as a result yields a different response for the same input, we thus say that the meter *tests* or senses the distribution $v(t)$, then the convolution integral takes the form

$$\int_{-\infty}^{\infty} v(t)\phi(t,t_0) \, dt = R[\phi(t,t_0)] \qquad (A.6)$$

Thus, for a fixed input V, the response R is a number depending on the system function $\phi(t,t_0)$.

If we interpret Eq. (A.6) as a conventional integral and if this integral equation is well-defined, then we say that the voltage source is defined by the ordinary function $v(t)$. But, as stated previously, it is possible that there does not exist an ordinary function satisfying Eq. (A.6). Because the response $R[\phi(t,t_0)]$ still exists, we must assume that there is a voltage source

V that causes this response and that a method of characterizing the source is by means of the distribution of Eq. (A.6).

The preceding discussion casts the theory of distributions in the form of physical measurements for ease of interpretation. Based on the defining relationship of Eq. (A.5), we now investigate the properties of a particular distribution: the δ function.

PROPERTIES OF IMPULSE FUNCTIONS

The impulse function $\delta(t)$ is a distribution assigning to the testing function $\phi(t)$ the number $\phi(0)$:

$$\int_{-\infty}^{\infty} \delta(t)\phi(t)\, dt = \phi(0) \tag{A.7}$$

It should be repeated that the relationship of Eq. (A.7) has no meaning as an integral, but the integral and the function $\delta(t)$ are defined by the number $\phi(0)$ assigned to the function $\phi(t)$.

We now describe the useful properties of impulse functions.

Sifting Property

The function $\delta(t - t_0)$ is defined by

$$\int_{-\infty}^{\infty} \delta(t - t_0)\phi(t)\, dt = \phi(t_0) \tag{A.8}$$

This property implies that the δ-function takes on the value of the function $\phi(t)$ at the time the δ-function is applied. The term *sifting property* arises in that if we let t_0 continuously vary, we can *sift* out each value of the function $\phi(t)$. This is the most important property of the δ-function.

Scaling Property

The distribution $\delta(at)$ is defined by

$$\int_{-\infty}^{\infty} \delta(at)\phi(t)\, dt = \frac{1}{|a|} \int_{-\infty}^{\infty} \delta(t)\phi\left(\frac{t}{a}\right) dt \tag{A.9}$$

where the equality results from a change in the independent variable. Thus, $\delta(at)$ is given by

$$\delta(at) = \frac{1}{|a|}\delta(t) \tag{A.10}$$

Product of a δ-function by an Ordinary Function

The product of a δ-function by an ordinary function $h(t)$ is defined by

$$\int_{-\infty}^{\infty} [\delta(t)h(t)]\phi(t) \, dt = \int_{-\infty}^{\infty} \delta(t)[h(t)\phi(t)] \, dt \qquad (A.11)$$

If $h(t)$ is continuous at $t = t_0$, then

$$\delta(t_0)h(t) = h(t_0)\delta(t_0) \qquad (A.12)$$

In general, the product of two distributions is undefined.

Convolution Property

The convolution of two impulse functions is given by

$$\int_{-\infty}^{\infty} \left[\int_{-\infty}^{\infty} \delta_1(\tau)\delta_2(t - \tau) \, d\tau \right] \phi(t) \, dt$$

$$= \int_{-\infty}^{\infty} \delta_1(\tau) \left[\int_{-\infty}^{\infty} \delta_2(t - \tau)\phi(t) \, dt \right] d\tau \qquad (A.13)$$

Hence,

$$\delta_1(t - t_1) * \delta_2(t - t_2) = \delta[t - (t_1 + t_2)] \qquad (A.14)$$

δ-functions as Generalized Limits

Consider the sequence $g_n(t)$ of distributions. If there exists a distribution $g(t)$ such that for every test function $\phi(t)$, we have

$$\lim_{n \to \infty} \int_{-\infty}^{\infty} g_n(t)\phi(t) \, dt = \int_{-\infty}^{\infty} g(t)\phi(t) \, dt \qquad (A.15)$$

then we say that $g(t)$ is the limit of $g_n(t)$

$$g(t) = \lim_{n \to \infty} g_n(t) \qquad (A.16)$$

We can also define a distribution as a generalized limit of a sequence $f_n(t)$ of ordinary functions. Assume that $f_n(t)$ is such that the limit

$$\lim_{n \to \infty} \int_{-\infty}^{\infty} f_n(t)\phi(t) \, dt$$

exists for every test function. This limit then is a number that depends on $\phi(t)$ and thus defines a distribution $g(t)$, where

$$g(t) = \lim_{n \to \infty} f_n(t) \qquad (A.17)$$

and the limiting operation is to be interpreted in the sense of Eq. (A.15). If Eq. (A.17) exists as an ordinary limit, then it defines an equivalent function if we assume that we can interchange the order of limit and integration in Eq. (A.15). It is from these arguments that the conventional limiting arguments, although awkward, are mathematically correct.

The δ-function can then be defined as a generalized limit of a sequence of ordinary functions satisfying

$$\lim_{n \to \infty} \int_{-\infty}^{\infty} f_n(t)\phi(t) \, dt = \phi(0) \tag{A.18}$$

If Eq. (A.18) holds, then

$$\delta(t) = \lim_{n \to \infty} f_n(t) \tag{A.19}$$

Each of these functions illustrated in Fig. A.1 satisfy Eq. (A.18) and define the δ-function in the sense of Eq. (A.19).

Another functional form of importance that defines the δ-function is

$$\delta(t) = \lim_{a \to \infty} \frac{\sin at}{\pi t} \tag{A.20}$$

Using Eq. (A.20), we can prove [Papoulis] that

$$\int_{-\infty}^{\infty} \cos(2\pi ft) \, df = \int_{-\infty}^{\infty} e^{j2\pi ft} \, df = \delta(t) \tag{A.21}$$

which is of considerable importance in evaluating particular Fourier transforms.

TWO-DIMENSIONAL IMPULSE FUNCTIONS

The two-dimensional impulse function $\delta(x,y)$ is a distribution assigning to the testing function $\phi(x,y)$ the number $\phi(0,0)$:

$$\int_{-\infty}^{\infty} \int_{-\infty}^{\infty} \delta(x,y)\phi(x,y) = \phi(0,0) \tag{A.22}$$

From this definition, the useful properties of two-dimensional impulse functions can be derived. In particular, the shifting property, which is key to developing the two-dimensional sampling theorem, is as follows:

$$\int_{-\infty}^{\infty} \int_{-\infty}^{\infty} \delta(x-x_0,y-y_0)h(x,y) \, dx \, dy = h(x_0,y_0) \tag{A.23}$$

REFERENCES

1. PAPOULIS, A. *The Fourier Integral and Its Applications*, 2d ed. New York: McGraw-Hill, 1984.

2. GUPTA, S. C. *Transform and State Variable Methods in Linear Systems*. New York: Wiley, 1966.

3. BRACEWELL, R. M. *The Fourier Transform and Its Applications*, 2d rev. ed. New York: McGraw-Hill, 1986.

4. LIGHTHILL, M. J. *An Introduction to Fourier Analysis and Generalized Function*. New York: Cambridge University Press, 1959.

5. ARSAC, J. *Fourier Transforms and the Theory of Distributions*. Englewood Cliffs, NJ: Prentice-Hall, 1966.

6. ZEMANIAN, A. H. *Distribution Theory and Transform Analysis*. New York: McGraw-Hill, 1965.

BIBLIOGRAPHY

This bibliography of FFT applications is categorized for ease of reference.
Classifications are as follows:

A. Biomedical Engineering
B. Array Processing
C. Deconvolution
D. Spectroscopy, Microscopy, Electrochemical Analysis
E. Antenna and Wave Propagation
F. Numerical Methods
G. FFT Hardware
H. Radar
 I. Instrumentation
J. General and Miscellaneous Applications
K. Spectral Estimation, Frequency Analysis
L. Time Domain Reflectometry, Delay Estimation
M. Network Analysis
N. Two-Dimensional Algorithms, Multidimensional Analysis
O. Sonics, Ultrasonics, Acoustics
P. Filter Design, Digital Filtering, Convolution, Correlation
Q. Communications
R. Speech Processing
S. Multichannel Filtering, Transmultiplexers

T. FFT Algorithms
U. Geophysical Processing
V. Mechanical Analysis, Fluid Mechanics, Structural Analysis
W. FFT Error Analysis
X. Image Processing, Optics
Y. Electrical Power Systems
Z. Simulation

ABBOUD, S., I. BRUDERMAN, AND D. SARDEH, "Frequency and Time Domain   A
Analysis of Airflow Breath Patterns in Patients with Chronic Obstructive
Airway Disease," *Comput. and Biomed. Res.*, Vol. 19, No. 3, pp. 266–
73, June 1986.

ABDELMALEK, N. N., T. KASVAND, AND J. P. CROTEAU, "Image Restoration   X
For Space Invariant Pointspread Functions," *Appl. Opt.*, Vol. 19, No.
7, pp. 1184–89, April 1980.

ABUSHAGUR, M. A. G., AND N. GEORGE, "Measurement of Optical Fiber   X
Diameter Using The Fast Fourier Transform," *Applied Optics*, Vol. 19,
No. 12, pp. 2031–33, June 1980.

ACOSTA, R. J., AND RICHARD Q. LEE, "Case Study of Sample-Spacing in   E
Planar Near-Field Measurement of High Gain Antennas," *NASA Tech
Memo 86872*, p. 9.

ACOSTA, R. J., "Compensation of Reflector Surface Distortions Using Con-   E
jugate Field Matching," *NASA Tech Memo 87198*, p. 9, June 1986.

ACOSTA, R. J., "Secondary Pattern Computation Of An Offset Reflector   E
Antenna," *NASA Tech Memo 87160*, p. 19, November 1985.

AGARWAL, R. C., AND J. W. COOLEY, "New Algorithms for Digital Con-   P
volution," *IEEE Trans. Acous. Speech and Signal Processing*, Vol.
ASSP-25, No. 5, pp. 392–410, October 1977.

AGRAWAL, Y. C. "CCD Chirp-Z FFT Doppler Signal Processor For Laser   H
Velocimetry," *Journal of Physics E: Scientific Instruments*, Vol. 17, No.
6, pp. 458–61, June 1984.

AGULLO, J., AND A. BARJAU, "Reflection Function r(t): A Matrix Approach   O
Versus FFT** Minus **1," *Journal of Sound and Vibration*, Vol. 106,
No. 2, pp. 193–201, April 1986.

AKIYAMA, TAKAO, "Pressure Estimation From Oscillatory Signals Obtained   O
Through BWR's Instrument Lines," *Journal of Dynamic Systems, Mea-
surement and Control, Transactions ASME*, Vol. 108, No. 1, pp. 80–85,
March 1986.

ALBA, D., AND G. R. MEIRA, "Instrumental Broadening Correction In Size   C
Exclusion Chromatography Through Fast Fourier Transform Techniques," *Journal of Liquid Chromatography*, Vol. 6, No. 13, pp. 2411–
31, 1983.

ALLAN, R. N., A. M. LEITE DA SILVA, A. A. ABU-NASSER, AND R. C. BUR-   Y
CHETT, "Discrete Convolution in Power System Reliability," *IEEE Trans.
Reliab.*, Vol. R-30, No. 5, pp. 452–56, December 1981.

ALLEN, G. H., "Programming an Efficient Radix-Four FFT Algorithm,"   T
Signal Processing, Vol. 6, No. 4, pp. 325–29, August 1984.

ALLEN, J. D. "Short-term Spectral Analysis, Synthesis, AND Modification   K
By Discrete Fourier Transform," *IEEE Trans. Acoust. Speech Signal
Process. ASSP-25*, pp. 235–38, 1977, corrections in *ASSP-25*, p. 589,
1977.

ALLEN, J. B., "Estimation of Transfer Functions Using the Fourier Trans-   C
form Ratio Method," *American Institute of Astronautics and Aeronautics
Journal*, Vol. 8, pp. 414–23, March 1970.

ALSOP, L. E., AND A. A. NOWROOZI, "Fast Fourier Analysis," *J. Geophys.*   U
Res., Vol. 70, No. 22, pp. 5482–83, November 1966.

ALTES, R. A., AND W. J. FAUST, "Unified Method of Broad-Band Echo   A
Characterization For Diagnostic Ultrasound," *IEEE Transactions on
Biomedical Engineering*, Vol. BME-27, No. 9, pp. 500–508, September
1980.

ANDERSON, G. L., "A Stepwise Approach To Computing The Multidimen-   N
sional Fast Fourier Transform of Large Arrays," *IEEE Transactions on
Acoustics, Speech, and Signal Processing*, Vol. ASSP-28, No. 3, pp. 280–
84, June 1980.

ANDREWS, H. C., AND B. R. HUNT, *Digitial Image Restoration*. Englewood   X
Cliffs, New Jersey: Prentice Hall, 1977.

ANDREWS, H. C., AND K. L. CASPARI, "A Generalized Technique for Spec-   K
tral Analysis," *IEEE Trans. on Computers*, Vol. C-19, No. 1, pp. 16–25,
January 1970.

ANON, "Battery Powered FFT Analysis," *Noise & Vibration Control*   G
Worldwide, Vol. 17, No. 2, pp. 60–63, February 1986.

ANSARI, R., "An Extension of the Discrete Fourier Transform," *IEEE*   T
Trans. Circuits & Syst., Vol. CAS-32, No. 6, pp. 618–19, June 1985.

AOKI, Y., AND A. BOIVIN, "Computer Reconstruction of Images from a   X
Microwave Hologram," *Proceedings of the IEEE*, Vol. 58, pp. 821–22,
May 1970.

AOKI, Y., "Optical and Numerical Reconstructions of Images from Sound-   X
wave Holograms," *IEEE Trans. Audio and Electroacoustics*, Vol. AU-
18, pp. 258–67, September 1970.

ARAMBEPOLA, B., AND P. J. W. RAYNER, "Discrete Transforms Over Po-   P
lynomial Rings With Applications in Computing Multidimensional Convolutions," *IEEE Trans. Acoust., Speech and Signal Proces.*, Vol.
ASSP-28, No. 4, pp. 407–14, August 1980.

ARIDGIDES, A., AND D. R. MORGAN, "Effects of Input Quantization in Float-   W
ing-Point Digital Pulse Compression," *IEEE Trans. Acoust., Speech &
Signal Process.*, Vol. ASSP-33, No. 2, pp. 434–35, April 1985.

ARNOT, N. R., G. G. WILKINSON, AND R. E. BURGE, "Applications of the   N

ICL Dap For Two-Dimensional Image Processing," *Computer Physics Communications*, Vol. 26, No. 3 & 4, June 1982.

ARNOTT, P. J., G. W. PFEIFFER, AND M. E. TAVEL, "Spectral Analysis of   A
Heart Sounds: Relationships Between Some Physical Characteristics and
Frequency Spectra of First and Second Heart Sounds In Normals and
Hypertensives," *Journal of Biomedical Engineering*, Vol. 6, No. 2, pp.
121–28, April 1984.

ARONSON, E. A., "Fast Fourier Integration of Piecewise Polynomial Func-   F
tions," *Proceedings of the IEEE*, Vol. 57, pp. 691–92, April 1969.

ARRIDGE, R. G. C., "Improvement of the High Frequency Content of Fast   K
Fourier Transforms," *Journal of Physics D: Applied Physics*, Vol. 17,
No. 6, pp. 1101–05, June 1984.

AZIMI-SADJADI, M. R., AND R. A. KING, "Two-Dimensional Block Proces-   N
sors-Structures and Implementations," *IEEE Trans. Circuits & Syst.*,
Vol. CAS-33, No. 1, pp. 42–50, January 1986.

BABIC, H., AND G. C. TEMES, "Optimum Low-Order Windows For Discrete   K
Fourier Transform Systems," *IEEE Trans. Acoust. Speech Signal Process.*, Vol. ASSP-24, pp. 512–17, 1976.

BABU, B. N. S., "Performance of an FFT-Based Voice Coding System in   R
Quiet and Noisy Environments," *IEEE Transactions on Acoustics,
Speech and Signal Processing*, Vol. ASSP-31, No. 5, pp. 1323–27, Oc-
tober 1983.

BAKHRAKH, L. D., AND O. S. LITVINOV, "On an Effective Method of Com-   E
puting the Coefficients in the Fourier Series of Functions of Sharply Di-
rected Antenna Diagrams," *Trans. In: Sov. Phys.-Dokl.*, Vol. 25, No. 9,
pp. 702–4, September 1980.

BALCOU, Y., "A Method To Increase The Accuracy of Fast Fourier Trans-   K
form Calculations For Rapidly Varying Functions," *IEEE Trans. Instrum. and Meas.*, Vol. IM-30, No. 1, pp. 38–40, March 1981.

BARGER, H. A., "Evaluation of Discrete Transforms For Use In Digital   R
Specch Recognition," *Comput. and Electr. Eng.*, Vol. 6, No. 3, pp. 183–
97, September 1979.

BASANO, L., AND P. OTTONELLO, "Real-Time FFT To Monitor Muscle Fa-   A
tigue," *IEEE Transactions on Biomedical Engineering*, Vol. 33, No. 11,
pp. 1049–51, 1986.

BECKER, W., R. JANSSEN, A. MUELLER-HELLMANN, AND H-C. SKUDELNY,   J
"Analysis of Power Converters For AC-Fed Traction Drives and Micro-
computer-Aided On-Line Optimization of Their Line Response," *IEEE
Transactions on Industry Applications*, Vol. IA-20, No. 3, pp. 605–14,
May–June 1984.

BELLAHSENE, B. E., J. W. HAMILTON, J. G. WEBSTER, AND P. BASS, "Im-   A
proved Method For Recording And Analyzing The Electrical Activity Of
The Human Stomach," *IEEE Transactions on Biomedical Engineering*,
Vol. T-BME-32, No. 11, pp. 911–15, November 1985.

BELLANGER, M. G., AND J. L. DAGUET, "TDM-FDM Transmultiplexer: Dig-   Q ital Polyphase and FFT," *IEEE Trans. Commun.*, Vol. COM-22, No. 9, pp. 1199–1205, September 1974.

BELOV, Y. I., AND D. E. EMEL'YANOV, "Properties of Fourier Transform of Near-zone Antenna Field Measured at a Nonspherical Surface," *Radiophys. & Quantum Electron.*, Vol. 29, No. 2, pp. 147–58, February 1986.   E

BELYI, A. A., A. 1. BOVBEL, AND V. I, MIKULOVICH, "Investigation of Algorithms of the Fast Fourier Transform, Base 4, with a Constant Structure," *Radio Engineering and Electronic Physics*, Vol. 25, No. 8, pp. 43–50, August 1980.   T

BENIGNUS, V. A., "Estimation of Coherence Spectrum of Non-Gaussian Time Series Populations," *IEEE Trans. Audio and Electroacoustics*, Vol. AU-17, pp. 198–201, September 1969.   K

BENIGNUS, V. A., "Estimation of the Coherence Spectrum and its Confidence Interval Using the Fast Fourier Transforms," *IEEE Trans. Audio and Electroacoustics*, Vol. AU-17, pp. 145–50, June 1969; Correction, Vol. 18, p. 320, September 1970.   K

BENNETT, J. C., "Fast Algorithm for Calculation of Radiation Integral and its Application to Plane-polar Near Field and Far Field Transformation," *Electron. Lett. (GB)*, Vol. 21, No. 8, pp. 343–44, April 1985.   E

BERGLAND, G. D., AND H. W. HALE, "Digital Real-time Spectral Analysis," *IEEE Trans. Electronic Computers*, Vol. EC-16, No. 2, pp. 180–85, April 1967.   K

BERGLAND, G. D., "A Fast Fourier Transform Algorithm for Real-valued Series," *Commun. ACM*, Vol. 11, No. 10, pp. 703–10, October 1968.   T

BERGLAND, G. D., "A Guided Tour Of The Fast Fourier Transform," *IEEE Spectrum*, Vol. 6, pp. 41–51, July 1969.   J

BERGLAND, G. D., "A Radix-eight Fast Fourier Transform Subroutine for Real-valued Series," *IEEE Trans. Audio and Electroacoustics*, Vol. 17, No. 2, pp. 138–44, June 1969.   T

BERGLAND, G. D., "The Fast Fourier Transform Recursive Equations for Arbitrary Length Records," *Math. Computation*, Vol. 21, pp. 236–38, April 1967.   T

BERGLAND, G. D., A Fast Fourier Transform Algorithm Using Base 8 Iterations," *Math. Comput.*, Vol. 22, pp. 275–79, April 1968.   T

BERNARD, R., D. VIDAL-MADJAR, F. BAUDIN, AND G. LAURENT, "Data Processing & Calibration For An Airborne Scatterometer," *IEEE Trans. On Geoscience & Remote Sensing*, Vol. GE-24, No. 5, September 1986.   H

BERROU, J.-L, AND R. A. WAGSTAFF, "Virtual Beams From An FFT Beamformer and Their Use to Access the Quality of a Towed-Array System," *Proceedings of ICASSP 82. IEEE International Conference on Acoustics, Speech and Signal Processing*, Vol. 2, pp. 811–14, Paris, France, May 1982.   B

BERRUT, J. P., "Fredholm Integral Equation Of The Second Kind For Conformal Mapping," *J. Comput Appl. Math*, Vol. 14, No. 1–2, pp. 99–110, February 1986.   F

BERTRAM, S., "Frequency Analysis Using The Discrete Fourier Trans-   K
form," *IEEE Trans. Audio Electroacoust. AU-18*, pp. 495–500, 1970.

BERTRAM, S., "On The Derivation Of The Fast Fourier Transform," *IEEE*   T
Trans. Audio Electroacoust., Vol. AU-18, pp. 55–58, March 1970.

BHUYAN, L. N., AND D. P. AGRAWAL, "Performance Analysis of FFT Al-   G
gorithms and Multiprocessor Systems," *IEEE Trans. Software Eng.*, Vol.
SE-9, No. 4, pp. 512–21, July 1983.

BINGHAM, C., M. D. GODFREY, AND J. W. TUKEY, "Modern Techniques of   K
Power Spectral Estimation," *IEEE Trans. Audio and Electroacoustics*,
Vol. AU-18, pp. 439–42, June 1967.

BINGHAM, C., M. D. GODFREY, AND J. W. TUKEY, "Modern Techniques Of   K
Power Spectrum Estimation," *IEEE Trans. Audio Electroacoust.*, Vol.
AU-17, pp. 56–66, June 1967.

BLANKEN, J. D., AND P. L. RUSTAN, "Selection Criteria For Efficient Im-   T
plementation of FFT Algorithms," *IEEE Transactions on Acoustics,*
Speech and Signal Processing, Vol. ASSP-30, No. 1, pp. 107–9, February
1982.

BLANKENSHIP, P. E., AND E. M. HOFSTETTER, "Digital Pulse Compression   H
Via Fast Convolution," *IEEE Trans. Acoust., Speech and Signal Pro-*
cess., Vol. ASSP-23, No. 2, pp. 189–201, April 1975.

BLUESTEIN, L. I., "A Linear Filter Approach to the Computation of the   T
Discrete Fourier Transform," Nerem Record, pp. 218–19, 1968.

BOGART, B. P., AND E. PARZEN, "Informal Comments on the Uses of Power   K
Spectrum Analysis," *IEEE Trans. Audio and Electroacoustics*, Vol. AU-
15, pp. 74–76, June 1967.

BOITE, R., H. LEICH, AND J. HANCQ, "An Efficient Computation Method   F
for the Bayard-Bode Relations with Applications," *1982 International*
Symposium on Circuits and Systems, Vol. 3, pp. 872–75, May 1982.

BOKHARI, S. A., AND N. BALAKRISHNAN, "Analysis of Cylindrical Antennas-   E
A Spectral Iteration Technique," *IEEE Trans. Antennas & Propag.*, Vol.
AP-33, No. 3, pp. 251–58, March 1985.

BOLD, GARY E. J., "Comparison Of The Time Involved In Computing Fast   T
Hartley And Fast Fourier Transforms," *IEEE Proceedings*, Vol. 73, No.
12, pp. 1863–64, December 1985.

BOLLINGER, K. E., AND R. GILCHRIST, "Voltage Regulator Models Using   Y
Automated Frequency Response Equipment," *IEEE Transactions on*
Power Apparatus and Systems, Vol. PAS-101, No. 8, pp. 2899–2905,
August 1982.

BOMAR, L. C., W. J. STEINWAY, S. A. FAULKNER, AND L. L. HARKNESS,   H
"CW Multi-Tone Radar Ranging Using DFT Techniques," *Conference*
Proceedings of the 13th European Microwave Conference, pp. 127–32,
Nurenberg, Ger., September 1983.

BONGIOVANNI, G., "VLSI Network For Variable Size FFT's," *IEEE Trans-*   G
actions on Computers, Vol. C-32, No. 8, pp. 756–60, August 1983.

BONGIOVANNI, G., P. CORSINI, AND G. FROSINI, "One Dimensional And Two   N
Dimensional Generalized Discrete Fourier Transform," *IEEE Trans.*
Acoust. Speech Signal Process., Vol. ASSP-24, pp. 97–99, 1976.

BONNEROT, G., AND M. BELLANGER, "Odd-time Odd-frequency Discrete   T
Fourier Transform For Symmetric Real-valued Series," *Proc. IEEE*, Vol.
64, pp. 392–93, 1976.

BORGIOLI, R. C., "Fast Fourier Transform Correlation Versus Direct Dis- P
crete Time Correlation," *Proceedings of the IEEE*, Vol. 56, pp. 1602–
1604, September 1968.

BORUP, D. T., AND O. P. GANDHI, "Calculation Of High-Resolution SAR   E
Distributions In Biological Bodies Using The FFT Algorithm And Con-
jugate Gradient Method," *IEEE Transactions on Microwave Theory and*
Techniques, Vol. MTT-33, No. 5, pp. 417–19, May 1985.

BORUP, D. T., AND O. P. GANDHI, "Fast-Fourier-Transform Method for   E
Calculation of SAR Distributions in Finely Discretized Inhomogeneous
Models of Biological Bodies," *IEEE Trans. Microwave Theory and*
Tech., Vol. MTT-32, No. 4, pp. 355–60, April 1984.

BOURQUIN, J. J., "Simulation by the Fast Fourier transform," *Conference* Z
Proceedings. 28th Midwest Symposium on Circuits and Systems, pp. 278–
82, August 1985.

BOVBEL, E. I., A. M. ZAITSEVA, AND V. I. MIKULOVICH, "Effective Algo-   T
rithms for the Fast Fourier Transform With a Mixed Base," *Radio Eng.*
and Electron. Phys., Vol. 27, No. 5, pp. 99–104, May 1982.

BOYER, A. L., AND E. C. MOK, "Brachytherapy Seed Dose Distribution   A
Calculation Employing the Fast Fourier Transform," *Med. Phys.*, Vol.
13, No. 4, pp. 525–29, August 1986.

BOYES, J. D., "Noise and Vibration Analysis of Reciprocating Machines," V
Noise and Vib. Control Worldwide (GB), Vol. 12, No. 3, pp. 90–92, April
1981.

BRASS, A., AND G. S. PAWLEY, "Two And Three Dimensional FFTs On   N
Highly Parallel Computers," *Parallel Comput*, Netherlands, Vol. 3, No.
2, pp. 167–84, May 1986.

BRAULT, J. W., AND O. R. WHITE, "The Analysis and Restoration of As-   C
tronomical Data via the Fast Fourier Transform," *Astronomy and As-*
trophysics, Vol. 13, No. 2, pp. 169–89, July 1971.

BRENNER, N. M., "Fast Fourier Transform of Externally Stored Data,"   T
IEEE Trans. Audio and Electroacoustics, Vol. AU-17, No. 2, pp. 128–
32, June 1969.

BRIGGS, W. L., L. B. HART, R. A. SWEET, AND A. O'GALLAGHER, "Multi-   T
processor FFT Methods," *SIAM J. Sci. & Stat. Comput.*, Vol. 8, No.
1, pp. S27–42, January 1987.

BRIGHAM, E. O., AND R. E. MORROW, "The Fast Fourier Transform," *IEEE*   J
Spectrum, Vol. 4, No. 12, pp. 63–70, December 1967.

BRIGHAM, E. O., *The Fast Fourier Transform*. Englewood Cliffs, New Jer- J
sey: Prentice Hall, 1974.

BRONZINO, J. D., "Quantitative Analysis Of The EEG-General Concepts   A
and Animal Studies," *IEEE Transactions On Biomedical Engineering*,
Vol. BME-31, No. 12, pp. 850–56, December 1984.

BROWN, B. L., W. J. STRONG, AND A. C. RENCHER, "Fifty Four Voices From   R
Two: The Effects of Simultaneous Manipulations of Rate, Mean Fundamental Frequency, and Variance of Fundamental Frequency on Rat-
ings of Personality From Speech," *J. Acoust. Soc. Am.*, Vol. 55, No. 2,
pp. 313–18, February 1974.

BROWN, T. D., R. H. GABEL, D. R. PEDERSEN, L. D. BELL, AND W. F. BLAIR,   A
"Some Characteristics Of Laminar Flow Velocity Spectra Detected By
A 20 MHZ Pulsed Ultrasound Doppler," *J. Biomech*, Vol. 18, No. 12,
pp. 927–38, 1985.

BRUCE, J. D., "Discrete Fourier Transforms, Linear Filters and Spectrum   P
Weighting," *IEEE Trans. Audio and Electroacoustics*, Vol. AU-16, No.
4, pp. 495–99, December 1968.

BUCCI, O. M., AND G. DIMASSA, "Exact Sampling Approach For Reflector   E
Antennas Analysis," *IEEE Trans. Antennas and Propag.*, Vol. AP-32,
No. 11, pp. 1259–62, November 1984.

BUCCI, O. M., AND G. DIMASSA, "Sampling Approach for Fast Computation   E
of Scattered Fields," *Electron. Lett.* (GB), Vol. 19, No. 1, pp. 15–17,
January 1983.

BUCCI, O. M., G. D'ELIA, G. FRANCESCHETTI, AND R. PIERRI, "Efficient   E
Computation of the Far Field of Parabolic Reflectors by Pseudo-sampling
Algorithm," *IEEE Trans. Antennas and Propag.*, Vol. AP-31, No. 6, pp.
931–37, November 1983.

BUIJS, H. L., "Fast Fourier Transformation of Large Arrays of Data,"   T
Applied Optics., Vol. 8, No. 1, pp. 211–12, January 1969.

BUIJS, H. L., "Implementation of a Fast Fourier Transform (FFT) For   X
Image Processing Applications," *IEEE Trans. Acoust. Speech Signal
Process.*, Vol. ASSP-22, pp. 420–24, 1974.

BUIJS, H. L., A. POMERLEAU, M. FOURNIER, AND W. G. TAM, "Implemen-   X
tation of a Fast Fourier Transform (FFT) For Image Processing Applications," *IEEE Trans. Acoust., Speech and Signal Process.*, Vol. ASSP-
22, No. 6, pp. 420–24, December 1974.

BURRUS, C. S., AND P. W. ESCHENBACHER, "An In-Place, In-Order Prime   T
Factor FFT Algorithm," *IEEE Transactions on Acoustics, Speech, and
Signal Processing*, Vol. ASSP-29, No. 4, pp. 806–17, August 1981.

BURRUS, C. S., "Comments on 'Selection Criteria For Efficient Implemen-   T
tation of FFT Algorithms'," *IEEE Trans. Acoust., Speech and Signal
Process.*, Vol. ASSP-31, No. 1, Pt. 1, p. 206, February 1983.

BURRUS, C. S., "Index Mappings For Multidimensional Formulation of the   T
FFT and Convolution," *IEEE Trans. Acoust. Speech Signal Process.*,
Vol. ASSP-25, pp. 239–42, June 1977.

BURRUS, C. S., "New Prime Factor FFT Algorithm," *Proceedings-   T
ICASSP, IEEE International Conference on Acoustics, Speech and Signal Processing*, Vol. 1, pp. 335–38, Atlanta, Georgia, March 1981.

BUSIGIN, A., "FFT Software For The IBM PC," *IEEE Micro*, Vol. 6, No.   T
1, p. 6, 1986.

BUTCHER, W. E., AND G. E. COOK, "Comparison of Two Impulse Response   C
Identification Techniques," *IEEE Trans. Automatic Control*, Vol. AC-
15, pp. 129–30, February 1970.

BUTZ, A. R., "FFT Length in Digital Filtering," *IEEE Trans. Comput.*,   P
Vol. C-28, No. 8, pp. 577–80, August 1979.

CAIN, A. B., J. H. FERZIGER, AND W. C. REYNOLDS, "Discrete Orthogonal   F
Function Expansions for Non-Uniform Grids Using the Fast Fourier
Transform," *J. Comput. Phys.*, Vol. 56, No. 2, pp. 272–86, November
1984.

CANDEL. S. M., "Dual Algorithms for Fast Calculation of the Fourier-Bessel   F
Transform," *IEEE Trans. Acoust., Speech and Signal Process.*, Vol.
ASSP-29, No. 5, pp. 963–72, October 1981.

CANDEL, S. M., "Fast Computation of Fourier-Bessel Transforms," *Pro-*   F
*ceedings of ICASSP 82. IEEE International Conference on Acoustics,
Speech and Signal Processing*, pp. 2076–79, Paris, France, May 1982.

CARAISCOS, C., AND B. LIU, "Two Dimensional FFT Using Mixed Time and   T
Frequency Decimations," *Proceedings of ICASSP 82. IEEE International Conference on Acoustics, Speech and Signal Processing*, Vol. 1,
pp. 24–27, Paris, France, May 1982.

CARLSON, D. A., "Time-Space Tradeoffs on Back-to-Back FFT Algo-   T
rithms," *IEEE Trans. Comput.*, Vol. C-32, No. 6, pp. 585–9, June 1983.

CARSON, C. T., "The Numerical Solution of Waveguide Problems by Fast   E
Fourier Transforms," *IEEE Trans. Microwave Theory*, Vol. MTT-16,
No. 11, pp. 955–58, November 1968.

CARTER, G. C., "Coherence And Its Estimation Via The Partitioned Mod-   K
ified Chirp-Z Transform," *IEEE Trans. Acoust. Speech Signal Process.*,
Vol. ASSP-23, pp. 257–63, 1975.

CARTER, G. C., "Receiver Operating Characteristics For A Linearly Thresh-   Q
olded Coherence Estimation Detector," *IEEE Trans. Acoust. Speech Signal Process.*, Vol. ASSP-25, pp. 90–92, 1977.

CARTER, G. C., C. H. KNAPP, AND A. H. NUTTALL, "Estimation Of The   K
Magnitude-Squared Coherence Function Via Overlapped Fast Fourier
Transform Processing," *IEEE Trans. Audio Electracoust.*, Vol. AU-21,
pp. 337–44, 1973.

CATEDRA, M. F., "Solution to Some Electromagnetic Problems Using Fast   E
Fourier Transform with Conjugate Gradient Method," *Electronic Letters
(GB)*, Vol. 22, No. 20, pp. 1049–51, September 1986.

CAWLEY, P., "Accuracy Of Frequency Response Function Measurements   I
Using FFT-Based Analyzers With Transient Excitation," *Journal of Vibration, Acoustics, Stress, and Reliability In Design*, Vol. 108, No. 1,
pp. 49–49, January 1986.

CAWLEY, P., "Reduction of Bias Error In Transfer Function Estimates   J
Using FFT-Based Analyzers," *Journal of Vibration, Acoustics, Stress,
and Reliability in Design*, Vol. 106, No. 1, pp. 29–35, January 1984.

CETIN, A. E., AND R. ANSARI, "Iterative Procedure for Designing Two-dimensional FIR Filters," *Electron. Lett. (GB)*, Vol. 23, No. 3, pp. 131–33, January 1987.

P

CHANDER, V., V. K. AATRE, AND K. RAMAKRISHMA, "2's Complement Error and Scaling in FFT Structures," *1984 IEEE International Symposium on Circuits and Systems Proceedings*, Vol. 1, pp. 280–83, May 1984.

W

CHANG, T. L., AND D. R. ELLIOTT, "Communication Channel Demultiplexing Via An FFT," *IEEE Conference Record of the Thirteenth Asilomar Conference on Circuits, Systems and Computers*, pp. 162–66, Pacific Grove, Ca., November 1979.

S

CHANOUS, D., "Synthesis of Recursive Digital Filters Using the FFT," *IEEE Trans. Audio and Electroacoustics*, Vol. 18, pp. 211–12, June 1970.

P

CHAUDHRY, F. I., S. C. DENNIS, AND J. B. HARNESS, "Spectral Analysis of Electrocardiogram Signals Of The Isolated Guinea Pig Heart For The Detection Of Arrhythmias," *Journal of Biomedical Engineering*, Vol. 4, No. 4, pp. 289–93, October 1982.

A

CHEN, W. H., C. H. SMITH, AND S. C. FRALICK, "A Fast Computational Algorithm For The Discrete Cosine Transform, *IEEE Trans. Commun.*, Vol. COM-25, pp. 1004–1009, 1977.

T

CHIANG, Y. C., AND A. A. SEIREG, "Acoustic Design Of Variably Segmented Pipes," *Comput Mech Eng.*, Vol. 3, No. 6, pp. 57–59, May 1985.

O

CHILDERS, D. G., *Modern Spectrum Analysis*. New York: IEEE Press, 1978.

K

CHORNIK, B., H. E. BISHOP, A. LE MOEL, AND C. LE GRESSUS, "Deconvolution in Auger Electron Spectroscopy," *Scanning Electron Microsc*, Pt. 1, pp. 77–88, 1986.

C

CHOWDARY, N., AND W. STEENAART, "Accumulation Of Product Roundoff Errors In Modified FFT's," *IEEE Transactions On Circuits and Systems*, Vol. CAS-33, No. 1, pp. 103–7, January 1986.

T

CHOWDARY, N. U., AND W. STEENAART, "A High Speed Two-dimensional FFT Processor," *ICASSP 84. Proceedings of the IEEE International Conference on Acoustics, Speech and Signal Processing*, 4.11/1-4 Vol. 1, San Diego, Ca., March 1984.

N

CHOWDHURY, S. K., AND A. K. MAJUMDER, "Digital Spectrum Analysis of Respiratory Sound," *IEEE Transactions On Biomedical Engineering*, Vol. BME-28, No. 11, pp. 784–88, November 1981.

A

CHOY, F. K., AND W. H. LI, "Frequency Component and Modal Synthesis Analysis of Large Rotor-bearing Systems with Base Motion Induced Excitations," *J. Franklin Inst.*, Vol. 323, No. 2, pp. 145–68, 1987.

V

CHRISTODOULOU, C. G., AND J. F. KAUFFMAN, "On the Electromagnetic Scattering From Infinite Rectangular Grids With Finite Conductivity," *IEEE Transactions on Antennas and Propagation*, Vol. AP-34, No. 2, pp. 144–54, February 1986.

E

CHU, S., AND C. S. BURRUS, "Prime Factor FTT Algorithm Using Distributed Arithmetic," *IEEE Transactions On Acoustics, Speech, and Signal Processing*, Vol. ASSP-30, No. 2, pp. 217–27, 1982.

T

CHU, W. T., "Architectural Acoustic Measurements Using Periodic Pseu-　　O
dorandom Sequences and FFT," *Journal of the Acoustical Society of
America*, Vol. 76, No. 2, pp. 475–78, August 1984.

CLARK, C., R. J. DOOLING, AND T. BUNNELL, "Analysis and Synthesis of　　J
Bird Vocalizations: An FFT-Based Software System," *Behav. Res.
Methods und Instrum*, Vol. 15, No. 2, pp. 251–53, April 1983.

CLAYTON, C., J. A. MCCLEAN, AND G. J. MCCARRA, "FFT Performance　　I
Testing of Data Acquisition Systems," *IEEE Transactions on Instrumentation and Measurement*, Vol. IM-35, No. 2, pp. 212–15, June 1986.

CLINE, C. L., AND H. J. SIEGEL, "Augmenting ADA for SIMD Parallel Pro-　　T
cessing," *IEEE Trans. On Software Engineering*, Vol. SE-11, No. 9, pp.
970–77, September 1985.

COBB, R. F., "Compute Lowpass Filter Responses For Special Driving　　Z
Functions," *EDN*, Vol. 29, No. 9, pp. 265–72, May 1984.

COBB, R. F., "Extend Basic FFT Capabilities to Handle Many Filter　　Z
Types," *EDN*, Vol. 29, No. 12, pp. 183–93, June 1984.

COBB, R. F., "Format Driving Functions To Simplify FFT Routines," *EDN*,　　Z
Vol. 29, No. 7, pp. 237–46, April 1984.

COBB, R. F., "Use Fast Fourier Transform Programs to Simplify, Enhance　　Z
Filter Analysis," *EDN*, Vol. 29, No. 5, pp. 209–18, March 1984.

COCHRAN, W. T., et al., "What is the Fast Fourier Transform?," *Proc.*　　J
IEEE, Vol. 55, pp. 1664–73, October 1967.

COOLEY, J. W., P. A. W. LEWIS, AND P. D. WELCH, "The Application of　　K
the Fast Fourier Transform Algorithm to the Estimation of Spectra and
Cross-spectra," *Journal of Sound Vibration*, Vol. 12, pp. 339–52, July
1970.

COOLEY, J. W., AND J. W. TUKEY, "An Algorithm For The Machine Cal-　　T
culation Of Complex Fourier Series," *Math. of Comput.*, Vol. 19, No.
90, pp. 297–301, 1965.

COOLEY, J. W., P. A. W. LEWIS, AND P. D. WELCH, "Historical Notes On　　J
The Fast Fourier Transform," *Proc. IEEE*, Vol. 55, pp. 1675–77, October
1967.

COOLEY, J. W., P. A. W. LEWIS, AND P. D. WELCH, "The Fast Fourier　　F
Transform Algorithm. Programming Considerations In The Calculation
Of Sine, Cosine and Laplace Transforms," *J. Sound Vib.*, Vol. 12, pp.
315–37, July 1970.

COOLEY, J. W., P. A. W. LEWIS, AND P. D. WELCH, "The Finite Fourier　　J
Transform," *IEEE Trans. Audio and Electroacoustics.*, Vol. 17, No. 2,
pp. 77–85, June 1969.

COOLEY, J. W., P. A. W. LEWIS, AND P. D. WELCH, "Application of the　　P
Fast Fourier Transforms to Computation of Fourier Integrals, Fourier
Series, and Convolution Integrals," *IEEE Trans. Audio and Electro-
coustics*, Vol. AU-15, No. 2, pp. 79–84, June 1967.

COOLEY, J. W., P. A. W. LEWIS, AND P. D. WELCH, "The Fast Fourier　　J

Transform And Its Applications," *IEEE Trans. Educ.*, Vol. E-12, No. 1, pp. 27–34, March 1969.

COOPER, D. E., AND S. C. MOSS, "Picosecond Optoelectronic Measurement Of The High-Frequency Scattering Parameters Of A GaAs FET," *IEEE J. Quantum Electron*, Vol. QE-22, No. 1, pp. 94–100, January 1986.   J

COREY, L. E., J. C. WEED, AND T. C. SPEAKE, "Modeling Triangularly Packed Array Antennas Using a Hexagonal FFT," *1984 International Symposium Digest, Antennas and Propagation*, Vol. 2, pp. 507–10, June 1984.   E

CORINTHIOS, M. J., K. C. SMITH, AND J. L. YEN, "A Parallel Radix-4 Fast Fourier Transform Computer," *IEEE Trans. Comput.*, Vol. C-24, No. 1, pp. 80–92, January 1975.   G

CORSINI, P., AND G. FROSINI, "Properties of the Multidimensional Generalized Discrete Fourier Transform," *IEEE Trans. Comput.*, Vol. C-28, No. 11, pp. 819–20, November 1979.   T

COZZENS, J. H., AND L. A. FINKELSTEIN, "Computing the Discrete Fourier Transform Using Residue Number Systems In A Ring Of Algebraic Integers," *IEEE Transactions On Information Theory*, Vol. IT-31, No. 5, pp. 580–88, September 1985.   T

CRAIG, A. D., AND P. D. SIMS, "Fast Integration Techniques For Reflector Antenna Pattern Analysis," *Electron. Lett. (GB)*, Vol. 18, No. 2, pp. 60–62, January 1982.   E

CRITINA, S., AND M. D'AMORE, "DFT-based Procedure for Transmission-line Transient Computation," *IEEE Proc. C (GB)*, Vol. 134, No. 2, pp. 138–43, March 1987.   Y

CUSHMAN, R. H., "Add the FFT to Your Box of Design Tools," *EDN*, Vol. 26, No. 18, pp. 83–88, September 16, 1981.   M

D'ALESSIO, T., "'Objective' Algorithm For Maximum Frequency Estimation In Doppler Spectral Analysers," *Medical & Biological Engineering & Computing*, Vol. 23, No. 1, pp. 63–68, January 1985.   A

D'ELIA, G., G. LEONE, AND R. PIERRI, "Analysis of Fresnel Field Of A Subreflector By An Optimal Use Of FFT," *Electronics Letters*, Vol. 20, No. 8, pp. 332–34, April 12, 1984.   E

D'INCAU, S., AND R. G. GARNER, "Application of Two-Channel FFT Analysis to Paper Machines," *Journal of Pulp and Paper Science*, Vol. 9, No. 2, pp. 38–42, June 1983.   J

DAISHIDO, T., K. ASUMA, H. OHARA, S. KOMATSU, AND K. NAGANE, "FFT Processor as a Digital Lens for Radio Patrol Camera in Astrophysics," *ICASSP 86 Proceedings, International Conf. on Acoust. Speech and Signal Processing*, Vol. 4, pp. 2855–57, April 1986.   L

DALY, R. H., AND J. M. GLASS, "Digital Signal Processing for RADAR," *Electron Prog*, Vol. 17, No. 1, pp. 24–30, Spring 1975.   H

DAMASKOS, N. J., R. T. BROWN, J. R. JAMESON, AND P. L. E. USLENGHI, "Transient Scattering By Resistive Cylinders," *IEEE Transactions on Antennas and Propagation*, Vol. AP-33, No. 1, pp. 21–25, January 1985.   E

DAME, L., AND F. VAKILI, "Ultraviolet Resolution of Large Mirrors Via   X
Hartmann Tests and Two-Dimensional Fast Fourier Transform Analysis," *Opt. Eng.*, Vol. 23, No. 6, pp. 759–65, November–December 1984.

DANIELL, G. J., "Improvement In The Evaluation Of Convolutions Of Real   J
Symmetric Signals Using The Fast Fourier Transform," *Signal Processing*, Vol. 10, No. 3, pp. 311–13, April 1986.

DANILENKO, A. I., AND M. O. KISCHENKOV, "A Method For The Rapid   E
Calculation Of The Radiation Patterns Of Nonequidistant Systems Of
Radiators," *Radioelectron Commun Syst.*, Vol. 27, No. 2, pp. 77–79,
1984.

DASH, P. K., AND D. K. PANDA, "Spectral Observations of Power Network   X
Signals for Digital Signal Processing," *Microprocess. and Microsyst.*
(GB), Vol. 8, No. 9, pp. 475–80, November 1984.

dE AMEIDA, R. A. R., AND J. MAZZUCCO, JR., "Fast Calculation of Far-Field   E
Patterns of Reflector Antennas With Fast Fourier Transform," *Conf Proc
Eur Microwave Conf 10th*, Warsaw, Pol, pp. 61–66, September 1980.

DEEL, J. C., "Use of Modal Analysis and Structural Modification Software   V
For The Measurement and Prediction of Machinery Dynamic Characteristics," *Noise & Vibration Control Worldwide*, Vol. 14, No. 1, pp. 7–
10, February 1983.

DEL RE, E., F. RONCONI, P. SALVI, AND P. SEMENZATO, "Comparison of   Q
Non-FFT Methods of TDM-FDM Transmultiplexing," *Alta Freq.* (Italy),
Vol. 51, No. 1, pp. 9–21, January–February 1982.

DELL'AGUILA, A., L. SALVATORE, "A Minicomputer Based System for Analysis of Inverter-fed Induction Motors," *Microcomput. Appl.*, Vol. 4, No.
1, pp. 22–28, 1985.

DELOTTO, I., AND D. DOTTI, "Two-dimensional Transforms by Mini-com-   N
puters Without Matrix Transposing," *Computer Graphics and Image
Processing*, Vol. 4, No. 3, pp. 271–78, September 1975.

DESPAIN, A. M., A. M. PETERSON, O. S. ROTHAUS, AND E. H. WOLD, "Fast   T
Fourier Transform Processors Using Gaussian Residue Arithmetic," *J.
Parallel Distrib Comput*, Vol. 2, No. 3, pp. 219–37, August 1985.

DESPAIN, A., "Very Fast Fourier Transform Algorithms Hardware For Im-   G
plementation," *IEEE Trans. Comput.*, Vol. C-28, pp. 333–41, 1979.

DIDERICH, R., "Calculating Chebyshev Shading Coefficients Via the Dis-   P
crete Fourier Transform," *IEEE Proceedings Letters*, Vol. 62, No. 10,
pp. 1395–96, October 1974.

DILLARD, G. M., "Recursive Computation Of The Discrete Fourier Trans-   H
form With Applications To A Pulse-Doppler Radar System," *Comput.
Electr. Eng.*, Vol. 1, pp. 143–52, 1973.

DILTS, G. A., "Computation of Spherical Harmonic Expansion Coefficients   F
Via FFTs," *J. Comput. Phys.*, Vol. 57, No. 3, pp. 439–53, February 1985.

"Discrete Fourier Transform Applied to Time Domain Signal Processing,"   J
IEEE Communications Magazine, Vol. 20, No. 3, pp. 13–22, May 1982.

DOERNBERG, J., H.-S., LEE, AND D. A. HODGES, "Full-Speed Testing of A/   J
D Converters," *IEEE J. Solid-State Circuits*, Vol. SC-19, No. 6, pp. 820–
27, December 1984.

DOUGHTY, S., "Steady-State Torsional Response With Viscous Damping,"   V
Journal of Vibration, Acoustics, Stress, and Reliability in Design, Vol.
107, No. 1, pp. 123–27, January 1985.

DOYLE, J. F., AND S. KAMLE, "Experimental Study Of The Reflection And   V
Transmission Of Flexural Waves At Discontinuities," *Journal of Applied
Mechanics, Transactions ASME*, Vol. 52, No. 3, pp. 669–73, September
1985.

DOYLE, J. F., "Further Developments In Determining The Dynamic Contact   V
Law," *Experimental Mechanics*, Vol. 24, No. 4, pp. 265–70, December
1984.

DRUBIN, M., "Computation of the Fast Fourier Transform Data Stored in   T
External Auxiliary Memory for Any General Radix ($r = 2$ exp. n, n
greater than or equal to 1.)," *IEEE Trans. on Computers.*, Vol. C-20,
No. 12, pp. 1552–58, December 1971.

DUBNER, H., AND J. ABATE, "Numerical Inversion of Laplace Transforms   F
By Relating Them to the Finite Fourier Cosine Transforms," *Journal of
the Association for Computing Machinery*, Vol. 15, No. 1, pp. 115–23,
January 1968.

DUCHOVIC, R. J., AND G. C. SCHATZ, "The FFT Method for Determining   F
Semiclassical Eigenvalues-application to Asymmetric Top Rigid Rotors,"
J. Chem. Phys., Vol. 84, No. 4, pp. 2239–46, February 1986.

DUHAMEL, P., AND H. HOLLMANN, "'Split Radix' FFT Algorithm," *Elec-*   T
tronics Letters, Vol. 20, No. 1, pp. 14–16, January 5, 1984.

DUHAMEL, P., AND H. HOLLMANN, "Existence of a 2**n FFT Algorithm   T
with A Number of Multiplications Lower Than 2**n** Plus **1," *Elec-*
tronics Letters, Vol. 20, No. 17, pp. 690–92, August 16, 1984.

DUHAMEL, P., "Implementation of 'Split-Radix' FFT Algorithms For Com-   T
plex, Real, AND Real-Symmetric Data," *IEEE Trans. Acoust., Speech &
Signal Process.*, Vol. ASSP-34, No. 2, pp. 285–95, April 1986.

DUMERMUTH, G. P., P. J. HUBER, B. KLEINER, AND T. GASSER, "Numerical   J
Analysis of Electroencephalographic Data," *IEEE Trans. Audio and
Electroacoustics*, Vol. AU-18, pp. 404–11, December 1970.

DUMOULIN, C. L., G. C. LEVY, AND F. A. L. ANET, "FFT Algorithm For   T
Virtually Stored Data," *Computers & Chemistry (GB)*, Vol. 5, No. 2–3,
pp. 125–38, 1981.

DUNHAM, W. R., J. A. FEE, L. J. HARDING, AND H. J. GRANDE, "Application   J
of Fast Fourier Transforms to EPR Spectra of Free Radicals In Solution,"
J. Magn. Resonance, Vol. 40, No. 2, pp. 351–59, August 1980.

DURAND, L. G., J. DE GUISE, G., CLOUTIER, R. GUARDO, AND M. BRAIS,   A
"Evaluation Of FFT-Based and Modern Parametric Methods For The
Spectral Analysis Of Bioprosthetic Valve Sounds," *IEEE Transactions
on Biomedical Engineering*, Vol. BME-33, No. 6, pp. 572–78, June 1986.

DURRANI, T. S., AND E. GOUTIS, "Optimisation Techniques for Digital Image X
Reconstruction From Their Projections," *IEE Proc. E (GB)*, Vol. 127,
No. 5, pp. 161–69, September 1980.

DURVASULA, S., B. PANDA, AND R. K. KAPANIA, "Impulse Excitation Tech- V
nique For Determining Vibration Characteristics Of Aeroelastic Models
of Tall Structures," *J Inst Eng India Part ND 2*, Vol. 61, May 1981.

DWYER, R. F., "Essential Limitations to Signal Detection and Estimation: P
An application to the arctic under ice environmental noise problem,"
Proc. IEEE, Vol. 72, No. 11, pp. 1657–60, November 1984.

EAKER, C. W., AND G. C. SHATZ, "Fourier Transform Methods for Cal- F
culating Action Variables and Semiclassical Eigen Values for Coupled
Oscillator Systems," *J. Chem. Phys.*, Vol. 81, No. 12, Pt. 2, pp. 5913–
19, December 1985.

EAST, K. A., AND T. D. EAST, "Computerized Acoustic Detection of Ob- A
structive Apnea," *Comput Methods Prog Biomed*, Vol. 21, No. 3, pp.
213–20, December 1985.

EDDY, D. R., AND F. J. BREMNER, "Computer Packages That Contain the J
FFT," *Behav. Res. Methods and Instrum.*, Vol. 15, No. 2, pp. 254–57,
April 1983.

EKLUNDH, J. O., "A Fast Computer Method for Matrix Transposing," *IEEE* N
Transactions Computers, Vol. C-21, No. 7, pp. 801–3, July 1972.

EL BANNA, M., AND M. EL NOKALI, "On The Transient Response Of A O
Circular Transducer," *Wave Motion*, Vol. 8, No. 3, pp. 235–41, May
1986.

EPSTEIN, G., "Recursive Fast Fourier Transforms," *AFIPS Proc. 1968 Fall* T
Joint Computer Conf., Vol. 33, Part 1, Washington D.C., pp. 141–43,
1968.

ERICKSON, B. K., "Programmed Calculation of Discrete Fourier Trans- T
forms," *IEEE Trans. Consum. Electron.*, Vol. CE-25, No. 5, pp. 772–
76, November 1979.

ERSOY, O., "On Relating Discrete Fourier, Sine, and Symmetric Cosine T
Transforms," *IEEE Trans. Acoust., Speech & Signal Process.*, Vol.
ASSP-33, No. 1, pp. 219–22, February 1985.

ESHLEMAN, R. L., "Machinery Diagnostics And Your FFT," *S. V. Sound* V
and Vibration, Vol. 17, No. 4, pp. 12–18, April 1983.

EVANS, J. C., AND P. H. MORGAN, "Simulation of Electron Spin Resonance D
Spectra By Fast Fourier Transform," *J. Magn. Resonance*, Vol. 52, No.
3, pp. 529–31, May 1983.

FABER, A. S., AND C. E. HO, "Wide-band Network Characterization by M
Fourier Transformation of Time Domain Measurements," *IEEE Journal*
of Solid State Circuits," Vol. SC-4, No. 4, pp. 231–35, August 1969.

FARMER, R. G., AND B. L. AGRAWAL, "State-of-the-art Techniques for Y
Power System Stabilizer Tuning," *IEEE Trans. on Power Apparatus and*
Systems, Vol. PAS-102, No. 3, pp. 699–709, March 1983.

FAVOUR, J. D., AND J. M. LEBRUN, "Transient Synthesis for Mechanical   V
Testing," *Instruments and Control Systems*, Vol. 43, pp. 125–27, September 1970.

FERRIE, J. F., C. W. NAWROCKI, AND G. C. CARTER, "Applications of the   O
Partitioned and Modified Chirp Z-Transform to Oceanographic Measurements," *IEEE Trans. Acoust., Speech and Signal Process.*, Vol.
ASSP-23, No. 2, pp. 243–44, April 1975.

FILHO, W. S., O. F. BRUM, AND R. B. PANERAI, "Spectral Analysis Of Electrical Impedance Measurements On The Lower Limbs," *IEEE Transactions on Biomedical Engineering*, Vol. BME-30, No. 7, pp. 387–92,   A
July 1983.

FISHCHER, D., G., GOLUB, O. HALD, C. LEIVA, AND O. WIDLUND, "On Four-   F
ier-Toeplitz Methods For Separable Elliptic Problems," *Math. Comput.*,
Vol. 28, No. 26, pp. 349–68, April 1974.

FORMAN, M. L., "Fast Fourier Transform Technique and Its Application   D
to Fourier Spectroscopy," *J. Opt. Soc. Am.*, Vol. 56, pp. 978–97, July
1966.

FORNBERG, B., "A Vector Implementation of the Fast Fourier Transform   T
Algorithm," *Math. Comput.*, Vol. 36, No. 153, pp. 189–91, January 1981.

FORNBERG, B., "On A Fourier Method For the Integration of Hyperbolic   F
Equations," *SIAM J. Numer. Anal.*, Vol. 12, No. 4, pp. 509–28, September 1975.

FORTES, J. M. P., AND R. SAPAIO-NETO, "A Fast Algorithm for Sorting and   Q
Counting Third-order Intermodulation Products," *IEEE Trans. Commun.*, Vol. COM-34, No. 12, pp. 1266–72, December 1986.

FOSKETT, C. T., "Noise and Finite Register Effects In Infrared Fourier   D
Transform Spectroscopy," *Appl. Spectrosc.*, Vol. 30, No. 5, pp. 531–
39, September–October 1976.

FOX, J., G. SURACE, AND P. A. THOMAS, "Self-Testing 2-MU m CMOS Chip   G
Set For FFT Applications," *IEEE Journal of Solid-State Circuits*, Vol.
SC-22, No. 1, pp. 15–19, February 1987.

FRASER, D., "Array Permutation by Index-Digit Permutation," *J. Assoc.*   T
Comput. Mach., Vol. 23, No. 2, pp. 298–309, April 1976.

FRASER, D., "BIT Reversal And Generalized Sorting Of Multidimensional   N
Arrays," *Signal Processing*, Vol. 9, No. 3, pp. 163–76, October 1985.

FRASER, D., "Incrementing a Bit-reversed Integer," *IEEE Trans. on Com-*   T
puters (Short Notes). Vol. C-18, p. 74, January 1969.

FREISCHLAD, K. R., AND C. L. KOLIOPOULOS, "Modal Estimation of a Wave   X
Front From Difference Measurements Using the Discrete Fourier Transform," *J. Opt. Soc., Am. A.*, Vol. 3, No. 11, pp. 1852–61, November
1986.

G-AE Subcommittee on Measurement Concepts, "What is the Fast Fourier   J
Transform?," *IEEE Trans. Audio and Electroacoustics*, Vol. AU-15, pp.
45–55, June 1967; also *Proceedings of the IEEE*, Vol. 55, pp. 1664–74,
October 1967.

GAMBARDELLA, G., "Time Scaling and Short-time Spectral Analysis,"   K
Acoustical Society of America Journal, Vol. 44, pp. 1745–47, December
1968.

GAN, R. G., K. F. EMAN, AND S. M. WU, "Extended FFT Algorithm for   T
ARMA Spectral Estimation," *IEEE Transactions on Acoustics, Speech,
und Signal Processing*, Vol. ASSP-32, No. 1, pp. 168–70, February 1984.

GAVRILOV, A. B., N. I. GORLOV, M. M. SIMONOV, AND Z. G. KURBANGALEEV,   L
"Methods For Increasing Accuracy Of Automated Pulse Reflectometry,"
Measurement Techniques, Vol. 28, No. 10, pp. 871–74, October 1985.

GECKINLI, N. C., AND D. YAVUZ, "Some Novel Windows and a Concise   K
Tutorial Comparison of Window Families," *IEEE Trans. Acoust. Speech
Signal Process.*, Vol. ASSP-26, pp. 501–7, 1978.

GENTLEMAN, W. M., AND G. SANDE, "Fast Fourier Transforms For Fun and   J
Profit," *AFIPS Proc. Fall Joint Comput. Conf.*, Washington, D.C., Vol.
29, pp. 563–78, 1966.

GENTLEMAN, W. M., "An Error Analysis of Goertzel's (Watt's) Method of   W
Computing Fourier Coefficients," *Computer J.*, Vol. 12, pp. 160–65, May
1969.

GENTLEMAN, W. M., "Matrix Multiplication and Fast Fourier Transforms,"   T
Bell Syst. Tech. J., Vol. 47, pp. 1099–1103, July–August 1968.

GHOSH, S. K., AND S. P. S. GUPTA, "X-Ray Diffraction Line Profile Of   J
Cold-Worked Hexagonal Alloys Zn-Ag: ETA And Epsilon Phase," *Metallurgical Transactions A*, Vol. 16A, No. 8, pp. 1427–35, August 1985.

GIANZERO, S., AND B. ANDERSON, "Integral Transform Solution To The Fun- U
damental Problem in Resistivity Logging," *Geophysics*, Vol. 47, No. 6,
pp. 946–56, June 1982.

GIBBS, J., "Windowing Boosts Performance of Dynamic Signal Analyzers,"   K
EDN, Vol. 26, No. 15, pp. 109–13, August 1981.

GIRGIS, A. A., AND F. M. HAM, "A New FFT-Based Digital Frequency   Y
Relay for Load Shedding," *IEEE Trans. Power Appar. and Syst.*, Vol.
PAS-101, No. 2, pp. 433–39, February 1982.

GIRGIS, A. A., AND F. M. HAM, "A Quantitative Study of Pitfalls In The   J
FFT," *IEEE Trans. Aerosp. and Electron. Syst.*, Vol. AES-16, No. 4,
pp. 434–39, July 1980.

GLASS, J. M., "An Efficient Method For Improving Reliability of a Pipeline   G
FFT," *IEEE Trans. Comput.*, Vol. C-29, No. 11, pp. 1017–20, November
1980.

GLASSMAN, J. A., "A Generalization of the Fast Fourier Transform," *IEEE*   T
Trans. Comput., Vol. C-19, pp. 105–16, February 1970.

GLISSON, T. H., AND C. I. BLACK, "On Digital Replica Correlation Algorithm   O
with Applications to Active Sonar," *IEEE Trans. Audio and Electroacoustics.*, Vol. AU-17, No. 3, pp. 190–97, September 1969.

GLISSON, T. H., C. I. BLACK, AND A. P. SAGE, "The Digital Computation   K
of Discrete Spectra Using the Fast Fourier Transform," *IEEE Trans.
Audio Electroacoust.*, Vol. AU-18, pp. 271–87, September 1970.

GOLD, B., AND T. BIALLY, "Parallelism in Fast Fourier Transform Hard-   G
ware," *IEEE Trans. Audio. Electroacoust.*, Vol. AU-21, 1973.

GOLDBERG, B. G., "A Continuous Recursive DFT Analyzer-The Discrete   K
Coherent Memory Filter," *IEEE Trans. Acoust., Speech and Signal Process.*, Vol. ASSP-28, No. 6, pp. 760–62, December 1980.

GONZALEZ, JR., R. R., AND W. J. KOH, "Spectral Analysis of Doppler Ul-   A
trasonic Flow Signals By a Personal Computer," *Computers in Biology
and Medicine*, Vol. 13, No. 4, pp. 281–86, 1983.

GOOD, I. J., "The Relationship Between Two Fast Fourier Transforms,"   T
IEEE Trans. Comput., Vol. C-20, pp. 310–17, March 1971.

GORDON, R., AND R. M. RANGAYYAN, "Geometric Deconvolution: A Meta-   C
Algorithm For Limited View Computed C Tomography," *IEEE Trans-
actions on Biomedical Engineering*, Vol. BME-30, No. 12, pp. 806–10,
December 1983.

GOTTLIEB, P., AND L. J. DE LORENZA, "Parallel Data Streams and Serial   T
Arithmetic for Fast Fourier Transform Processors," *IEEE Trans. Acoust.
Speech and Signal Process.*, Vol. ASSP-22, No. 2, pp. 111–17, April
1974.

GOTWALS, J. K., "Performing the Fast Fourier Transform With a Personal   T
Computer," *1985 ACM SIGSMALL Symposium on Small Systems*, pp.
142–54, May 1985.

GRACE, O. D., "Two Finite Fourier Transforms for Bandpass Signals,"   S
IEEE Trans. Audio and Electroacoustics, Vol. AU-18, pp. 501–2, December 1970.

GRACHEV, I. D., M. K. H. SALAKHOV, AND I. S. FISHMAN, "Efficient Al-   D
gorithms for the Processing of Multidimensional Spectroscopic Data Arrays,"*Comput. Enhanced Spectrosc. (GB)*, Vol. 2, No. 1, pp. 1–12, January–March 1984.

GRANSER, H., "Deconvolution Of Gravity Data Due To Lateral Density   C
Distributions," *Geoexploration*, Vol. 23, No. 4, pp. 537–47, December
1985.

GRAY, D. A., "Applications Of Parametric And Non-Parametric Spectral   O
Estimation Techniques To Passive Sonar Data," *Journal of Electrical
and Electronics Engineering*, Australia, Vol. 5, No. 2, pp. 112–19, June
1985.

GREENSPAN, R. L., P. H. ANDERSON, "Channel Demultiplexing by Fourier   S
Transform Processing," *EASCON 74 Proceedings*, pp. 360–72, 1974.

GRIGORYAN, A. M., AND M. M. GRIGORYAN, "Discrete Two-dimensional   N
Fourier Transforms in a Tensor Representation: New Orthogonal Functions," *Optoelectron. Instrum. & Data Process.*, No. 1, pp. 22–28, 1986.

GROGINSKY, HERBERT L., "An FFT Chart-letter, Fast Fourier Transform   J
Processors; Chart Summarizing Relations Among Variables," *Proceedings of the IEEE*, Vol. 58, pp. 1782–84, October 1970.

GUENDEL, L., "Novel High-speed Fourier-Vector Processor," *Signal Pro-   G
cessing*, Vol. 9, No. 2, pp. 107–20, September 1985.

GUESSOUM, A., AND R. M. MERSEREAU, "Fast Algorithms for the Multidimensional Discrete Fourier Transform," *IEEE Trans. Acoust., Speech & Signal Process.*, Vol. ASSP-34, No. 4, pp. 937–43, August 1986.   T

GUSEV, V. G., AND G. V. LOSKUTOVA, "Use of the Bivariate Fast Fourier Transform Algorithm For Process of Information Received From a Linear Array Antenna," *Radio Engineering and Electronic Physics*, Vol. 27, No. 12, pp. 78–82, December 1982.   B

GUTKNECHT, M. H., "Numerical Experiments on Solving Theodorsen's Integral Equation For Conformal Maps With the Fast Fourier Transform and Various Nonlinear Interative Methods," *SIAM J. Sci. and Stat. Comput.*, Vol. 4, No. 1, pp. 1–30, March 1983.   F

GUTKNECHT, M. H., "Solving Theodorsen's Integral Equation For Conformal Maps With The Fast Fourier Transform and Various Nonlinear Iterative Methods," *Numer. Math.* (Germany) Vol. 36, No. 4, pp. 405–29, 1981.   F

HAAS, W. H., AND C. S. LINDQUIST, "Transitional FFT-Based Filters For Delay Estimation," *Conf Rec Asilomar Conf Circuits Syst Comput 14th* November 1980.   L

HAGER, W. W., "A Modified Fast Fourier Transform for Polynomial Evaluation and the Jenkins-Traub Algorithm," *Numer. Math.* (Germany), Vol. 50, No. 3, pp. 253–61, 1987.   F

HAIRER, E., C. LUBRICH, AND M. SCHLICHTE, "Fast Numerical Solution of Nonlinear Volterra Convolution Equations," *SIAM J. Sci. and Stat. Comput.*, Vol. 6, No. 3, pp. 532–41, July 1985.   F

HALBERG, L. I., AND K. E. THIELE, "Extraction of Blood Flow Information Using Doppler-Shifted Ultrasound," *Hewlett-Packard Journal*, Vol. 37, No. 6, pp. 35–40, June 1986.   A

HALL, J. F., "FFT Algorithm For Structural Dynamics," *Earthquake Engineering & Structural Dynamics*, Vol. 10, No. 6, pp. 797–811, November–December 1982.   V

HALLIWELL, G., "Using Fourier Transform Analysers To Investigate Component Vibration," *Nuclear Engineering International (GB)*, Vol. 25, No. 296, pp. 42–45, March 1980.   V

HAM, F. M., AND A. A. GIRGIS, "Measurement Of Power Frequency Fluctuations Using The FFT," *IEEE Trans Ind Electron*, Vol. IE-32, No. 3, pp. 199–204, August 1985.   J

HANNA, M. T., "Two-dimensional Filtering of Sensor Array Data," *1986 IEEE International Symposium on Circuits and Systems*, Vol. 1, pp. 17–20, May 1986.   N

HARALICK, R. M., "A Storage Efficient Way to Implement The Discrete Cosine Transform," *IEEE Trans. Comput.*, Vol. C-25, No. 7, pp. 764–65, July 1976.   T

HARDING, J. C., JR., D. A., WADE, R. A. MARINO, E. G. SAUER, AND S. M. KLAINER, "A Pulsed NQR-FFT Spectrometer For Nitrogen-14," *J. Magn. Resonance*, Vol. 36, No. 1, pp. 21–33, October 1979.   D

HARMANSSON, B., D. YEVICK, AND A. T. FRIBERG, "Optical Coherence Calculations with the Split-step Fast Fourier Transform Method," *Appl. Opt.*, Vol. 16, pp. 2645–47, August 1986. X

HARRIS, D. B., J. H. MCCLELLAN, D. S. K. CHAN, AND H. W. SCHUESSLER, "Vector Radix Fast Fourier Transform," *1977 IEEE Int. Conf. on Acoustics, Speech and Sig. Proc. Rec.*, pp. 548–51, May 1977. N

HARRIS, F. J., "The Discrete Fourier Transform Applied To Time Domain Signal Processing," *IEEE Communications Magazine*, Vol. 20, No. 3, pp. 13–22, May 1982. J

HARRIS, J. H., "On the Use of Windows for Harmonic Analysis with the Discrete Fourier Transform," *Proc. IEEE*, Vol. 66, pp. 51–83, January 1978. K

HARRISON, R. E., "Calculator Program Speeds Through 32 FFT Points," *Electronic Design*, Vol. 28, No. 10, pp. 233–36, May 1980. T

HARTWELL, J. W., "A Procedure for Implementing the Fast Fourier Transform on Small Computers," *IBM Journal of Research and Development*, Vol. 15, pp. 355–63, September 1971. T

HASHMATY-MANESH, D., AND S. C. TAM, "Application of Winograd's Fast Fourier Transform (FFT) to the Calculation of Diffraction Optical Transfer Function (OTF)," *Proc. SPIE Int. Soc. Opt. Eng.*, Vol. 369, pp. 692–95, September 1983. X

HASSANEIN, H., AND M. RUDKO, "On the Use of Discrete Cosine Transform in Cepstral Analysis," *IEEE Trans. Acoust., Speech and Signal Process.*, Vol. ASSP-32, No. 4, pp. 922–25, August 1984. J

HAUBRICH, R. A., AND J. W. TUKEY, "Spectrum Analysis of Geophysical Data," *Proc. IBM Scientific Computing Symp. on Environmental Sciences*, pp. 115–28, 1967. K

HAVRILLA, J. J., "A Technique for Determining the Effects of Aircraft Components on the Pattern of Radar Antennas Before They are Built," *Proceedings of the IEEE 1984 National Aerospace and Electronics Conference*, Vol. 1, pp. 304–8, May 1984. E

HAYASHI, K., K. K. DHAR, K. SUGAHARA, AND K. HIRANO, "Design of High-Speed Digital Filters Suitable For Multi-DSP Implementation," *IEEE Transactions On Circuits and Systems*, Vol. CAS-33, No. 2, pp. 202–17, February 1986. P

HAYKIN, S. S., AND C. M. THORSTEINSON, "Decision-directed Delay-lock Loop Using Fast Fourier Transform Crosscorrelation," *Proc. Inst. Electr. Eng. (GB)*, Vol. 121, No. 4, pp. 245–49, April 1974. L

HEARN, T. C., J. MAZUMDAR, AND L. J. MAHAR, "First Heart Sound Spectra in Relation to Anterior Mitral-Leaflet Closing Velocity," *Medical & Biological Engineering & Computing*, Vol. 20, No. 4, pp. 466–72, July 1982. A

HEIDEMAN, M. T., AND C. S. BURRUS, "On the Number of Multiplications Necessary To Compute A Length-2**N DFT," *IEEE Transactions on Acoustics, Speech, and Signal Processing*, Vol. ASSP-34, No. 1, pp. 91–95, February 1986. T

HEIDEMAN, M. T., D. H. JOHNSON, AND C. S. BURRUS, "Gauss And The T
History Of The Fast Fourier Transform," *IEEE ASSP Mag.* (USA), Vol.
1, No. 4, pp. 14–21, October 1984.

HELMS, H. D., "Fast Fourier Transform Method of Computing Difference P
Equations and Simulating Filters," *IEEE Trans. Audio and Electroacoustics,* Vol. AU-15, No. 2, pp. 85–90, June 1967.

HELMS, H. D., "Nonrecursive Digital Filters: Design Methods for Achiev- P
ing Specifications on Frequency Response," *IEEE Trans. Audio Electroacoust.*, Vol. AU-16, No. 3, pp. 336–42, September 1968.

HENERY, R. J., "Solution of Wiener-Hopf Integral Equations Using the Fast F
Fourier Transform," *J. Inst. Math. and Appl.* (GB), Vol. 13, No. 1, pp.
89–96, February 1974.

HENRICI, P., "Fast Fourier Methods In Computational Complex Analysis," F
Swiss Federal Inst. of Technol., Zurich, Switzerland, SIAM Rev., Vol.
21, No. 4, pp. 481–527, October 1979.

HERMAN, A. J., R. M. ANANIA, J. H. CHUN, C. A. JACEWITZ, AND R. E. F. U
PEPPER, "Fast Three-Dimensional Modeling Technique and Fundamen-
tals of Three-Dimensional Frequency-Domain Migration," *Geophysics*,
Vol. 47, No. 12, pp. 1627–44, December 1982.

HERMANSSON, B., AND D. YEVICK, "Numerical Analyses of the Modal Ei- X
genfunctions of Chirped and Unchirped Multiple-stripe-geometry Laser
Arrays," *J. Op. Soc. Am. A,* Vol. 4, No. 2, pp. 379–90, February 1987.

HERMANSSON, B., D. YEVICK, AND A. T. FRIBERG, "Optical Coherence Cal- X
culations With The Split-Step Fast Fourier Transform Method," *Appl.
Opt.*, Vol. 25, No. 16, pp. 2645–47, August 1986.

HEUTE, U., AND P. VARY, "Digital Filter Bank With Polyphase Network S
and FFT Hardware: Measurements and Applications," *Signal Processing*, Vol. 3, No. 4, pp. 307–19, October 1981.

HEUTE, U., "Impact Of FFT Coefficient Errors On Polyphase Filter W
Banks," *Signal Processing*, Vol. 7, No. 2, pp. 119–33, October 1984.

HEUTE, U., "Results of a Deterministic Analysis of FFT Coefficient Er- W
rors," *Signal Processing*, Vol. 3, No. 4, pp. 321–31, October 1981.

HIGGINS, R. J., "Fast Fourier Transform: An Introduction With Some Min- J
icomputer Experiments," *AM. J. Phys.* (USA), Vol. 44, No. 8, pp. 766–
73, August 1976.

HINICH, M. J., AND C. S. CLAY, "The Application of the Discrete Fourier U
Transform in the Estimation of Power Spectra, Coherence and Bispectra
of Geophysical Data," *Review of Geophysics,* Vol. 6, pp. 347–62, August
1968.

HIROSAKI, B., "An Orthogonally Multiplexed QAM System Using the Dis- Q
crete Fourier Transform," *IEEE Trans. Commun.*, Vol. COM-29, No. 7,
pp. 982–89, July 1981.

HITCHEN, M., J. B. HARNESS, AND A. J. MEARNS, "Thermal Entrainment A
Device for Cardiovascular Investigation," *Journal of Medical Engineering & Technology*, Vol. 4, No. 4, pp. 179–82, July 1980.

HODGE, A. J., W. J. ADELMAN, R. B. WALTZ, AND C. L. TYNDALE, "Analysis   A
of Periodic Structure In Model Subcellular Macromolecular Arrays By
Fourier Processing Of Single Line Video Signals In Scanning Transmission Electron Microscopy," *IEEE Transactions on Biomedical Engineering*, Vol. BME-29, No. 6, pp. 439–47, June 1982.

HOLM, W. A., AND J. D. ECHARD, "FFT Signal Processing For Non-Co-   H
herent Radar Systems," *Proceedings of ICASSP 82. IEEE International
Conference on Acoustics, Speech and Signal Processing*, Vol. 1, pp. 363–
66, Paris, France, May 1982.

HOLZER, F., AND R. REIBOLD, "Numerical Analysis of Ultrasonic Trans-   O
ducer Vibrations From Optically Measured Beam Profiles," *Acustica*,
Vol. 60, No. 3, pp. 236–43, May 1986.

HONMA, H., AND M. SAGAWA, "Improving The Accuracy And Error Anal-   W
ysis In Floating-Point FFT Computation," *Electronics and Communications in Japan*, Vol. 67, No. 11, pp. 9–18, November 1984.

HORNSBY, J. S., "Full-Wave Analysis of Microstrip Resonator and Open-   E
Circuit End Effect," *IEEE Proceedings, Part H: Microwaves, Optics and
Antennas*, Vol. 129, No. 6, pp. 338–41, December 1982.

HORTA, L. G., AND JER-NAN JUANG, "Identifying Approximate Linear   J
Models for Simple Nonlinear Systems," *Journal of Guidance, Control,
and Dynamics*, Vol. 9, No. 4, pp. 385–90, July–August 1986.

HOWARD, A. Q. JR., "On Approximating Fourier Integrals Transforms By   U
Their Discrete Counterparts in Certain Geophysical Applications," *IEEE
Trans. Antennas and Propag.*, Vol. AP-23, No. 3, pp. 264–66, March
1975.

HOWARD, S. J., "Method For Continuing Fourier Spectra Given By The   K
Fast Fourier Transform," *J. Opt. Soc. Am.*, Vol. 71, No. 1, pp. 95–98,
January 1981.

HOYER, E. A., AND W. R. BERRY, "An Algorithm for the Two-Dimensional   N
FFT," *1977 IEEE Int. Conf. on Acoustics, Speech and Sig. Proc., Rec.*,
pp. 552–55, May 1977.

HSI-PING, L., AND D. D. KOSLOFF, "Numerical Evaluation of the Hilbert   F
Transform by the Fast Fourier Transform (FFT) Technique," *Geophys.
J. R. Astron. Sco. (GB)*, Vol. 67, No. 3, pp. 791–99, December 1981.

HSU, Y. C., "A Method Of Zip For Tracking Target Velocity and Increasing   H
Velocity Resolution," *1983 IEEE International Symposium of Circuits
and Systems*, Vol. 1, pp. 101–4, Newport Beach, CA, May 1983.

HUANG, T. S., "Two-Dimensional Windows," *IEEE Trans. Audio and Elec-   N
troacoustics*, Vol. AU-20, No. 1, pp. 88–89, March 1972.

HUNG, C. C., AND R. MITTRA, "Secondary Pattern and Focal Region Dis-   E
tribution of Reflector Antennas Under Wide-Angle Scanning," *IEEE
Trans. Antennas and Propag.*, Vol. AP-31, No. 5, pp. 756–63, September
1983.

HUNT, B. R., "Digital Image Processing," *Proc. IEEE*, Vol. 63, No. 4, pp.   X
693–708, April 1975.

IKEDA, N., T. AKOI, H. KOGA, K. TAEKIM, AND K. KIDO, "Recognition of   R
Vowels Using the Local Peaks in FFT Spectrum," *J. Acoust. Soc. Jpn.*
(Japan), Vol. 41, No. 12, pp. 886–90, December 1985.

IKEDA, Y., AND M. NORIGOE, "New Realization of DFT Applied to CCITT   S
No. 5 Telephone Signaling System," *IEEE J. Sel. Areas Commun.*, Vol.
SAC-2, No. 2, pp. 334–39, March 1984

IMAI, M., AND S. INUKUCHI, "Frequency Identification By Complex Spec-   R
trum (Speech)," *ICASSP 86 Proceedings. IEEE-IECEJ-ASJ Interna-*
tional Conference on Acoustics, Speech and Signal Processing (Cat. No.
86CH2243-4), Vol. 1, pp. 117–20, Tokyo, Japan, April 1986.

INADA, H., "Backscattered Short Pulse Response of Surface Waves From   J
Dielectric Spheres," *Appl. Opt.*, Vol. 13, No. 8, pp. 1928–33, August
1974.

JAIN, A. K., AND J. JASIULEK, "A Class of FFT Based Algorithms For Linear F
Estimation and Boundary Value Problems," *Proceedings of the 1981*
Joint Automatic Control Conference, FA-6B/1–5, Vol. 2, Charlottesville,
VA, June 1981.

JAIN, A. K., AND J. JASIULEK, "Fast Fourier Transform Algorithms For   F
Linear Estimation, Smoothing, And Riccati Equations," *IEEE Trans-*
actions on Acoustics, Speech, and Signal Processing, Vol. ASSP-31, No.
6, pp. 1435–46, December 1983.

JAIN, A. K., "A Fast Karhunen-Loeve Transform For A Class of Random   F
Processes," *IEEE Trans. Commun.*, Vol. COM-24, No. 9, pp. 1023–29,
September 1976.

JAMES, D. V., "Quantization Errors in the Fast Fourier Transform," *IEEE* W
Trans. Acoust., Speech and Signal Process., Vol. ASSP-23, No. 3, pp.
277–83, June 1975.

JAMIESON, L. H., P. T. MUELLER, JR., AND H. J. SIEGEL, "FFT Algorithms   T
For SIMD Parallel Processing Systems," *J. Parallel and Distrib. Com-*
put., Vol. 3, No. 1, pp. 48–71, March 1986.

JANSEN, B. H., J. R. BOURNE, AND J. W. WARD, "Autoregressive Estimation A
of Short Segment Spectra For Computerized EEG Analysis," *IEEE*
Transactions on Biomedical Engineering, Vol. BME-28, No. 9, pp. 630–
38, September 1981.

JENKE, L. M., "Application of Digital Fourier Analysis in Processing Dy-   V
namic Aerodynamic Heating Measurements," *AIAA J.*, Vol. 17, No. 6,
pp. 641–42, June 1979.

JENKINS, G. M., AND D. G. WATTS, *Spectral Analysis and its Applications.* K
San Francisco: Holden-Day, 1968.

JENKINS, W. K., "Inherent Phase Distortion In Rectangular Format FFT   H
Processing Algorithms For Synthetic Aperture Radar," *Proc of the Int*
Symp on Network Theory, pp. 101–5, 1979.

JESSHOPE, C. R., "The Implementation of Fast Radix 2 Transforms on Array   T
Processors," *IEEE Trans. Comput.*, Vol. C-29, No. 1, pp. 20–28, January
1980.

JOHNSON, D. G., AND J. I. SEWELL, "Improved Z Plane Polynomial Inter-   M
polative Analysis of Switched-Capacitor Networks," *IEEE Trans. Circuits and Syst.*, Vol. CAS-31, No. 7, pp. 666–68, July 1984.

JOHNSON, H. W., AND C. S. BURRUS, "On The Structure of Efficient DFT   T
Algorithms," *IEEE Trans. Acoust., Speech & Signal Process.*, Vol.
ASSP-33, No. 1, pp. 248–54, February 1985.

JOHNSON, L. R., AND A. K. JAIN, "Efficient Two-Dimensional FFT Algo-   N
rithm," *IEEE Transactions On Pattern Analysis and Machine Intelli-
gence*, Vol. PAMI-3, No. 6, pp. 698–701, November 1981.

JOHNSON, M. M., "Direct Application of the Fast Fourier Transform to   X
Open Resonator Calculations," *Appl. Opt.*, Vol. 13, No. 10, pp. 2326–
28, October 1974.

JOHNSON, S. A., Y. ZHOU, M. K. TRACY, M. J. BERGGREN, AND F. STRENGER,   O
"Inverse Scattering Solutions by a Sinc Basis, Multiple Source, Moment
Method. III. Fast Algorithms," *Ultrason. Imaging*, Vol. 6, No. 1, pp.
103–16, January 1984.

JOHNSTON, J. A., "Generating Multipliers For A Radix-4 Parallel FFT al-   T
gorithm," *Signal Processing*, Vol. 6, No. 1, pp. 61–66, January 1984.

JOHNSTON, J. A., "Input/Output Memory For Digital Convolution Via The   T
Parallel Pipeline Fast Fourier Transformer," *Signal Processing*, Vol. 10,
No. 2, pp. 193–99, March 1986.

JOHNSTON, J. A., "Parallel Pipeline Fast Fourier Transformer," *IEEE Pro-   H
ceedings, Part F* (GB): Communications, Radar and Signal Processing,
Vol. 130, No. 6, pp. 564–72, October 1983.

JONES, R. H., "A Reappraisal of the Periodogram in Spectral Analysis,"   K
Technometrics, Vol. 7, No. 4, pp. 531–42, November 1965.

JONES, W. R., "Precision FFT Correlation Techniques for Nondeterministic   L
Waveform," *IEEE EASCON Conv. Record*, pp. 375–80, October 1974.

KABAL, P., AND B. SAYAR, "Performance of Fixed-point FFTs: Rounding   W
and Scaling Considerations," *ICASSP 86 Proceedings. International
Conference on Acoustics, Speech and Signal Processing*, Vol. 1, pp. 221–
24, *Tokyo*, Japan, April 1986.

KAHANER, D. K., "Matrix Description of the Fast Fourier Transform,"   T
IEEE Trans. Audio Electroacoust., Vol. AU-18, No. 4, pp. 442–50, December 1970.

KAMANGAR F. A., AND K. R. RAO, "Fast Algorithms for the 2D-discrete   N
Cosine Transform," *IEEE Trans. Comput.*, Vol. C-31, pp. 899–906, 1982.

KANDA, M., "Effects of Resistive Loading Of 'TEM' Horns," *IEEE Trans-   E
actions On Electromagnetic Compatibility*, Vol. EMC-24, No. 2, Pt. 2,
pp. 245–55, May 1982.

KANDA, M., "Transients in a Resistively Loaded Linear Antenna Compared   F
With Those In A Conical Antenna and A TEM Horn," *IEEE Trans. Antennas and Propag.*, Vol. AP-28, No. 1, pp. 132–36, January 1980.

KANEKO, T., AND H. YAMAUCHI, "Addressing Technique for Bit-reversal   G

Transfer," *Trans. Inst. Electron. and Commun. Eng.* (Japan), Part (D),
Vol. J69D, No. 7, pp. 1124–26, July 1986.

KANEKO, T. K., AND B. LIU, "Accumulation of Round-off Errors in Fast   W
Fourier Transforms," *J. Assoc. Comput. Mach.*, Vol. 17, No. 4, pp. 637–
54, October 1970.

KARLSSON, L., "Numerical Analysis of Damped Transient Beam Vibration   V
by Use of Fourier Transforms," *Int. J. Numer. Methods Eng.* (GD), Vol.
21, No. 4, pp. 683–89, April 1985.

KATAYAMA, H., "Diversity Becoming More Apparent In Selection Of FFT   I
Analyzers," *JEE, Journal of Electronic Engineering*, Vol. 22, No. 218,
pp. 54–58, February 1985.

KATYL R. H., "FFT Calculation of Magnetic Fields In Air Coils," *IEEE*   J
Transactions on Magnetics, Vol. MAG-16, No. 3, pp. 545–49, May 1980.

KATYL, R. H., "FFT Calculation of Two-Step Semiconductor Impurity Dif-   J
fusion," *IEEE Transactions on Electron Devices*, Vol. ED-27, No. 5, pp.
991–93, May 1980.

KAY, S. M., AND S. L. MARPLE, JR., "Spectrum Analysis—A Modern Per-   K
spective," *Proc. IEEE*, Vol. 69, pp. 1380–1419, 1981.

KEMERAIT, R. C., AND D. G. CHILDERS, "Signal Detection and Extraction   K
By Cepstrum Techniques," *IEEE Trans. Informat. Theory*, IT-18, No.
6, pp. 745–59, November 1972.

KENNEDY, P. D., "FFT Signal Processing for Noncoherent Airborne Ra-   H
dars," *Proceedings of the 1984 IEEE National Radar Conference*, pp.
79–83, Atlanta, GA, March 1984.

KERLEY, L. M., "Teaching Concepts of Data Structures Via the Fast Four- T
ier Transform," *SIGCSE Bull.*, Vol. 18, No. 3, pp. 26–35, September
1986.

KEYS, R. G., "An Algorithm For Computing The Nth Roots of Unity in   T
Bit-Reversed Order," *IEEE Trans. Acoust., Speech and Signal Process.*,
Vol. ASSP-28, No. 6, pp. 762–63, December 1980.

KILPATRICK, D., J. V. TYBERG, AND W. W. PARMLEY, "Blood Velocity Mea-   A
surement By Fiber Optic Laser Doppler Anemometry," *IEEE Transactions on Biomedical Engineering*, Vol. BME-29, No. 2, pp. 142–45,
February 1982.

KIM, C. E., AND M. G. STRINTZIS, "High Speed Multidimensional Convo-   P
lution," *IEEE Trans. Pattern Anal. and Mach. Intell.*, Vol. PAMI-2, No.
3, pp. 269–73, May 1980.

KIN-CHUE, N. G., "On the Accuracy of Numerical Fourier Transforms,"   F
J. Comput. Phys., Vol. 16, No. 4, pp. 396–400, December 1974.

KING, R. E., "Digital Image Processing in Radioisotope Scanning," *IEEE* A
Trans. Bio-Med. Eng., BME-21, No. 5, pp. 414–16, September 1974.

KIRCHNER, P. D., W. J. SCHAFF, G. N. MARACAS, L. F. EASTMAN, T. I.   D
CHAPPELL, AND C. M. RANSOM, "The Analysis of Exponential and Non-
exponential Transients in Deep-Level Transient Spectroscopy," *J. Appl.
Phys.*, Vol. 52, No. 11, pp. 6462–70, November 1981.

KITAI, K., AND K. SIEMENS, "Discrete Fourier transform via Walsh transform," *IEEE Trans. Acoust. Speech Signal Process.*, Vol. ASSP-27, p. 288, 1979.

KNIGHT, W. R., AND R. KAISER, "A Simple Fixed-Point Error Bound For The Fast Fourier Transform," *IEEE Trans. Acoust., Speech and Signal Process.*, Vol. ASSP-27, No. 6, pp. 615–20, December 1979.

KOGA, H., T. AOKI, N. IKEDA, K. T. KIM, AND K. KIDO, "Recognition Of Vowels In Spoken Words Using Local Peaks In FFT Spectra," *Mem. Tohoku Inst. Technol. Ser. I* (Japan), No. 6, pp. 47–55, March 1986.

KOLBA, D. P., AND I. W. PARKS, "A Prime Factor FFT Algorithm Using High-speed Convolution," *IEEE Trans. Acoust. Speech Signal Process.*, Vol. ASSP-25, pp. 281–94, 1977.

KONVALINKA, I. S., "Iterative Nonparametric Spectrum Estimation," *IEEE Trans. Acoust., Speech and Signal Process.*, Vol. ASSP-32, No. 1, pp. 59–69, February 1984.

KRON, D. G., AND J. J. LAMBIOTTE, JR., "Computing the Fast Fourier Transform on a Vector Computer," *Math. Comput.*, Vol. 33, No. 147, pp. 977–92, July 1979.

KOSLOFF, D. D., AND E. BAYSAL, "Forward Modeling By a Fourier Method," *Geophysics*, Vol. 47, No. 10, pp. 1402–12, October 1982.

KRAUSE, L. O., "Digital Dechannelizer and Detoner for 14-Channel 8-Ary FSK," *Sixteenth Asilomar Conference on Circuits, Systems and Computers*, pp. 457–60, Pacific Grove, CA, November 1983.

KRYTER, R. C., "Application of the Fast Fourier Transform Algorithm to On-line Reactor Diagnosis," *IEEE Trans. Nuclear Sci.*, Vol. 16, pp. 210–17, February 1969.

KUMARESAN, R., AND P. K. GUPTA, "A Prime Factor FFT Algorithm With Real Valued Arithmetic," *Proceedings of the IEEE*, Vol. 73, No. 7, pp. 1241–43, July 1985.

LACKOFF, M. R., AND L. R. LEBLANC, "Frequency-Domain Seismic Deconvolution Filtering (For Swallow Water Profiles)," *J. Acoust. Soc. Am.*, Vol. 57, No. 1, pp. 151–59, January 1975.

LADD, S. A., "Software Solution To Vibration Analysis," *Mechanical Engineering*, Vol. 108, No. 2, pp. 73–75, February 1986.

LAM, P. T. C., S. W. LEE, AND R. ACOSTA, "Secondary Pattern Computation Of An Arbitrarily Shaped Main Reflector," *NASA Tech memo 87162*, p. 119, November 1985.

LAM, P. T., S. LEE, C. C. HUNG, AND R. ACOSTA, "Strategy For Reflector Pattern Calculation: Let The Computer Do The Work," *IEEE Trans. Antennas & Propag.*, Vol. AP-34, No. 4, pp. 592–95, April 1986.

LARSEN, T., AND G. DYRIK, "Fast Fourier Transforms Using a Microcomputer," *Electronics and Wireless World*, Vol. 91, No. 1595, pp. 80–82, September 1985.

LAWRENCE, N. B., AND J. D. MOORE, "Performance Results of An MTI/

FFT CFAR Radar Signal Processor," *Proc SOUTHEASTCON Reg 3 Conf '81*, pp. 21–25, Huntsville, Ala., April 1981.

LEBLANC, L. R., "Narrow-band Sampled Data Techniques for Detection Via the Underwater Acoustic Communications Channel," *IEEE Trans. Commun. Tech.*, Vol. 17, pp. 481–88, August 1969.   O

LEE, H., AND G. WADE, "Resolution For Images From Fresnel or Fraunhofer Diffraction Using The FFT," *IEEE Transactions on Sonics and Ultrasonics*, Vol. SU-29, No. 3, pp. 151–56, May 1982.   O

LEE, H. S., H. MORI, AND H. AISO, "Parallel Processing FFT For VLSI Implementation," *Transactions of the Institute of Electronics and Communication Engineers of Japan*, Section E, Vol. E68, No. 5, pp. 284–91, May 1985.   G

LEE, L.-S., Y-P HARN, AND Y-C CHEN, "A Simple Sample Value Scrambler Using FFT Algorithms for Secure Voice Communications," *NTC '80. IEEE 1980 National Telecommunications Conference*, 49.4/1–5, Houston, TX, November–December 1980.   Q

LEE, R. Q., AND R. ACOSTA, "Numerical Method For Approximating Antenna Surfaces Defined By Discrete Surface Points," *NASA Tech Memo 87125*, p. 19, 1985.   E

LEE, W. H., "Sampled Fourier Transform Hologram Generated By Computer," *Applied Optics*, Vol. 9, pp. 639–43, March 1970.   X

LEGOFF, H., A. RAMADANE, AND P. LEGOFF, "Modeling Of Simultaneous Heat And Mass Transfer In Gas-Liquid Absorption In A Laminar Falling Film," *International Journal of Heat and Mass Transfer*, Vol. 28, No. 11, pp. 2005–17, November 1985.   F

LEHTINEN, M. S., AND P. R. GOTHONI, "System For Measuring Tremor Intensity In Rats," *IEEE Trans Biomed Eng.*, Vol. BME-32, No. 8, pp. 549–53, August 1985.   A

LEWIS, R. L., "Numerical Computation of the Far-Field Excited by a Prescribed Aperture Distribution along Perpendicular Plane Cuts Using the FFT: Program Verification by Comparison with Exact Expressions for Uniformly-Excited Rectangular or Circular Apertures," *1980 International Symposium Digest. Antennas and Propagation*, pp. 287–8, Quebec, Canada, June 1980.   E

LI, K. K., G. ARJAVALINGAM, A. DIENES, AND J. R. WHINNERY, "Propagation of Picosecond Pulses on Microwave Striplines," *IEEE Transactions on Microwave Theory and Techniques*, Vol. MTT-30, No. 8, pp. 1270–73, August 1982.   E

LIANG, C. S., AND R. CLAY, "Computation of Short-pulse Response From Radar Targets—An Application of the Fast Fourier Transform Technique," *Proceedings of the IEEE*, Vol. 58, No. 1, pp. 169–71, January 1970.   H

LIM, J. S., AND N. A. MALIK, "A New Algorithm For Two-Dimensional Maximum Entropy Power Spectrum Estimation," *IEEE Trans. Acoust., Speech and Signal Process.*, Vol. 29, No. 3, Pt. 1, pp. 401–13, June 1981.   K

LIN, D. X., AND R. D. ADAMS, "Determination Of The Damping Properties   V
Of Structures By Transient Testing Using Zoom-FFT," *Journal of Physics E* (GB): Scientific Instruments, Vol. 18, No. 2, pp. 161–65, February
1985.

LINDERMAN, R. W., P. M. CHAU, W. H. KU, AND P. P. REUSENS, "CUSP: G
A 2-MU M CMOS Digital Signal Processor," *IEEE Journal of Solid-State
Circuits*, Vol. SC-20, No. 3, pp. 761–69, June 1985.

LIPSHITZ, S. P., T. C. SCOTT, AND J. VANDERKOOY, "Increasing The Audio   I
Measurement Capability Of FFT Analyzers By Microcomputer Postpro-
cessing," *Journal of the Audio Engineering Society*, Vol. 33, No. 9, pp.
626–48, September 1985.

LIU, B., AND A. PELED, "A New Hardware Realization of High-speed Fast   G
Fourier Transformers," *IEEE Trans. Acoust. Speech Signal Process.*,
Vol. ASSP-23, pp. 543–47, 1975.

LIU, B., AND F. MINTZER, "Calculation of Narrow-band Spectra By Direct   T
Decimation," *IEEE Trans. Acoust. Speech Signal Process.*, Vol. ASSP-
26, pp. 529–34, 1978.

LIU, B., AND T. KANEKO, "Roundoff Error in Fast Fourier Transforms (De-   T
cimation in Time)," *Proc. IEEE*, Vol. 63, No. 6, pp. 991–92, June 1975.

LIU, B., *Digital Filters and the Fast Fourier Transform*, Stroudsburg, Penn-   J
sylvania: Dowden, Hutchinson, and Ross, 1975.

LOO, C., "Calculations of Intermodulation Noise Due to Hard and Soft   Q
Limiting of Multiple Carriers," *IEEE International Conf. on Communications*, IEEE, New York, June 1974.

LOPRESTI, P. V., AND H. L. SURI, "Fast Algorithm for the Estimation of   K
Autocorrelation Functions," *IEEE Trans. Acoust. Speech Signal Process*, Vol. ASSP-22, pp. 449–53, 1974.

LORD, A., "Understanding Dual-Channel FFT Measurements," *Noise and*   V
Vib. Control Worldwide (GB), Vol. 12, No. 6, pp. 241–45, September
1981.

LUCHINI, P., "Two-Dimensional Numerical Integration Using a Square   F
Mesh," *Computer Physics Communications*, Vol. 31, No. 4, pp. 303–10,
March 1984.

MACTAGGART, I. R., AND M. A. JACK, "Radix-2 FFT Butterfly Processor   G
Using Distributed Arithmetic," *Electronics Letters*, Vol. 19, No. 2, pp.
43–44, January 20, 1983.

MAKHOUL, J., "A Fast Cosine Transform in One and Two Dimensions,"   T
IEEE Trans. Acoust. Speech Signal Process., Vol. ASSP-28, No. 1, pp.
27–34, February 1980.

MALIK, N. H., S. M. E. HAQUE, AND W. SHEPHERD, "Analysis And Per-   J
formance Of Three-Phase Phase-Controlled Thyristor AC Voltage Con-
trollers," *IEEE Trans Ind Electron*, Vol. IE-32, No. 3, pp. 192–99, August
1985.

MARBLE, A. E., J. W. ASHE, D. H. K. TSANG, D. BELLIVEAU, AND D. N.   A
SWINGLER, "Assessment of Algorithms Used to Compute The Fast Four-

ier Transform of Left Ventricular Pressure On A Microcomputer," *Medical & Biological Engineering & Computing*, Vol. 23, No. 2, pp. 190–94, March 1985.

MARKEL, J. D., "FFT Pruning," *IEEE Trans. Audio Electroacoust.* AU-19, No. 4, pp. 305–11, 1971.   T

MARTINSON, L. W., AND R. J. SMITH, "Digital Matched Filtering With Pipeline Floating Point Fast Fourier Transforms (FFT's)," *IEEE Trans. Acoust., Speech and Signal Process.*, Vol. ASSP-23, No. 2, pp. 222–34, April 1975.   H

MARUHN, J. A., T. A. WELTON, AND C. Y. WONG, "Remarks on the Numerical Solution of Poisson's Equation for Isolated Charge Distributions," *J. Comput. Phys.*, Vol. 20, No. 3, pp. 326–35, March 1976.   F

MATHEWS, J. D., "Incoherent Scatter Radar Probing Of The 60-100KM Atmosphere and Ionosphere," *IEEE TRANS on Geoscience and Remote Sensing*, Vol. GE-24, No. 5, pp. 765–76, September 1986.   H

MCCLARY, W. K., "Fast Seismic Inversion," *Geophysics*, Vol. 48, No. 10, pp. 1371–72, October 1983.   C

MCCLELLAN, J. H. "Multidimensional Spectral Estimation," *Proceedings of the IEEE*, Vol. 70, No. 9, pp. 1029–39, September 1982.   K

MCDOUGAL, J. R., L. C. SURRATT, AND L. F. STOOPS, "Computer Aided Design of Small Superdirective Antennas Using Fourier Integral and Fast Fourier Transform Techniques," *SWIEECO Record*, pp. 421–25, 1970.   E

MEHALIC, M. A., P. L. RUSTAN, AND G. P. ROUTE, "Effects of Architecture Implementation on DFT Algorithm Performance," *IEEE Trans. Acoust., Speech & Signal Process.*, Vol. ASSP-33, No. 3, pp. 684–93, June 1985. G

MEL'KANOVICK, A. F., L. M. KUSHKULEI, AND I. I. ARBIT, "Examination Of The Spectral And Time Characteristics Of Information Signals Of Ultrasonic Flaw Detectors," *Soviet Journal of Nondestructive Testing*, Vol. 21, No. 5, pp. 300–306, May 1985.   O

MENCARAGLIA, F., AND V. NATALE, "Method of Computing Fast Cosine Transforms," *Infrared Physics*, Vol. 24, No. 6, pp. 551–53, November 1984.   T

MENSA, D. L., "Wideband Radar Cross Section Diagnostic Measurements," *IEEE Transactions on Instrumentation and Measurement,* Vol. IM-33, No. 3, pp. 206–14, September 1984.   H

MEQUIO, C., R. H. COURSANT, AND P. PESQUE, "Simulation of the Acousto-Electric Response Of Piezoelectric Structures By Means of a Fast Fourier-Transform Algorithm," *Acta Electronica*, Vol. 25, No. 4, pp. 311–23, 1983.   O

MERHAUT, J., "Impulse Measurement of Horn-type Loudspeaker Drivers," *J. Audio Eng. Soc.*, Vol. 34, No. 4, pp. 245–54, April 1986.   O

MERMELSTEIN, PAUL, "Computer-generated Spectrogram Displays for On-line Speech Research," *IEEE Trans. Audio and Electroacoustics*, Vol. AU-19, pp. 44–47, March 1971.   R

MERSEREAU, R. M., AND T. C. SPEAKE, "A Unified Treatment of Cooley-   N

Tukey Algorithms For the Evaluation Of The Multidimensional DFT,"
IEEE Trans. Acoust. Speech and Signal Process., Vol. ASSP-29, No. 5,
pp. 1011–18, October 1981.

METZ, L. S., AND O. O. GANDHI, "Numerical Calculations of the Potential   F
Due to an Arbitrary Charge Density Using The Fast Fourier Transform,"
Proc. IEEE, Vol. 62, No. 7, pp. 1031–32, July 1974.

MEYER, J. U., AND M. INTAGLIETTA, "Measurement Of The Dynamics Of   A
Arteriolar Diameter," *Annals Of Biomedical Engineering*, Vol. 14, No.
2, pp. 109–17, April 1986.

MIAN, G. A., AND A. P. NAINER, "A Fast Procedure To Design Equiripple   P
Minimum-Phase Fir Filters," *IEEE Trans. Circuits and Syst.*, Vol. CAS-
29, No. 5, pp. 327–31, May 1982.

MILLER, T. J. E., AND P. J. LAWRENSON, "Penetration Of Transient Magnetic   J
Fields Through Conducting Cylindrical Structures With Particular Ref-
erence To Superconducting A. C. Machines," *Proc. Inst. Electr. Eng.*
(GB), Vol. 123, No. 5, pp. 437–43, May 1976.

MILUTINOVIC, V., J. A. B. FORTES, AND L. H. JAMIESON, "A Multimicro-   G
processor Architecture for Real-time Computation of a Class of DFT Algorithms," *IEEE Trans. Acoust., Speech & Signal Process.*, Vol. ASSP-
34, No. 5, pp. 1301–1309, October 1986.

MITCHELL, L. D., "Improved Methods For The Fast Fourier Transform   K
(FFT) Calculation Of The Frequency Response," *ASME Pap 81-DET-8
for MEET*, p. 3, September 1981.

MITTRA, R., AND W. L. KO, "New Techniques for Efficient Pattern Com-   E
putation of Aperture and Reflector Antennas," *Electron. Lett.* (GB), Vol.
16, No. 14, pp. 549–51, July 3, 1980.

MITTRA, R., W. L. KO, AND M. S. SHESHADRI, "A Novel Technique for the   E
Computation of Secondary Patterns of Reflector Antennas," *Second In-
ternational Conference on Antennas and Propagation*, 1/481-5, Part 1,
Heslington, York, England, April 1981.

MIYA, K., M. UESAKA, AND F. C. MOON, "Finite Element Analysis of Vi-   V
bration of Torodial Field Coils Coupled With LaPlace Transform," *Trans.
ASME J. Appl. Mech.*, Vol. 49, No. 3, pp. 594–600, September 1982.

MIYAKAWA, H., H. HARASHIMA, K. WATANABE, M. KISHI, AND K. OHYAMA,   J
"Detection of Multipath Reflected Waves By Two-Antenna Method,"
Journal of the Institute of Television Engineers of Japan, Vol. 36, No.
2, pp. 126–31, February 1982.

MIYAKAWA, Y., N. MIKI, AND N. NAGAI, "Adaptive Identification of a Time-   R
varying ARMA Speech Model," *IEEE Trans. on Acoustics, Speech and
Signal Processing*, Vol. ASSP-34, No. 3, pp. 423–33, June 1986.

MIZUSHINA, S., Y. XIANG, AND T. SUGIURA, "Large Waveguide Applicator   A
For Deep Regional Hyperthermia," *IEEE Transactions on Microwave
Theory and Techniques*, Vol. MTT-34, No. 5, pp. 644–48, May 1986.

MOHAN, M. V., AND V. V. RAO, "Error Analysis of Adpcm-FET," *IEEE*   W
Trans. Acoust. Speech and Signal Process., Vol. ASSP-27, No. 4, pp.
424–26, August 1979.

MOHAN, R., AND C. CHEN-SHOU, "Use of Fast Fourier Transforms in Cal- A culating Dose Distributions for Irregularly Shaped Fields for Three-dimensional Treatment Planning," *Med. Phys.*, Vol. 14, No. 1, pp. 70–77, January–February 1987.

MOHARIR, P. S., "Extending The Scope Of Golub's Method Beyond Com- F plex Multiplication," *IEEE Trans Comput*, Vol. C-34, No. 5, pp. 484–87, May 1985.

MOKRY, M., AND L. H. OHMAN, "Application Of The Fast Fourier Trans- V form To Two-Dimensional Wind Tunnel Wall Interference," *Journal of Aircraft*, Vol. 17, No. 6, pp. 402–8, June 1980.

MONTPETIT, M. J., M. NACHMAN, AND L. G. DURAND, "Application of Nu- A merical Methods for Feature Extraction From Phonopneumograms," *J. Clin. Eng.*, Vol. 10, No. 4, pp. 339–45, October–December 1985.

MOORER, J. A., "Algorithm Design for Real-time Audio Signal Processing," T *ICASSP 84, Proceedings of the IEEE International Conference on Acoustics, Speech and Signal Processing*, 12B.3/1–4, Vol. 1, San Diego, CA, March 1984.

MORGERA, S. D., AND R. SANKAR, "Digital Signal Processing for Precision L Wide-swath Bathymetry," *IEEE J. Oceanic Eng.*, Vol. OE-9, No. 2, pp. 73–84, April 1984.

MORITANI, T., H. TANAKA, T. YOSHIDA, C. ISHII, T. YOSHIDA, AND M. A SHINDO, "Relationship Between Myo-Electric Signals and Blood Lactate During Incremental Forearm Exercise," *American Journal of Physical Medicine*, Vol. 63, No. 3, pp. 122–32, June 1984.

MUIR, R. A., D. M. CHABRIES, AND R. W. CHRISTIANSEN, "Frequency Do- O main Compensation for Failed Elements in Linear Sonar Arrays," *ICASSP 86 Proceedings, IEEE-IECEJ-ASJ International Conference on Acoustics, Speech and Signal Processing*, Vol. 3, pp. 1865–68, Tokyo, Japan, April 1986.

MUNSON, D. C. JR., "Floating Point Error Bound in the Prime Factor FFT," W *ICASSP 80 Proceedings, Part 1. IEEE International Conference on Acoustics, Speech and Signal Processing*, pp. 69–72, Denver, CO, 1980.

MUNSON, D. C., JR., AND B. LIU, "Floating Point Roundoff Error In The W Prime Factor FFT," *IEEE Transactions on Acoustics, Speech and Signal Processing*, Vol. ASSP-29, No. 4, pp. 877–82, August 1981.

MURO, M., A. NAGATA, K. MURAKAMI, AND T. MORITANI, "Surface EMG A Power Spectral Analysis of Neuromuscular Disorders During Isometric and Isotonic Contractions," *American Journal of Physical Medicine*, Vol. 61, No. 5, pp. 244–54, October 1982.

NACCARATO, D. F., AND Y. T. CHIEN, "A Direct Two-Dimensional FFT N With Applications In Image Processing," *Proceedings of the 1979 IEEE Computer Society Conference on Pattern Recognition and Image Processing*, pp. 233–38, Chicago, IL, August 1979.

NAGAI, K., "Fourier Domain Reconstruction of Synthetic Focus Acoustic O Imaging System," *Proceedings of the IEEE*, Vol. 72, No. 6, pp. 748–49, June 1984.

NAGAI, K., "Measurement of Time Delay Using the Time Shift Property   L
of the Discrete Fourier Transform (DFT)," *IEEE Transactions on Acoustics, Speech and Signal Processing*, Vol. ASSP-34, No. 4, pp. 1006–8,
August 1986.

NAGAI, K., "Pruning the Decimation-in-time FFT Algorithm with Fre-   T
quency-shift," *IEEE Trans. on Acoustics, Speech and Signal Processing*,
Vol. 34, No. 4, pp. 1008–10, 1986.

NANDAGOPAL, D., J. MAZUMDAR, R. E. BOGNER, AND E. GOLDBLATT, "Spec-   A
tral Analysis of Second Heart Sound in Normal Children by Selective
Linear Prediction Coding," *Medical & Biological Engineering & Computing*, Vol. 22, No. 3, pp. 229–39, May 1984.

NARASIMHA, M. J., AND A. M. PETERSON, "Design of a 24-channel Trans-   Q
multiplier," *IEEE Trans. Acoust Speech Signal Process.*, Vol. ASSP-27,
pp. 752–62, 1979.

NARASIMHA, M. J., AND A. M. PETERSON, "On the Computation of the Dis-   T
crete Cosine Transform," *IEEE Trans. Commun.*, Vol. COM-26, pp.
934–36, 1978.

NARASIMHAN, M. S., AND M. KARTHIKEYAN, "Evaluation of Fourier Trans-   E
form Integrals Using FFT With Improved Accuracy and Its Applications," *IEEE Transactions on Antennas and Propagation*, Vol. AP-32,
No. 4, pp. 404–8, April 1984.

NAWAB, H., AND J. H. MCCLELLAN, "Bounds On the Minimum Number of   T
Data Transfers in WFTA and FFT Programs," *IEEE Trans. Speech and
Signal Process.*, Vol. ASSP-27, No. 4, pp. 394–8, August 1979.

NEBESNY, K. W., AND N. R. ARMSTRONG, "Deconvolution of Auger Electron   C
Spectra For Lineshape Analysis and Quantitation using a Fast Fourier
Transform Algorithm," *J. Electron Spectrosc. & Relat. Phenom.* (Neth-
erlands), Vol. 37, No. 4, pp. 355–73, March 1986.

NEILL, T. B. M., "Nonlinear Analysis of a Balanced Diode Modulator,"   M
Electronics Letters (GB), Vol. 6, No. 5, pp. 125–28, March 5, 1970.

NESBET, R. K., AND D. C. CLARY, "Fourier Transform Method For The   F
Classical Trajectory Problem," *J. Chem. Phys.*, Vol. 71, No. 3, pp. 1372–
79, August 1, 1979.

NESTER, W. H., "The Fast Fourier Transform and the Butler Matrix," *IEEE T
Trans. Antennas Propagations*, Vol. AP-16, p. 360, (correspondence),
May 1968.

NEWLAND, D. E., *Introduction to Random Vibrations and Spectral Anal-   K
ysis*, London England: Longman, 1975.

NGUYEN, D. T., K. SWANN, AND J. R. MCMILLAN, "Microprogrammed Dig-   R
ital Filter-Bank For Real-Time Spectral Analysis of Speech," *Journal of
Electrical and Electronics Engineering*, Australia, Vol. 4, No. 3, pp. 219–
26, September 1984.

NOBILE, A., AND V. ROBERTO, "Efficient Implementation of Multidimen-   T
sional Fast Fourier Transforms On A Cray X-MP," *Computer Physics
Communications*, Vol. 40, No. 2–3 pp. 189–201, June 1986.

NORTON, S. J., AND M. LINZER, "Reconstructing Spatially Incoherent RAN- O
DOM Sources In The Nearfield: Exact Inversion Formulas For Circular
and Spherical Arrays," *Journal of the Acoustical Society of America*,
Vol. 76, No. 6, pp. 1731–37, December 1984.

NUGENT, S. T., AND J. P. FINLEY, "Spectral Analysis of Periodic AND Normal A
Breathing In Infants," *IEEE Transactions on Biomedical Engineering*,
Vol. BME-30, No. 10, pp. 672–75, October 1983.

NUSSBAUMER, H. J., AND P. QUANDALLE, "Computation of Convolutions T
and Discrete Fourier Transforms," *IBM J. Res. Develop.*, Vol. 22, pp.
134–44, 1978.

NUSSBAUMER, H. J., AND P. QUANDALLE, "Fast Computation of Discrete T
Fourier Transforms Using Polynomial Transforms," *IEEE Trans. Acoust.
Speech Signal Process.*, Vol. ASSP-27, pp. 169–81, 1979.

NUSSBAUMER, H. J., "Fast Computation of Discrete Fourier Transforms," T
IBM Tech. Disclosure Bull., Vol. 22, No. 1, pp. 149–50, June 1979.

NUSSBAUMER, H. J., *Fast Fourier Transform and Convolution Algorithms*. J
Berlin, Germany and New York: Springer-Verlag, 1981.

NUSSBAUMER, H. J., "New Polynomial Transform Algorithms For Multi- P
dimensional DFTS and Convolutions," *IEEE Trans. Acoust., Speech and
Signal Process.*, Vol. ASSP-29, No. 1, pp. 74–84, February 1981.

NUSSBAUMER, H. J., "Polynomial Transform Implementation of Digital Fil- P
ter Banks," *IEEE Trans. Acoust., Speech and Signal Process.*, Vol.
ASSP-31, No. 3, pp. 616–22, June 1983.

NUTTALL, A. H., "Generation of Dolph-Chebyshev Weights Via A Fast P
Fourier Transform," *Proc. IEEE*, Vol. 62, No. 10, p. 1396, October 1974.

NUTTALL, A. H., "Some Windows with Very Good Sidelobe Behavior," K
IEEE Trans. Acoust. Speech Signal Process., Vol. ASSP-29, pp. 84–87,
1981.

NUTTALL, ALBERT, H., "Alternate Forms for Numerical Evaluation of Cu- F
mulative Probability Distributions Directly from Characteristic Functions," *Proceedings of the IEEE*, Vol. 58, pp. 1872–73, November 1970.

NWACHUKWU, E. O., "Address Generation In An Array Processor," *IEEE G
Transactions On Computers*, Vol. C-34, No. 2, pp. 170–73, February
1985.

O'LEARNY, G. C., "Nonrecursive Digital Filtering Using Cascade Fast P
Fourier Transformers," *IEEE Trans. Audio AND Electroacoustics*, Vol.
AU-18, pp. 177–83, June 1970.

OKINO, M., AND Y. HIGASHI, "Measurement of Seabed Topography By Mul- O
tibeam Sonar Using CFFT," *IEEE J. Oceanic Eng.*, Vol. OE-11, No. 4,
pp. 474–9, October, 1986.

OMER, W., "Faster Fourier Transforms," *Electron. and Wireless World* J
(GB), Vol. 92, No. 1605, pp. 57–58, July 1986.

ONO, T., "FFT Analyzers Draw More Data From a Single Input Signal," I
JEE, Journal of Electronic Engineering, Japan, Vol. 20, No. 200, pp. 86–
88, August 1983.

ONO, T., "Field-Oriented Faster FFT Analyzers Wanted," *JEE, Journal of Electronic Engineering*, Vol. 23, No. 230, pp. 52–55, February 1986.   I

ONOE, M., "A Method for Computing Large-scale Two-dimensional Transform Without Transposing Data Matrix," *Proc. IEEE*, Vol. 63, No. 1, pp. 196–97, 1975.   N

OPPENHEIM, A. V., AND C. J. WEINSTEIN, "A Bound on the Output of a Circular Convolution with Application to Digital Filtering," *IEEE Trans. Audio and Electroacoustics*, Vol. AU-17, pp. 120–24, June 1969.   P

OPPENHEIM, A. V., AND C. J. WEINSTEIN, "Effects of Finite Register Length in Digital Filtering and the Fast Fourier Transform," *Proc. IEEE*, Vol. 60, No. 8, pp. 957–76, August 1972.   W

OPPENHEIM, A. V., AND R. W. SCHAFER, *Digital Signal Processing*. Englewood Cliffs, New Jersey: Prentice Hall, 1975.   J

OPPENHEIM, A. V., "Speech Spectrogram Using the Fast Fourier Transform," *IEEE Spectrum*, Vol. 7, pp. 57–62, August 1970.   R

OPPENHEIM, A. V., D. JOHNSON, AND K. STEIGLITZ, "Computation of Spectra with Unequal Resolution Using the Fast Fourier Transform," *Proc. IEEE*, Vol. 59, pp. 299–301, February 1971.   K

ORFANIDIS, S. J., AND T. G. MARSHALL, "Two-Dimensional Transforms Of The Sample National Television Systems Committee (NTSC) Color Video Signal," *Optical Engineering*, Vol. 20, No. 3, pp. 417–20, June 1981.   X

OSAKA, T., AND Y. YATSUDA, "Study on Time-Dependence of the Oxygen Evolution Reaction On Nickel by FFT Impedance Measurement," *Electrochimica Acta*, Vol. 29, No. 5, pp. 677–81, May 1984.   J

OSINSKII, L. M., AND O. V. GLUSHKO, "Fast Fourier Transform Pipeline Schemes with Arbitrary Overlap of Input Data Arrays," *Radioelectron. & Commun. Syst.*, Vol. 29, No. 1, pp. 34–39, 1986.   G

OVSYANIK, V. P., L. S. KOVALENKO, AND A. N. VOVCHINSKII, "Investigation Of Temporal And Spectral Characteristics Of Some Signals Of Artifacts And Background Bioclectrical Activity," *Biomedical Engineering*, Vol. 19, No. 1, pp. 4–7, January–February 1985.   A

PAPOULIS, A., "A New Algorithm In Spectral Analysis AND BAND-Limited Extrapolation," *IEEE Trans. Circuits AND Syst.*, Vol. CAS-22, No. 9, pp. 735–42, September 1975.   K

PARAMANAND, S., AND P. RAMAKRISHNAN, "Powder Signature—A Strategy For Powder Characterization By Fourier Analysis," *International Journal of Powder Metallurgy and Powder Technology*, Vol. 21, No. 2, pp. 111–18, April 1985.   J

PARKER, R., AND S. A. T. STONEMAN, "On The Use Of Fast Fourier Transforms When High Frequency Resolution Is Required," *Journal of Sound and Vibration*, Vol. 104, No. 1, pp. 75–79, January 1986.   O

PARMENTER, W. W., AND R. G. CHRISTIANSEN, "Recovery of Modal Information From a Beam Undergoing Random Vibration," *Trans. ASME Ser. B.*, Vol. 96, No. 4, pp. 1307–13, November 1974.   V

PASTERKAMP, H., R. FENTON, F. LEAHY, AND V. CHERNICK, "Spectral Anal-   A
ysis of Breath Sounds In Normal Newborn Infants," *Medical Instrumentation*, Vol. 17, No. 5, pp. 355–57, September–October 1983.

PEARL, J., "On Coding and Filtering Stationary Signals By Discrete Fourier   Q
Transforms," *IEEE Trans. Informat. Theory*, Vol. IT-19, pp. 229–32,
1973.

PEARSON, A. E., AND F. C. LEE, "On The Identification Of Polynomial   F
Input-Output Differential Systems," *IEEE Trans. Autom. Control*, Vol.
AC-30, No. 8, pp. 778–82, August 1985.

PEARSON, A. E., AND F. C. LEE, "Parameter Identification Of Linear Dif-   F
ferential Systems Via Fourier Based Modulating Functions," *Control
Theory Adv. Technol.*, Vol. 1, No. 4, pp. 239–66, December 1985.

PEDERSEN, J. E., "Fast Dedicated Microprocessor For Real-Time Fre-   A
quency Analysis of Ultrasonic Blood-Velocity Measurements," *Medical
& Biological Engineering & Computing*, Vol. 20, No. 6, pp. 681–86, November 1982.

PEI, S-C, AND E-F HUANG, "In-Order, Partially In-Place Mixed Radix FFT   T
Algorithm," *IEEE Transactions on Acoustics, Speech, and Signal Processing*, Vol. ASSP-31, No. 5, pp. 1314–17, October 1983.

PEI, S-C, AND J-L WU, "Split-Radix Fast Hartley Transform," *Electronics   T
Letters*, Vol. 22, No. 1, pp. 26–27, January 1986.

PEI, SOO-CHANG, AND SHEN-TAN LU, "Design of Minimum-Phase Fir Digital   P
Filters by Differential Cepstrum," *IEEE Transactions on Circuits and
Systems*, Vol. CAS-33, No. 5, pp. 570–76, May 1986.

PELED, A., AND S. WINOGRAD, "TDM-FDM Conversion Requiring Reduced   Q
Computational Complexity," *IEEE Trans. Commun.*, Vol. COM-26, pp.
707–19, 1978.

PERERA, W. A., AND P. J. W. RAYNER, "Optimal Design of Discrete Coef-   T
ficient DFTs for Spectral Analysis, Extension To Multiplierless FFTs,"
IEEE Proceedings, Part G: Electronic Circuit and Systems, Vol. 133,
No. 1, pp. 8–18, February 1986.

PETERS, W. N., "Applications of the Two-Dimensional Fast Fourier Trans-   Z
form For Optical Systems Analysis," *Proceedings of the Society of
Photo-Optical Instrumentation Engineers*, Vol. 193, pp. 70–77, August
1979.

PICKERING, W. M., "On The Solution Of Poisson Equation On A Regular   F
Hexagonal Grid Using FFT Methods," *Journal of Computational Physics*, Vol. 64, No. 2, pp. 320–33, 1986.

PINKOWITZ, D., "Fast Fourier Transform Speeds Signal-to-noise Analysis   I
for A/D Converter," *Digital Design*, Vol. 16, No. 6, pp. 64–66, May 1986.

POLGE, R. J., AND B. K. BHAGAVAN, "Efficient Fast Fourier Transform   T
Programs for Arbitrary Factors With One Step Loop Unscrambling,"
IEEE Trans. Comput., Vol. C-25, No. 5, pp. 534–39, May 1976.

POLGE, R. J., AND E. R. MCKEE, "Extension of Radix-2 Fast Fourier Trans-   T
form (FFT) Program to Include a Prime Factor," *IEEE Trans. Acoust
Speech Signal Process.*, Vol. ASSP-22, No. 5, pp. 388–89, October 1974.

POLGE, R. J., B. K. BHAGAVAN, AND J. M. CARSWELL, "Fast Computational   T
Algorithms For BIT Reversal," *IEEE Trans. Comput.*, Vol. C-23, No.
1, pp. 1–9, January 1974.

POMERLEAU, A., "Real-Data FFT Algorithm For Image Processing Appli-   N
cations," *Canadian Electrical Engineering Journal*, Vol. 8, No. 2, pp.
65–72, April 1983.

POMERLEAU, A., H. L. BUIJS, AND M. FOURNIER, "A Two-pass Fixed Point   W
Fast Fourier Transform Error Analysis," *IEEE Trans. Acoust Speech
Signal Process.*, Vol. ASSP-25, pp. 582–85, 1977.

POMERLEAU, A., M. FOURNIER, AND H. L. BUIJS, "On the Design of a Real   G
Time Modular FFT Processor," *IEEE Trans. Circuits Syst.*, Vol. CAS-
23, pp. 630–33, 1976.

PORTNOFF, M. R., "Implementation of the Digital Phase Vocoder Using The   R
Fast Fourier Transform," *IEEE Trans. Acoust., Speech* AND Signal Process., Vol. ASSP-24, No. 3, pp. 243–48, June 1976.

PORTNOFF, M. R., "Time-frequency Representation of Digital Signals and   K
Systems Based on Short-time Fourier Analysis," *IEEE Trans. Acoust.
Speech Signal Process.*, Vol. ASSP-28, No. 1, pp. 55–69, February 1980.

PRABHU, A. V., V. K. AATRE, T. SOUMINI, AND S. A. KARIPEL, "Frequency   T
Zooming Techniques For High Resolution Spectrum Analysis," *Defense
Science Journal* (India), Vol. 35, No. 3, pp. 281–85, July 1985.

PRAKASH, S., AND V. V. RAO, "A New Radix-6 FFT Algorithm," *IEEE*   T
Trans. Acoust. Speech and Signal Process., Vol. ASSP-29, No. 4, pp.
939–41, August 1981.

PRAKASH, S., AND V. V. RAO, "Fixed-Point Error Analysis of Radix-4 FFT,"   W
Signal Processing, Vol. 3, No. 2, pp. 123–33, April 1981.

PRAKASH, S., AND V. V. RAO, "Vector Radix FFT Error Analysis," *IEEE*   W
Transactions on Acoustics, Speech, and Signal Processing, Vol. ASSP-
30, No. 5, pp. 808–11, October 1982.

PRASAD, K. P., AND P. SATYANARAYANA, "Fast Interpolation Algorithm   T
Using FFT," *Electronics Letters* (GB), Vol. 22, No. 4, pp. 185–87, February 1986.

PRESCOTT, J., AND R. L. JENKINS, "An Improved Fast Fourier Transform,"   T
IEEE Trans. Acoust Speech Signal Process., Vol. ASSP-22, No. 3, pp.
226–27, June 1974.

PREUSS, R. D. "Very Fast Computation of the Radix-2 Discrete Fourier   T
Transform," *IEEE Transactions on Acoustics, Speech, and Signal Processing*, Vol. ASSP-30, No. 4, pp. 595–607, August 1982.

PRICE, E. V., "The Fast Fourier Transform on a Digital Image Processor-   X
Implementation and Applications," *Architecture and Algorithms for Digital Image Processing*, San Diego, CA, August 1983.

PRIDHAM, R. G., AND R. E. KOWALCZK, "Use of FFT Subroutine in Digital   P
Filter Design Program," *Proceedings of the IEEE* (Letters), Vol. 57, p.
106, January 1969.

PRIESTLEY, B., "Fast Fourier-Transform in Basic," *Electronic Engineering*, T
Vol. 58, No. 711, pp. 33–34, 1986.

PROKOP'EV, A. I., "Use of Fast Fourier Transform in Calculation of Potential J
and Field Distributions in Gallium Arsenide Charge-coupled Devices in
the Absence of Mobile Charge," *Radioelectron. & Commun. Syst.*, Vol.
28, No. 6, pp. 62–66, 1985.

QUIRK, M., AND B. LIU, "On Narrow-BAND Spectrum Calculation By Direct K
Decimation," *ICASSP 81. Proceedings of the 1981 IEEE International
Conference on Acoustics, Speech and Signal Processing*, Vol. 1, pp. 85–
88, Atlanta, GA, March 1981.

RAABE, H. P., "Fast Beamforming with Circular Receiving Arrays," *IBM* B
J. Res. and Dev., Vol. 20, No. 4, pp. 398–408, July 1976.

RABINER, L. R., AND B. GOLD, *Theory and Application of Digital Signal* J
Processing. Englewood Cliffs, N.J.: Prentice Hall, 1975.

RABINER, L. R., R. W. SCHAFER, AND C. M. RADER, "The Chirp-Z Transform T
Algorithm," *IEEE Trans. Audio. Electroacoust.*, Vol. AU-17, No. 2, pp.
86–92, June 1969.

RADCLIFFE, C. J., AND C. D. MOTE, JR., "Identification AND Control of Ro- V
tating Disk Vibration," *Trans. ASME J. Dyn. Syst. Meas. and Control*,
Vol. 105, No. 1, pp. 39–45, March 1983.

RADER, C. M., AND N. M. BRENNER, "A New Principle For Fast Fourier T
Transformation," *IEEE Trans. Acoust., Speech and Signal Process.*,
Vol. ASSP-24, No. 3, pp. 268–70, June 1976.

RADER, C. M., "An Improved Algorithm for High Speed Autocorrelation T
with Applications to Spectral Estimation," *IEEE Trans. Audio and Electroacoustics*, Vol. 18, pp. 439–41, December 1970.

RADER, C. M., "Discrete Fourier Transforms When the Number of Data T
Samples is Prime," *Proc. IEEE* (Letters), Vol. 56, pp. 1107–1108, June
1968.

RAJALA, S. A., A. N. RIDDLE, AND W. E. SNYDER, "Application of the One- X
Dimensional Fourier Transform for Tracking Moving Objects In Noisy
Environments," *Comput. Vision, Graphics and Image Process.*, Vol. 21,
No. 2, pp. 280–93, February 1983.

RAJAONA, R. D., AND P. SULMONT, "A Method of Spectral Analysis Applied K
to Periodic and Pseudoperiodic Signals," *J. Comput. Phys.*, Vol. 61, No.
1, pp. 186–93, October 1985.

RAMIREZ, R. W., "Fast Fourier Transform Makes Correlation Simpler," P
Electronics, Vol. 48, No. 13, pp. 98–103, June 1975.

RAMIREZ, R. W., "The FFT: Fundamentals and Concepts," Englewood J
Cliffs, NJ: Prentice Hall, 1985.

RAO, P. N., AND G. BOOPATHY, "Analysis and Classification of Commu- Q
nication Signals," *Defense Science Journal* (India), Vol. 35, No. 3, pp.
367–74, July 1985.

READ, R., AND J. MEEK, "Digital Filters with Poles Via FFT," *IEEE Trans.* P
Audio and Electroacoustics, Vol. AU-19, pp. 322–23, December 1971.

REDDY, B. R. S., AND I. S. N. MURTHY, "ECG Data Compression Using A
Fourier Descriptor," *IEEE Trans. Biomed. Eng.*, Vol. BME-33, No. 4,
pp. 428–34, April 1986.

REDDY, D. C., K. S. RAO, AND K. J. R. MURTY, "Waveform Analysis For A
the Detection Of Airways Obstruction In Man," *Medical & Biological
Engineering & Computing*, Vol. 22, No. 6, pp. 481–85, November 1984.

REDDY, N. S., AND M. N. S. SWAMY, "Resolution of Range and Doppler H
Ambiguities in Medium PRF Radars in Multiple-target Environment,"
*ICASSP 84, Proceedings of the IEEE International Conference on Acous-
tics, Speech and Signal Processing*, Vol. 3, pp. 47.6/1–4, March 1984.

REDDY, R. S., I. S. N. MURTHY, AND P. C. CHATTERJEE, "Rhythm Analysis A
Using Vectorcadiograms," *IEEE Transactions on Biomedical Engineering*, Vol. BME-32, No. 2, pp. 97–104, February 1985.

REDDY, V. U., AND M. SUNDARAMURTHY, "Effect of Correlation Between T
Truncation Errors on Fixed-Point Fast Fourier Transform Error Analysis," *IEEE Transactions on Circuits and Systems*, Vol. CAS-27, No. 8,
pp. 712–16, August 1980.

REDINBO, G. R., AND K. K. RAO, "Expediting Factor-type Fast Finite Field T
Transform Algorithms," *IEEE Trans. Inf. Theory*, Vol. IT-32, No. 2, pp.
186–94, March 1986.

REED, F. A., AND P. L. FEINTUCH, "A Comparison of LMS Adaptive Can- Q
cellers Implemented In The Frequency Domain And The Time Domain,"
IEEE Trans. Circuits and Syst., Vol. CAS-28, No. 6, pp. 610–15, June
1981.

REED, I. S., T. K. TRUONG, B. BENJAUTHRIT, AND C. WU, "A Fast Algorithm T
For Computing A Complex-Number Theoretic Transform For Long Sequences," *IEEE Trans. Acoust., Speech and Signal Process.*, Vol. ASSP-
29, No. 1, pp. 122–24, February 1981.

RENDERS, H., J. SCHOUKENS, AND G. VILAIN, "High-Accuracy Spectrum K
Analysis of Sampled Discrete Frequency Signals By Analytical Leakage
Compensation," *IEEE Transactions on Instrumentation and Measure-
ment*, Vol. IM-33, No. 4, pp. 287–92, December 1984.

RENNIE, L. J., "The Tap III Beamforming System," *IEEE; San Diego Sec- B
tion of the Marine Technol. Soc. Oceans '79*, pp. 6–13, San Diego, CA,
September 1979.

REQUICHA, A. A. G., "Direct Computation of Distribution Functions From F
Characteristic Functions Using the Fast Fourier Transform," *Proceedings of the IEEE*, Vol. 58, No. 7, pp. 1154–55, July 1970.

RESCH, F. J., AND R. ABEL, "Spectral Analysis Using Fourier Transform K
Techniques," *Int. J. Numer. Methods Eng.* (GB), Vol. 9, No. 4, pp. 869–
902, 1975.

RIAD, S. M., AND R. B. STAFFORD, "Impulse Response Evaluation Using C
Frequency Domain Optimal Compensation Deconvolution," *23rd Midwest Symposium on Circuits AND Systems*, pp. 521–5, 1980.

RIBLET, G. P., "Use Of The FFT To Speed Analysis Of Planar Symmetrical E

3- AND 5-Ports By The Integral Equation Method," *IEEE Transactions on Microwave Theory and Techniques*, Vol. MTT-33, No. 10, pp. 1073–75, October 1985.

RICHARDS, T. L., AND K. ATTENBOROUGH, "Accurate FFT Based Hankel   O
Transforms for Predictions of Outdoor Sound Propagation," *J. Sound and Vib.* (GB), Vol, 109, No. 1, pp. 157–67, August 1986.

RIEDEL, N. K., D. A, McANINCH, C. FISHER, AND N. B. GOLDSTEIN, "Signal   G
Processing Implementation For An IDM PC-Based Workstation," *IEEE Micro*, Vol. 5, No. 5, pp. 52–67, October 1985.

RITTGERS, S. E., W. W. PUTNEY, AND R. W. BARNES, "Real-Time Spectrum   A
Analysis and Display of Directional Doppler Ultrasound Blood Velocity Signals," *IEEE Transactions on Biomedical Engineering*, Vol. BME-27, No. 12, pp. 723–28, December 1980.

ROBERTS, K. B., P. D. LAWRENCE, AND A. EISEN, "Dispersion of the So-   A
matosensory Evoked Potential (SEP) in Multiple Sclerosis," *IEEE Transactions on Biomedical Engineering*, Vol. BME-30, No. 6, pp. 360–64, June 1983.

ROBINSON, E. A., "Historical Perspective of Spectrum Estimation," *Pro-   E
ceedings of the IEEE*, Vol. 70, No. 9, pp. 885–907, September 1982.

ROBINSON, E. A., T. S. DURRANI, AND L. G. PEARDON, *Geophysical Signal   U
Processing*. Englewood Cliffs, NJ: Prentice Hall, 1986.

RODDY, D., "A Method of Using Simpson's Rule in the DFT," *IEEE Trans.   F
Acoust. Speech and Signal Process.*, Vol. ASSP-29, No. 4, pp. 936–37, August 1981.

ROSTE, T., O. HAABERG, AND T. A. RAMSTAD, "A Radix-4 FFT Processor   S
For Application in a 60-Channel Transmultiplexer Using TTL Technology," *IEEE Trans. Acoust., Speech and Signal Process.*, Vol. ASSP-27, No. 6, Pt. 2, pp. 746–51, December 1979.

ROTHWEILLER, J. H., "Implementation of the In-Order Prime Factor Trans-   T
form For Variable Sizes," *IEEE Transactions on Acoustics, Speech, and Signal Processing*, Vol. ASSP-30, No. 1, pp. 105–7, February 1982.

ROWLANDS, R. O., "The Odd Discrete Fourier Transform," *Proc. IEEE Int.   T
Conf. Acoust. Speech Signal Process.*, Philadelphia, Pennsylvania, pp. 130–33, 1976.

RUDNICK, P., "Digital Beamforming in the Frequency Domain," *Journal of   B
the Acoustical Society of America*, Vol. 46, No. 5, pp. 1089–90, November 1969.

RUSSEL, R. F., D. P. GAINES, AND F. W. SEDENQUIST, "Improved Radar   H
Range Resolution Using Frequency Agility and the Fast Fourier Transform," *Conference Proceedings of IEEE Southeastcon 84*, pp. 261–65, April 1984.

RUSSEL, R. F., F. W. SEDENQUIST, AND D. P. GAINES, "Frequency Agile/   H
Polarimetric Radar-simulation and Testing," *Proceedings of the 1984 IEEE National Radar Conference*, pp. 58–62, Atlanta, GA, March 1984.

RUSSELL, R. F., AND F. W. SEDENQUIST, "Digital Simulation of Polarimetric Z
Radars," *Simulation* (USA), Vol. 43, No. 5, pp. 242–46, November 1984.

SABLIK, M. J., R. E. BEISSNER, AND A. CHOY, "An Alternative Numerical E
Approach for Computing Eddy Currents: Case of the Double-layered
Plate," *IEEE Trans. Magn.*, Vol. MAG-20, No. 3, pp. 500–506, May
1984.

SADJADI, F. A., J. J. HWANG, E. L. HALL, AND M. J. ROBERTS, "Measure- L
ment of Two Phase Flow Velocities Using Image Correlation," *IEEE
Proceedings of the 5th International Conference on Pattern Recognition*,
pp. 386–92, Miami Beach, FL, December 1980.

SAID, S. M., AND K. R. DIMOND, "Improved Implementation of FFT Al- G
gorithm on a High-Performance Processor," *Electronics Letters*, Vol. 20,
No. 8, pp. 347–49, April 12, 1984.

SAKAMOTO, T., T. SUGIMOTO, AND M. NAKAMURA, "Clutter Rejection Signal H
Processor Using High Speed FFT For Radar System," *Noise and Clutter
Rejection in Radars and Imaging Sensors. Proceedings of the 1984 In-
ternational Symposium*, pp. 518–21, Tokyo, Japan, October, 1984.

SAKURAI, K., K. KOGA, AND T. MURATANI, "Speech Scrambler Using the R
Fast Fourier Transform Technique," *IEEE Journal on Selected Areas in
Communications*, Vol. SAC-2, No. 3, pp. 434–42, May 1984.

SALA, K. L., R. W. YIP, R. LESAGE, "Application of Fast Fourier Transform D
and Convolution Techniques To Picosecond Continuum Spectroscopy,"
Applied Spectroscopy, Vol. 37, No. 3, pp. 273–79, May–June 1983.

SANKARAN, R., K. A. MURALEEDHARAN, AND K. P. P. PILLAI, "Transient Y
Performance of Linear Induction Machines Following Reconnection of
Supply," *Proc. Inst. Electr. Eng.* (GB), Vol. 126, No. 10, pp. 979–83,
October 1979.

SARKAR, T. K., E. ARVAS, AND S. M. RAO, "Application Of FFT And The E
Conjugate Gradient Method For The Solution Of Electromagnetic Radiation From Electrically Large And Small Conducting Bodies," *IEEE
Transactions on Antennas and Propagation*, Vol. AP-34, No. 5, pp. 635–
40, May 1986.

SAZONOV, N. A., "Quantization Noise for Digital Signal Processing By Har- H
monic Analysis in an Aperture-synthesis Radar System," *Telecommun.
and Radio Eng. Part 2*, Vol. 40, No. 4, pp. 59–62, April 1985.

SCHAFER, R. W., AND L. R. RABINER, "Design and Simulation of a Speech R
Analysis-synthesis System Based on Short-time Fourier Analysis," *IEEE
Trans. Audio Electroacoust.*, Vol. AU-21, pp. 165–74, 1973.

SCHAFFER, J. P., E. J. SHAUGHNESSY, AND P. L. JONES, "The Deconvolution C
Of Doppler-Broadened Positron Annihilation Measurements Using Fast
Fourier Transforms and Power Spectral Analysis," *Nuclear Instruments
& Methods In Physics Research, Section B*: (Netherlands) Beam Inter-
actions with Materials and Atoms, Vol. 233, No. 1, pp. 75–79, Sep-
tember–October 1984.

SCHAFFER, J. P., AND P. L. JONES, "An Evaluation of the Fast Fourier-   C
Transform Power Spectrum Deconvolution Method As Applied To Doppler-Broadened Positron-Annihilation Spectra of High-Purity Alumi-
num," *Journal of Physics F-Metal Physics*, (GB), Vol. 16, No. 11, pp.
1885–96, November 1986.

SCHEUERMAN, H., AND II. GOCKLER, "A Comprehensive Survey of Digital   Q
Transmultiplexing Methods," *Proc. IEEE*, Vol. 69, No. 69, pp. 1419–50,
November 1981.

SCHLEHER, D. C. "Numerical Evaluation of Logarithmic Receiver Thresh-   Y
olds," *Electron. Lett.* (GB), Vol. 16, No. 23, pp. 875–76, November 6,
1980.

SCHREIER, P. G., "PC-based Spectrum Analysis Packages Reach Toward   G
DSP-Chip Performance," *Electron. Test*, Vol. 8, No. 8, pp. 39–44, August 1985.

SCHUTTE, J., "New Fast Fourier Transform Algorithm For Linear System   D
Analysis Applied In Molecular Beam Relaxation Spectroscopy," *Rev.
Sci. Instrum.*, Vol. 52, No. 3, pp. 400–404, March 1981.

SEBERN, M. J., J. D. HORGAN, R. C. MEADE, C. M. KRONENWETTER, P. P.   X
RUETZ, AND EN-LIN YEH, "Minicomputer Enhancement of Scintillation
Camera Images Using Fast Fourier Transform Techniques," *J. Nucl.
Med.*, Vol. 17, No. 7, pp. 647–52, July 1976.

SERDA, L. A., "Fast Two-Dimensional Discrete Fourier Transform,"   N
Trans. in: Radioelectron. and Commun. Syst., Vol. 26, No. 7, pp. 15–
19, 1983.

SEVERUD, L. K., M. J. ANDERSON, AND D. A. BARTA, "Seismic Damping   V
Factors Of Small Bore Piping As Influenced By Insulation and Support
Elements," *Journal of Pressure Vessel Technology, Transactions of the
ASME*, Vol. 107, No. 2, pp. 142–47, May 1985.

SHAARAWI, A. M., AND S. M. RIAD, "Computing the Complete FFT of a   F
Step-like Waveform," *IEEE Trans. Instrum. & Meas.*, Vol. IM-35, No.
1, pp. 91–92, March 1986.

SHANKARA REDDY, B. R., AND I. S. N. MURTHY, "ECG Data Compression   A
Using Fourier Descriptors," *IEEE Transactions on Biomedical Engineering*, Vol. BME-33, No. 4, pp. 428–34, April 1986.

SHAYG, S. A., AND Y. H. HAN, "Rayleigh Spectrometer," *Rev. Sci. In-   D
strum.*, Vol. 45, No. 2, pp. 280–85, February 1974.

SHCHERBAKOV, M. A., "Identification of Discrete Nonlinear Systems With   J
PseudoRANDom Inputs," *Soviet Automatic Control*, Vol. 16, No. 4, pp.
16–26, July–August 1983.

SHEROV, E. M., AND V. A. MAMONTOV, "Interpolation with the Aid of Fast   D
Fourier Transforms in Fourier Spectroscopy," *Trans. J. Appl. Spectro.*,
Vol. 24, No. 6, June 1976.

SHIOJIRI, E., AND Y. FUJII, "Transmission Capability of An Optical Fiber   E
Communication System Using Index Nonlinearity," *Applied Optics*, Vol.
24, No. 3, pp. 358–60, February 1985.

SHIRLEY, R. S., "Application of a Modified Fast Fourier Transform to Cal-   J
culate Human Operator Describing Functions," *IEEE Trans. Man-Ma-
chine Systems*, Vol. MMS-10, pp. 140–44, December 1969.

SHUNI C., AND C. S. BURRUS, "A Prime Factor FFT Algorithm Using Dis-   T
tributed Arithmetic," *IEEE Trans. Acoust. Speech and Signal Process.*,
Vol. ASSP-30, No. 2, pp. 217–27, April 1982.

SILBERBERG, M., "Improving the Efficiency of Laplace-transform Inversion   M
for Network Analysis," *Electronics Letters* (GB), Vol. 6, No. 4, pp. 105–
6, February 19, 1970.

SILVERMAN, H. F., "A High-quality Digital Filterbank for Speech Recog-   R
nition Which Runs in Real Time on a Standard Microprocessor," *IEEE
Trans. Acoust., Speech & Signal Process.*, Vol. ASSP-34, No. 5, pp.
1064–73, October 1986.

SILVERMAN, H. F., A. E. PEARSON, "On Deconvolution using the Discrete   C
Fourier Transform," *IEEE Transactions on Audio and Electroacoustics*,
Vol. AU-21, No: 2, pp. 112–18, April 1973.

SINGHAL, K., "Interpolation Using the Fast Fourier Transform," *Proceed-   F
ings of the IEEE*, Vol. 60, No. 12, p. 1558, December 1972.

SINGLETON, R. C., AND T. C. POULTER, "Spectral Analysis of the Call of   K
the Male Killer Whale," *IEEE Trans. Audio and Electroacoustics*, Vol.
AU-15, No. 2, pp. 104–13, June 1967.

SINGLETON, R. C., "A Method for Computing the Fast Fourier Transform   T
with Auxiliary Memory and Limited High-speed Storage," *IEEE Trans.
Audio Electroacoust.*, Vol. AU-15, pp. 91–98, June 1967.

SINGLETON, R. C., "Algol Procedures for the Fast Fourier Transform,"   T
Commun. ACM, Vol. 11, No. 11, pp. 773–76, Algorithm 338, November
1968.

SINGLETON, R. C., "An Algol Procedure for the Fast Fourier Transform   T
with Arbitrary Factors," *Commun. ACM*, Vol. 11, pp. 776–79, Algorithm
339, November 1968.

SINGLETON, R. C., "An Algorithm for Computing the Mixed Radix Fast   T
Fourier Transform," *IEEE Trans. Audio Electroacoust.*, Vol. AU-17,
No. 2, pp. 93–103, June 1969.

SINGLETON, R. C., "On Computing the Fast Fourier Transform," *Commun.*   T
ACM. Vol. 10, pp. 647–54, October 1967.

SINHA, B., J. DATTAGUPTA, AND A. SEN, "Improvement In The Speed Of   T
FFT Processors Using Segmented Memory And Parallel Arithmetic
Units," *Signal Processing*, Vol. 8, No. 2, pp. 267–74, April 1985.

SKARJUNE, R., "Deviation and Implementation of an Efficient Fast Fourier   T
Transform Algorithm (EFFT)," *Comput. & Chem.* (GB), Vol. 10, No.
4, pp. 241–51, 1986.

SKINNER, D. P., "Pruning The Decimation In-Time FFT Algorithm," *IEEE*   T
Trans. Acoust., Speech and Signal Process., Vol. ASSP-24, No. 2, pp.
193–94, April 1976.

SKOLLERMO, G., "A Fourier Method for the Numerical Solution of Poisson's   F
Equation," *Math. Comput.*, Vol. 29, No. 131, pp. 697–711, July 1975.

SKORMIN, V., "Frequency Approach to Mathematical Modeling of a Nu-   V
clear Power Plant Piping System," *Journal of Vibration, Acoustics,
Stress, and Reliability in Design*, Vol. 107, No. 1, pp. 106–11, January
1985.

SLOANE, E. A., "Comparison of Linearly and Quadratically Modified Spec-   K
tral Estimates of Gaussian Signals," *IEEE Trans. Audio and Electroacoustics*, Vol. AU-17, pp. 133–37, June 1969.

SLOATE, H., "Matrix Representations for Sorting and the Fast Fourier   T
Transform," *IEEE Trans. Circuits and Syst.*, Vol. CAS-21, No. 1, pp.
109–16, January 1974.

SMITH, D. E., "The Acquisition of Electrochemical Response Spectra By   D
On-Line Fast Fourier Transform Data Processing In Electrochemistry,"
Anal. Chem., Vol. 48, No. 2, pp. 221A–40, February 1976.

SMITH, D. E., "The Enhancement of Electroanalytical Data By On-Line   J
Fast Fourier Transform Data Processing In Electrochemistry," *Anal.
Chem.*, Vol. 48, No. 6, pp. 517–26, May 1976.

SMITH, J. R., R. MCLEAN, AND J. R. ROBBIE, "Assessment of Hydroturbine   Y
Models for Power-Systems Studies," *IEEE Proc. C* (GB), Vol. 130, No.
1, pp. 1–7, January 1983.

SMITH, W. W., "Zipping Through FFTs, Software Tools Turn PCs into   G
Signal Processors," *Electron. Des.*, Vol. 33, No. 7, pp. 175–81, March
1985.

SODERSTRAND, M. A., T. G. JOHNSON, AND G. A. CLARK, "Hardware Re-   P
alizations of Frequency-sampling Adaptive Filters," *1986 IEEE International Symposium on Circuits and Systems*, Vol. 3, pp. 900–903, May
1986.

SOOHOO, J., AND G. E. MEVERS, "Cavity Mode Analysis Using the Fourier   E
Transform Method," *Proc. IEEE*, Vol. 62, No. 12, pp. 1721–23, December 1974.

SORENSEN, H. V., M. T. HEIDEMAN, AND C. S. BURRUS, "On Computing   T
the Split-Radix FFT," *IEEE Transactions on Acoustics, Speech, and
Signal Processing*, Vol. ASSP-34, No. 1, pp. 152–56, February 1986.

SOUMEKH, M., "Image Reconstruction Techniques in Tomographic Imaging   X
Systems," *IEEE Trans. Acoust., Speech & Signal Process.*, Vol. ASSP-
34, No. 4, pp. 952–62, August 1986.

SPEAKE, T. C., AND R. M. MERSEREAU, "Evaluation Of Two-Dimensional   N
Discrete Fourier Transforms Via Generalized FFT Algorithms," *Pro-
ceedings—ICASSP, IEEE International Conference on Acoustics,
Speech and Signal Processing*, Vol. 3, pp. 1006–9, Atlanta, Georgia,
March 1981.

Special Issue on Fast Fourier Transforms and Its Application to Digital   K
Filtering and Spectral Analysis, *IEEE Trans. Audio Electroacoust.*, Vol.
AU-15, 1967.

Special Issue on Fast Fourier Transforms, *IEEE Trans. Audio Electroacoust.*, Vol. AU-17, June 1969.

Special Issue on TDM-FDM Conversion, *IEEE Transactions on Communications*, Vol. Com-30, No. 7, pp. 489–741, May 1978.

Special Issue on Transmultiplexers, *IEEE Transactions on Communications*, Vol. Com-30, No. 7, pp. 1457–1656, July 1982.

Special Issue on Two-Dimensional Digital Signal Processing, *IEEE Trans. Comput.*, Vol. C-21, 1972.

SPYRAKOS, C. C., AND D. E. BESKOS, "Dynamic Response of Frameworks By Fast Fourier Transform," *Computers and Structures*, Vol. 15, No. 5, pp. 495–505, 1982.

SREENIVAS, T. V., AND P. V. S. RAO, "High-Resolution Narrow-BAND Spectra By FFT Pruning," *IEEE Transactions on Acoustics, Speech, and Signal Processing*, Vol. ASSP-28, No. 2, pp. 254–57, April 1980.

STANGHAN, C. J., AND B. M. MACDONALD, "Electrical Characterization of Packages for High-Speed Integrated Circuits," *IEEE Transactions on Components, Hybrids and Manufacturing Technology*, Vol. CHMT-8, No. 4, pp. 468–73, December 1985.

STANLEY, W. D., AND S. J. PETERSON, "Fast Fourier Transforms on Your Home Computer," *Byte*, Vol. 3, No. 12, pp. 14, 16, 18, 20, 22, 24–25, December 1978.

STASINSKI, R., "Comments on 'Bounds on the Minimum Number of Data Transfer in WFTA AND FFT Programs'," *IEEE Trans. Acoust., Speech and Signal Process.*, Vol. ASSP-32, No. 6, pp. 1255–57, December 1984.

STEARNS, S. D., "Tests of Coherence Unbiasing Methods," *IEEE Trans. Acoust. Speech and Signal Process.*, Vol. ASSP-29, No. 2, pp. 321–23, April 1981.

STEFFEN, P. L., "Exact Calculation of the Impulse Response of Quarter-Plane Filters in a Finite Region By Means of the DFT," *IEEE Trans. Acoust., Speech and Signal Process.*, Vol. ASSP-30, No. 4, pp. 608–12, August 1982.

STEPHANISHEN, P. R., AND H. W. CHEN, "Nearfield Pressures and Surface Intensity for Cylindrical Vibrators," *J. Acoust. Soc. Am.*, Vol. 76, No. 3, pp. 942–48, September 1984.

STEPHANISHEN, P. R., AND K. C. BENJAMIN, "Forward and Backward Projection of Acoustic Fields Using FFT Methods," *Journal of the Acoustical Society of America*, Vol. 71, No. 4, pp. 803–12, April 1982.

STIGALL, P. D., R. E. ZIEMER, AND L. HUDEC, "Performance Study of 16-Bit Microcomputer-Implemented FFT Algorithms," *IEEE Micro*, Vol. 2, No. 4, pp. 61–66, November 1982.

STIGALL, P. D., R. E. ZIEMER, AND V. T. PHAM, "Performance Study of a Microcomputer-Implemented FSK Receiver," *IEEE Micro*, Vol. 1, No. 1, pp. 43–51, February 1981.

STOCKHAM, T. G., "High Speed Convolution and Correlation," *AFIPS*

Proc. Spring Joint Comput. Conf., Washington, DC.: Spartan, Vol. 28, pp. 229–33, 1966.

STONE, H. C., "Parallel Processing with the Perfect Shuffle," *IEEE Trans.*   T
on Computers. Vol. C-20, pp. 153–61, February 1971.

STRADER, N R., II, "Effects of Subharmonic Frequencies on DFT Coef-   K
ficients," *Proc. IEEE*, Vol. 68, No. 2, pp. 285–86, February 1980.

STRANG, G., "Proposal For Toeplitz Matrix Calculations," *Studies in Ap-*   F
plied Mathematics, Vol. 74, No. 2, pp. 171–76, April 1986.

STRUZINSKI, W. A., AND E. D. LOWE, "Performance Comparison of Four   O
Noise Background Normalization Schemes Proposed For Signal Detec-
tion Systems," *Journal of the Acoustical Society of America*, Vol. 76,
No. 6, pp. 1738–42, December 1984.

SUDHAKAR, R., R. C., AGARWAL, AND S. C. DUTTA ROY, "Fast Computation   T
of Fourier Transform at Arbitrary Frequencies," *IEEE Trans. Circuits
and Syst.*, Vol. CAS-28, No. 10, pp. 972–80, October 1981.

SUGITA, M., "FFT Spectrum Analyzers Find Wide Applications In Acous-   I
tic, Noise And Vibration Analysis," *JEE, Journal of Electronic Engineering*, Vol. 23, No. 230, pp. 56–58, February 1986.

SUNDARAMURTHY, M., AND V. U. REDDY, "Some Results in Fixed Point Fast   W
Fourier Transform Error Analysis," *IEEE Trans. Comput.*, Vol. C-26,
pp. 305–8, 1977.

SUOBANK, D. W., A. P. YOGANATHAN, E. C. HARRISON, AND W. H. COR-   A
CORAN, "Quantitative Method For the In Vitro Study Of Sounds Produced
By Prosthetic Aortic Heart Valves. Part I. Analytical Considerations,"
Medical & Biological Engineering & Computing, Vol. 22, No. 1, pp. 32–
39, January 1984.

SUZUKI, Y., T. SONE, AND K. I. KIDO, "A New FFT Algorithm of Radix 3,   T
6 AND 12," *IEEE Transactions on Acoustics, Speech, and Signal Processing*, Vol. ASSP-34, No. 2, pp. 380–83, April 1986.

SVERDLIK, M. B., "Matrix Interpretation AND Computational Efficiency of   T
FFT [Fast Fourier Transform] Algorithms," *Radio Engineering and Electronic Physics*, Vol. 29, No. 2, pp. 60–68, February 1984.

SVERDLIK, M. B., "Matrix Interpretation Of The FFT Algorithm For Mu-   T
tually Simple Factors," *Ratio Engineering and Electronic Physics*, Vol.
28, No. 10, pp. 36–43, October 1983.

SWARTZLANDER, E. JR., D. V. SATISH CHANDRA, H. T. NAGLE, JR., AND S.   W
A. STARKS, "Sign/Logarithm Arithmetic For FFT Implementation,"
IEEE Transactions on Computers, Vol. C-32, No. 6, pp. 526–34, June
1983.

SWARTZLANDER, E., "Systolic FFT Processors," *Systolic Arrays. First In-*   G
ternational Workshop, pp. 133–40, 1987.

SWARTZLANDER, E. E., JR., W. K. W. WENDELL, AND S. J. JOSEPH, "Radix   G
4 Delay Commutator For Fast Fourier Transform Processor Implementation," *IEEE Journal of Solid-State Circuits*, Vol. SC-19, No. 5, pp.
702–9, October 1984.

SWARZTAUBER, P. N., "FFT Algorithms For Vector Computers," *Parallel Comput*, Vol. 1, No. 1, pp. 45–63, August 1984. T

SWARZTAUBER, P. N., "Symmetric FFTs," *Math. Comput.*, Vol. 47, No. 175, pp. 323–46, July 1986. T

SWORD, C. K., AND M. SIMAAN, "Estimation of Mixing Parameters For Cancellation Of Discretized Eddy Current Signals Using Time And Frequency Domain Techniques," *Journal of Nondestructive Evaluation*, Vol. 5, No. 1, pp. 27–35, March 1985. E

SYSOYEV, V. U., "Convolution Of A Multifrequency Signal Using Truncated FFT Algorithms," *Telecommunications and Radio Engineering*, Vol. 38–39, No. 10, pp. 94–96, October 1984. T

SZAPIEL, S. "Point-spread Function Computation: Analytic End Correction in the Quasi-digital Method," *J. Opt. Soc. Am. A.*, Vol. 4, No. 4, pp. 625–8, April 1987. X

SZIKLAS, E. A., AND A. E. SIEGMAN, "Diffraction Calculations Using Fast Fourier Transform Methods," *Proc. IEEE*, Vol. 62, No. 3, pp. 410–12, March 1974. F

TAKAMURA, H., Y. OHTA, AND T. MATSUMOTO, "Symbolic Analysis of Linear Networks Using Matrix Partition Method and FFT Algorithm," *Electronics and Communications in Japan*, Vol. 65, No. 11, pp. 19–28, November 1982. M

TANG, D. T., AND D. LI, "Time Interval Damage Potential Of Seismic Testing Waveforms," *Journal of Pressure Vessel Technology, Transactions of the ASME*, Vol. 107, No. 4, pp. 373–79, November 1985. V

TAYLOR, C. W., K. Y. LEE, AND D. P. DAVE, "Automatic Generation Control Analysis with Governor Deadband Effects," *IEEE Trans. Power Appar. and Syst.*, Vol. PAS-98, No. 6, pp. 2030–36, November –December 1979. Y

TAYLOR, F. J., A. S. RAMNARAYAN, AND J. WASSERMAN, "Non-Invasive Aneurysm Detection Using Digital Signal Processing," *Journal of Biomedical Engineering*, Vol. 5, No. 3, pp. 201–10, July 1983. A

TAYLOR, F. J., G. PAPADOURAKIS, A. SKAVANTZOS, AND A. STOURAITIS, "A Radix-4 FFT Using Complex RNS Arithmetic," *IEEE Transactions On Computers*, Vol. C-34, No. 6, pp. 573–76, June 1985. T

TAYLOR, T. D., R. S. HIRSH, AND M. M. NADWORNY, "Comparison of FFT, Direct Inversion and Conjugate Gradient Methods For Use in Pseudo-Spectral Methods," *Computers & Fluids*, Vol. 12, No. 1, pp. 1–9, 1984. V

TEGOPOULOS, J. A., AND E. E. KRIEZIS, "Eddy Currents in Linear Conducting Media," *Studies in Electrical and Electronic Engineering*, Vol. 16, Eddy Currents in Linear Conduct Media. Elsevier, Amsterdam, Neth AND New York, NY, USA, p. 304, 1985. E

TEMES, G. C., "A Worst Case Error Analysis for the FFT," *IEEE Int. Symp. Circuits Syst.*, Munich, West Germany, pp. 98–101, 1976. W

TEMPERTON, C., "Self-Sorting Mixed-Radix Fast Fourier Transforms (Numerical Weather Prediction)," *J. Comput. Phys.*, Vol. 52, No. 1, pp. 1–23, October 1983. J

TETEWSKY A. K., "Accelerate Your PC's Arithmetic With An Array Pro-   G
cessor," *EDN*, Vol. 30, No. 22, pp. 155–64, October 1985.

THEILHEIMER, F., "A Matrix Version of the Fast Fourier Transform," *IEEE*   T
Trans. Audio Electroacoust., Vol. AU-17, No. 2, pp. 158–61, June 1969.

THRANE, N. B., AND N. D. KJAER, "Frequency Analysis Using Zoom-FFT,"   K
Noise & Vibration Control Worldwide, (GB), Vol. 12, No. 1, pp. 13–15,
January–February 1981.

TILLOTSON, T. C., AND E. O. BRIGHAM, "Simulation with the Fast Fourier   Z
Transform," *Instruments and Control Systems.* Vol. 42, pp. 169–71, Sep
tember 1969.

TIWARI, P. K., M. IBRAHIM, AND O. P. N. CALLA, "Implementation and   H
Testing of FFT Hardware through Microprocessor," *Journal of the In-*
stitution of Engineers (India), Part ET: Electronics & Telecommunication
Engineering Division, Vol. 66, Pt. 4, pp. 57–61, May 1986.

TOLIMIERI, R., AND S. WINOGRAD, "Computing The Ambiguity Surface,"   H
IEEE Trans. Acoust., Speech & Signal Process., Vol. ASSP-33, No. 5,
pp. 1239–45, October 1985.

TOM, V. T., "Adaptive Filter Techniques for Digital Image Enhancement,"   X
Proc. SPIE Int. Soc. Opt. Eng., Vol. 528, pp. 29–42, January 1985.

TORTOLI, P., G. MANES, AND C. ATZENI, "Velocity Profile Reconstruction   O
using Ultrafast Spectral Analysis Of Doppler Ultrasound," *IEEE Trans-*
actions On Sonics and Ultrasonics, Vol. SU-32, No. 4, pp. 555–61, July
1985.

TRAN-THONG, AND B. LIU, "Fixed-point Fast Fourier Transform Error Anal-   W
ysis," *IEEE Trans. Acoust. Speech Signal Process.*, Vol. ASSP-24, pp.
563–73, 1976.

TRAN-THONG, AND B. LIU, "Accumulation of Roundoff Errors in Floating   W
Point FFT," *IEEE Trans. Circuits Syst.*, Vol. CAS-24, pp. 132–43, 1974.

TRETTER, S. A., "Tracking the Frequency Translation of a Sum of Or-   Q
thogonal Sinusoids," *IEEE Trans. Aerosp. & Electron. Syst.*, Vol. AES-
22, No. 2, pp. 211–14, March 1986.

TRIDER, R. C., "A Fast Fourier Transform (FFT) Based Sonar Signal Pro-   O
cessor," *1976 IEEE International Conference on Acoustics, Speech and*
Signal Processing, pp. 389–93, April 1976.

TSANG, S. H. L., M. W. BENSON, AND R. H. GRANBERG, "Open and Blocked   E
Distributed Air Transmission Lines By the Fast Fourier Transform
Method," *Journal of Dynamic Systems, Measurement* AND Control,
Trans. ASME, Vol. 107, No. 3, pp. 213–19, September 1985.

TSENG, B. D., G. A. JULLIEN, AND W. C. MILLER, "Implementation of FFT   G
Structures Using the Residue Number System," *IEEE Trans. Comput.*,
Vol. C-28, No. 11, pp. 831–45, November 1979.

TSENG, B. D., W. C. MILLER, AND G. A. JULLIEN, "Analysis of Quantization   T
Error in a Rom Oriented FFT Processor," *25th Midwest Symposium on*
Circuits and Systems, pp. 6–8, 1982.

TSENG, F-I, AND T. K. SARKAR, "Detection of Branch Points By Modified   J

FFT," *IEEE Transactions on Geoscience and Remote Sensing*, Vol. GE-21, No. 4, pp. 468–72, October 1983.

TSENG, F-I, AND T. K. SARKAR, "Experimental Determination of Resonant   E Frequencies By Transient Scattering From Conducting Spheres and Cylinders," *IEEE Transactions On Antennas and Propagation*, Vol. AP-32, No. 9, pp. 914–18, September 1984.

TSENG, F., AND T. K. SARKAR, "Enhancement of Poles in Spectral Analysis," *IEEE Trans. Geosci. and Remote Sensing*, Vol. GE-20, No. 2, pp.   K 161–68, April 1982.

TSENG, F. I., T. K. SARKAR, AND D. D. WEINER, "A Novel Window for   K Harmonic Analysis," *IEEE Trans. Acoust., Speech and Signal Process.*, Vol. ASSP-29, No. 2, pp. 177–88, April 1981.

TUFTS, D. W., H. S. HERSEY, AND W. E. MOSIER, "Effects of FFT Coefficient Quantization on Bin Frequency Response," *Proc. IEEE*, Vol. 60,   W pp. 146–47, 1972.

TWOGOOD, R. E., AND M. P. EKSTROM, "An Extension Of Eklundh's Matrix   X Transposition Algorithm and Its Application In Digital Image Processing," *IEEE Trans. Comput.*, Vol. C-25, No. 9, pp. 950–52, September 1976.

UENO, M., "A Systematic Design Formulation for Butler Matrix Applied   E FFT Algorithm," *IEEE Trans. Antennas and Propag.*, Vol. AP-29, No. 3, pp. 496–501, May 1981.

ULRIKSSON, B., "Conversion Of Frequency-Domain Data To The Time Domain," *Proceedings of the IEEE*, Vol. 74, No. 1, pp. 74–77, January 1986.   T

ULRIKSSON, B., "Synthesis Procedure For Designing 90 Degree Directional   E Couplers With a Large Number of Sections," *IEEE Transactions On Microwave Theory and Techniques*, Vol. 30, No. 8, pp. 1216–19, August 1982.

ULRIKSSON, B., "Time Domain Reflectometer Using A Semiautomatic Network Analyzer And The Fast Fourier Transform," *IEEE Transactions*   L *on Microwave Theory and Techniques*, Vol. MTT-29, No. 2, pp. 172–74, February 1981.

UMAPATHI, REDDY, V., AND M. SUNDARAMURTHY, "New Results in Fixed-   W Point Fast Fourier Transform Error Analysis," *1976 IEEE International Conference on Acoustics, Speech and Signal Processing*, pp. 120–25, Philadelphia, PA, April 1976.

VAN DER AUWERAER, H., AND R. SNOEYS, "FFT Implementation Alternatives   G in Advanced Measurement Systems," *IEEE Micro*, Vol. 7, No. 1, pp. 39–49, February 1987.

VARG, P., AND U. HEUTE, "A Short-time Spectrum Analyzer with Poly-   K phase-network and DFT," *Signal Process.*, Vol. 2, pp. 55–65, 1980.

VASUDEVAN, N., AND A. K. MAL, "Response Of An Elastic Plate To Lo-   V calized Transient Sources," *Journal of Applied Mechanics, Transactions ASME*, Vol. 52, No. 2, pp. 356–62, June 1985.

VEENKANT, R. L., "A Serial Minded FFT," *IEEE Trans. Audio, Electro-   T acoust.*, Vol. AU-20, pp. 180–85, 1972.

VERLY, J. G., AND T. PELI, "Circular Harmonic Analysis of PSF's Corresponding to Separable Polar-Coordinate Frequency Responses With Emphasis On Fan Filtering," *IEEE Transactions on Acoustics, Speech, and Signal Processing*, Vol. ASSP-33, No. 1, pp. 300–307, February 1985.   P

VERNET, J. L., "Real Signals Fast Fourier Transform: Storage Capacity and Step Number Reduction By Means of an Odd Discrete Fourier Transform," *Proc. IEEE*, Vol. 59, No. 10, pp. 1531–32, October 1971.   T

VETTERLI, M., AND A. LIGTENBERG, "A Discrete Fourier-cosine Transform Chip," *IEEE J. Sel. Areas Commun.*, Vol. SAC-4, No. 1, pp. 49–61, January 1986.   G

"Vibration Monitor Simplifies Machinery," *Diagnostics Diesel and Gas Turbine Worldwide*, Vol. 16, No. 6, July–August 1984.

VISHNYAKOV, Y. M., AND G. A. KUKHAREV, "Fast Fourier Transform Procedures For The Processing Of Two-Dimensional Signals Without Transposition Operations," *Programming* AND Computer Software, Vol. 8, No. 3, pp. 124–28, May–June 1982.   N

VITYAZEV, V. V., AND A. I. STEPASHKIN, "Synthesis of a Digital Filter-Demodulator Based On The Double Fast Fourier Transform," *Telecommunications and Radio Engineering*, Vol. 35/36, No. 7, pp. 51–54, July 1982.   P

VLASENKO, V. A., AND Y. M. LAPPA, "A Matrix Approach to the Construction of Fast Multidimensional Discrete Fourier Transform Algorithms," *Radioelectron. & Commun. Syst.*, Vol. 29, No. 1, pp. 87–90, 1986.   T

WACKERSREUTHER, G., "On Two-Dimensional Polyphase Filter Banks," *IEEE Transactions on Acoustics, Speech, and Signal Processing*, Vol. ASSP-34, No. 1, pp. 192–99, February 1986.   N

WALLACH, Y., AND A. SHIMOR, "Alternating Sequential-parallel Versions of the FFT," *IEEE Trans. on Acoustics, Speech and Signal Processing*, Vol. ASSP-28, No. 2, pp. 236–42, April 1980.   T

WALTERS, L. C., "Interpolation In FFTs," *J Inst Electron Radio Eng.* (GB), Vol. 55, No. 11–12, pp. 415–19, November–December 1985.   J

WEALE, J. R., "Use of FFT In Microstrip Capacitance Calculations," *Electronics Letters*, Vol. 21, No. 3, p. 86, January 1985.   E

WEBB, R., "Frequency Domain Instrumentation Techniques For The Condition Monitoring of Rotating Machinery," *Noise and Vib. Control Worldwide* (GB), Vol. 14, No. 8, pp. 215–19, October 1983.   V

WEBBER, C. L., JR., "A C-language Program for the Computation of Power Spectra on a Laboratory Microcomputer," *Comput. Methods and Programs Biomed.* (Netherlands), Vol. 22, No. 3, pp. 285–91, June 1986.   T

WEE, W. G., AND TSUNG-TAO-HSIEH, "An Application Of The Projection Transform Technique In Image Transmission," *IEEE Trans. Syst., Man and Cybern.*, Vol. SMC-6, No. 7, pp 486–93, July 1976.   X

WEINSTEIN, C. J., "Roundoff Noise in Floating Point Fast Fourier Transform Computation," *IEEE Trans. Audio Electroacoust.*, Vol. AU-17, No. 3, pp. 209–15, September 1969.   W

WELCH, P. D., "A Fixed Point Fast Fourier Transform Error Analysis," W
IEEE Trans. Audio Electroacoust., Vol. AU-17, pp. 151–57, June 1969.

WELCH, P. D., "The Use of Fast Fourier Transform for the Estimation of K
Power Spectra: A Method Based on Time Averaging Over Short, Modified Periodograms," *IEEE Trans. Audio Electroacoust.*, Vol. AU-15, pp.
70–73, June 1967.

WELLACH, Y., AND A. SHINOR, "Alternating Sequential-parallel Versions of T
the FFT," *IEEE Trans. on Acoustics, Speech and Signal Processing*,
Vol. ASSP-28, No. 2, pp. 236–42, April 1980.

WHELCHEL, J. E. W., AND D. F. GUINN, "FFT Organizations for High Speed T
Digital Filtering," *IEEE Trans. Audio and Electroacoustics*, Vol. AU-
18, No. 2, pp. 159–68, June 1970.

WHITE, P. H., "Application of the Fast Fourier Transform to Linear Dis- J
tributed System Response Calculations," *Acoustical Society of America
Journal*, Vol. 46, Pt. 2, pp. 273–74, July 1969.

WHITE, S. A., "A Simple FFT Butterfly Arithmetic Unit," *IEEE Trans. G
Circuits and Syst.*, Vol. CAS-28, No. 4, pp. 352–55, April 1981.

WILKEN, W., AND J. WEMPEN, "An FFT-Based, High Resolution Measuring O
Technique with Application to Outdoor Ground Impedance at Grazing
Incidence," *Noise Control Eng. J.*, Vol. 27, No. 2, pp. 52–60, Sep-
tember–October 1986.

WILLEY, T., R. CHAPMAN, H. YOHO, T. S. DURRANI, AND D. PREIS, "Systolic C
Implementations For Deconvolution, DFT and FFT," *IEEE Proceedings, Part F, Communications, Radar, and Signal Processing* (GB): Vol.
132, No. 6, pp. 466–72, October 1985.

WILLIAMS, E. G., AND J. D. MAYNARD, "Numerical Evaluation Of The Ray- O
leigh Integral For Planar Radiators Using The FFT," *Journal of the
Acoustical Society of America*, Vol. 72, No. 6, pp. 2020–30, December
1982.

WILLIAMS, E. G., "Numerical Evaluation Of The Radiation From Unbaf- O
fled, Finite Plates Using The FFT," *Journal of the Acoustical Society of
America*, Vol. 74, No. 1, pp. 343–47, July 1983.

WILLIAMS, J. R., "Fast Beamforming Algorithm," *Journal of the Acoustical
Society of America*, Vol. 44, No. 5, pp. 1454–55, 1968.

WILLIS, H. L., AND J. V. AANSTOOS, "Some Unique Signal Processing Ap- Y
plications In Power System Planning," *IEEE Trans. Acoust., Speech and
Signal Process.*, Vol. ASSP-27, No. 6, Pt. 2, pp. 685–97, December 1979.

WITTE, H., S. GLASER, AND M. ROTHER, "New Spectral Detection and Elim- A
ination Test Algorithms of ECG and EOG Artifacts In Neonatal EEG
Recordings," *Medical & Biological Engineering & Computing*, Vol. 25,
No. 2, pp. 127–30, March 1987.

WITTIG, L. E., AND A. K. SINHA, "Simulation Of Multicorrelated RANDOM F
Processes Using The FFT Algorithm," *J. Acoust. Soc. AM.*, Vol. 58,
No. 3, pp. 603–34, September 1975.

WOLD, E. H., AND A. M. DESPAIN, "Pipeline AND Parallel-Pipeline FFT G

Processors For VLSI Implementations," *IEEE Transactions on Computers*, Vol. C-33, No. 5, pp. 414–26, May 1984.

WOLINSKI, K., "Analysis of Errors in Mixed Fast Fourier Transform Algorithms With Decimation in Frequency For Fixed Point Arithmetic," *Proceedings of ICASSP 82. IEEE International Conference on Acoustics, Speech and Signal Processing,* pp. 2089–93, 1982.   W

WOOD, S. L., AND M. MORF, "A Fast Implementation of a Minimum Variance Estimator For Computerized Tomography Image Reconstruction," *IEEE Trans. Biomed. Eng.*, Vol. BME-28, No. 2, pp. 56–68, February 1981.   A

YAHYA, R-S, "Microwave Holography of Large Reflector Antennas-Simulation Algorithms," *IEEE Transactions on Antennas and Propagation*, Vol. AP-33, No. 11, pp. 1194–1203, October 1985.   E

YAMAGUCHI, T., AND N. ARAKAWA, "Effects of Finite Kernel Word Length In Signal Processing," *ICASSP 84. Proceedings of the IEEE International Conference on Acoustics, Speech and Signal Processing*, 30.12/1–4 Vol. 2, San Diego, CA, March 1984.   W

YARLAGADDA, R., J. B. BEDNAR, AND T. L. WATT, "Fast Algorithms For 1p Deconvolution," *IEEE Trans. Acoust., Speech & Signal Process.*, Vol. ASSP-33, No. 1, pp. 174–82, February 1985.   C

YEH, C., AND F. MANSHADI, "On Weakly Guiding Single-Mode Optical Waveguides," *Journal of Lightwave Technology*, Vol. LT-3, No. 1, pp. 199–205, February 1985.   E

YEH, H.-G., "Power Spectrum Estimation By Using Digital Frequency Tracking Filter," *Proceedings of the 1986 American Control Conference*, Vol. 3, pp. 1642–44, June 1986.   K

YEH, M., J. L. MELSA, AND D. L. COHN, "A Direct FFT Scheme For Interpolation, Decimation, Amplitude Modulation and Single Side Band Modulation," *Sixteenth Asilomar Conference on Circuits, Systems and Computers*, pp. 437–41, Pacific Grove, CA, November 1983.   Q

YEVICK, D., AND B. HERMANSSON, "Band Structure Calculation With The Split-Step Fast Fourier Transform Technique," *Solid State Communications*, Vol. 54, No. 2, pp. 197–99, April 1985.   J

YEVICK, D., AND B. HERMANSSON, "New Approach To Perturbed Optical Waveguides," *Opt. Lett.*, Vol. 11, No. 2, pp. 103–5, February 1986.   X

YEW, C. H., AND C. S. CHEN, "Study of Linear Wave Motions Using FFT AND Its Potential Application To Non-Destructive Testing," *International Journal of Engineering Science*, (GB), Vol. 18, No. 8, pp. 1027–36, 1980.   V

YING, S. P., AND E. E. DENNISON, "Vibration Diagnosis For Turbine-Generators," *Noise & Vibration Control Worldwide*, Vol. 12, No. 2, pp. 50–52, March 1981.   K

YIP, P., "Some Aspects of the Zoom Transform," *IEEE Trans. Comput.*, Vol. C-25, No. 3, pp. 287–96, March 1976.   T

YLITALO, J., E. ALASARELA, A. TAURIANINEN, K. TERVOLA, AND J. KOIVI-KANGAS, "Three-dimensional Ultrasound C-scan Imaging Using Holo-   O graphic Reconstruction," *IEEE Trans. Ultrason., Ferroelectr. and Freq. Control*, Vol. UFFC-33, No. 6, pp. 731–39, November 1986.

YOKOTA, Y., M. TOMITA, H. HASHIMOTO, AND H. ENDOH, "Construction Of An On-Line System For FFT Processing AND Analysis Of Atomic Resolution Microscopic Images And Its Applications," *Ultramicroscopy*, Vol. 6, No. 4, pp. 313–21, 1981.   X

YONG, A., AND M. JUANATEY, "Microprogrammable Peripheral Unit And The FFT," *Software & Microsystems*, Vol. 4, No. 2, pp. 35–39, April 1985.   G

YONG, CHING LIM, "An Interpolation Technique for Computing the DFT of a Sparse Sequence," *IEEE Trans. Acoust., Speech & Signal Process.*, Vol. ASSP-33, No. 6, pp. 1456–60, December 1985.   T

YOST, M., F. J. BREMNER, R. J. HELMER, AND M.-E. C. CHINO, "The Effect of Smoothing Functions on Data Obtained from a FFT," *Behav. Res. Methods Instrum. & Comput.*, Vol. 18, No. 2, pp. 263–66, April 1986.   A

YOST, R. A., "On Nonuniform Windowing M-Ary FSK Data in a DFT-Based Detector," *IEEE Trans. Commun.*, Vol. COM-28, No. 12, pp. 2014–19, December 1980.   Q

ZELENKEVICH, V. M., V. A. KAPLUN, AND A. B. TEREKHOVICH, "Automating The Design of Antenna-Radome Radiating Systems," *Telecommunications and Radio Engineering*, Vol. 37-38, No. 8, pp. 58–60, August 1983.   E

ZHONG, L., AND WEN-HONG CHIN, "A Novel Method for Evaluating the Resolution and the Wavefront of Plane Diffraction Grating with FFT," *Proc. SPIE Int. Soc. Opt. Eng.*, Vol. 599, pp. 297–302, 1986.   X

ZIEMER, R. E. "Computer Evaluation Of A Broadband M-ary Signaling Scheme and FFT-Processing Receiver For Data Transmission In HF Channels," *1976 International Conference on Communications*, pp. 44/1–6, Philadelphia, PA, June 1976.   S

INDEX